D1740129

PERCEPTION OF
COMPLEX SMELLS AND TASTES

PERCEPTION OF COMPLEX SMELLS AND TASTES

Edited by

David G. Laing
CSIRO Division of Food Processing,
Sydney, Australia

William S. Cain
John B. Pierce Foundation Laboratory,
New Haven, Connecticut, USA

Robert L. McBride
CSIRO Division of Food Processing,
Sydney, Australia

Barry W. Ache
C. V. Whitney Laboratory and
Departments of Zoology and Neuroscience,
University of Florida, Florida, USA

ACADEMIC PRESS
Harcourt Brace Jovanovich, Publishers
Sydney   San Diego   New York   Berkeley
Boston   London   Tokyo   Toronto

ACADEMIC PRESS AUSTRALIA
30-52 Smidmore Street, Marrickville, NSW 2204

United States Edition published by
ACADEMIC PRESS INC.
1250 Sixth Avenue
San Diego, California 92101-4311

United Kingdom Edition published by
ACADEMIC PRESS, INC.(LONDON) LTD.
24/28 Oval Road, London NW1 7DX

Copyright © 1989 by
ACADEMIC PRESS AUSTRALIA

All rights reserved. No part of this publication may be
reproduced or transmitted in any form or by any means,
electronic or mechanical, including photocopy, recording,
or any information storage and retrieval system, without
permission in writing from the publisher.

National Library of Australia Cataloguing-in-Publication Data

---

Perception of complex smells and tastes.

Bibliography.
Includes index.
ISBN 0 12 042990 X.

1. Chemical senses. 2. Smell. 3. Taste. I. Laing, David G. (David
George), date— .

591.1'826

Library of Congress Catalog Card Number: 88-071452

Transferred to digital printing 2006

Contents

Contributors

The numbers in parentheses indicate the pages on which the authors' contributions begin.

Barry W. Ache (101), C. V. Whitney Laboratory and Department of Zoology, University of Florida, 9505 AIA South, St Augustine, Florida, USA 32086.

R. Patrick Akers (49), The Worcester Foundation for Experimental Biology, 222 Maple Avenue, Shrewsbury, Massachusetts, USA D1545.

Jelle Atema (83), Boston University Marine Program, Marine Biological Laboratory, Woods Hole, Massachusetts, USA 02543.

Ann M. Belcher (1), Monell Chemical Senses Center, 3500 Market Street, Philadelphia, Pennsylvania, USA 19104; German Primate Center, Kellnerweg 4, 34 Gottingen, Federal Republic of Germany.

Paola Borroni (83), Boston University Marine Program, Marine Biological Laboratory, Woods Hole, Massachusetts, USA 02543.

Joseph G. Brand (115), Monell Chemical Senses Center, 3500 Market Street, Philadelphia, Pennsylvania, USA 19104; Veterans Administration Medical Center, Philadelphia, Pennsylvania, USA 19104; Department of Biochemistry, University of Pennsylvania School of Dental Medicine, Philadelphia, Pennsylvania, USA 19104

Bruce P. Bryant (115), Monell Chemical Senses Center, 3500 Market Street, Philadelphia, Pennsylvania, USA 19104.

Robert H. Cagan (115), Colgate Palmolive Company, Research and Development Division, Piscataway, New Jersey, USA 08854.

William S. Cain (173), John B. Pierce Foundation Laboratory, 290 Congress Avenue, New Haven, Connecticut, USA 06519.

William E. S. Carr (27), C. V. Whitney Laboratory and Department of Zoology, University of Florida, 9505 AIA South, St Augustine, Florida, USA 32086.

Cees De Graaf (245), Department of Food Science and Department of Market Research, The Netherlands Agricultural University, De Dreijen 12, 6703BD Wageningen, The Netherlands.

Peter C. Daniel (65), Department of Biology, Georgia State University, Atlanta, Georgia, USA 30303.

Charles D. Derby (65), Department of Biology, Georgia State University, Atlanta, Georgia, USA 30303.

Melvin P. Enns (285), Department of Psychology, St Lawrence University, Canton, New York, USA 13617.

Gisela M. Epple (1), Monell Chemical Senses Center, 3500 Market Street, Philadelphia, Pennsylvania, USA 19104; German Primate Center, Kellnerweg 4, 34 Gottingen, Federal Republic of Germany.

Jacqueline B. Fine-Levy (65), Department of Biology, Georgia State University, Atlanta, Georgia, USA 30303.

Marion E. Frank (127), Department of Biostructure and Function, University of Connecticut Health Center, Farmington, Connecticut, USA 06032.

Jan E. R. Frijters (245), Department of Food Science and Department of Market Research, The Netherlands Agricultural University, De Dreijen 12, 6703BC Wageningen, The Netherlands.

Marie-Nadia Girardot (65), Department of Biology, Georgia State University, Atlanta, Georgia, USA 30303.

Richard A. Gleeson (27), C. V. Whitney Laboratory and Department of Zoology, University of Florida, 9505 AIA South, St Augustine, Florida, USA 32086.

Katherine L. Greenfield (1), Monell Chemical Senses Center, 3500 Market Street, Philadelphia, Pennsylvania, USA 19104; Department of Chemistry, University of Pennsylvania, Philadelphia, Pennsylvania, USA 19104.

Linda Handrich (83), Boston University Marine Program, Marine Biological Laboratory, Woods Hole, Massachusetts, USA 02543.

David E. Hornung (285), Department of Biology, St Lawrence University, Canton, New York, USA 13617.

Bruce Johnson (83), Boston University Marine Program, Marine Biological Laboratory, Woods Hole, Massachusetts, USA 02543.

D. Lynn Kalinoski (115), Monell Chemical Senses Center, 3500 Market Street, Philadelphia, Pennsylvania, USA 19104.

Jan H. A. Kroeze (225), Psychological Laboratory, Utrecht University, Sorbonnelaan 16, 3584 CA Utrecht, The Netherlands.

I. Küderling (1), German Primate Center, Kellnerweg 4, 34 Gottingen, Federal Republic of Germany.

Paul Laffort (205), Laboratoire de Physiologie de la Chimioreception (UA1190), CNRS, 91190 Gif-sur-Yvette, France.

David G. Laing (189), CSIRO Division of Food Research, PO Box 52, North Ryde, New South Wales, Australia 2113.

Harry T. Lawless (297), Product Evaluation Department, S.C. Johnson & Son, Racine, Wisconsin, USA 53403.

Robert L. McBride (265), CSIRO Division of Food Research, PO Box 52, North Ryde, New South Wales, Australia 2113.

K. Nordstrom (1), Department of Dermatology, University of Pennsylvania, Philadelphia, Pennsylvania, USA 19104.

Robert J. O'Connell (49), The Worcester Foundation for Experimental Biology, 222 Maple Avenue, Shrewsbury, Massachusetts, USA 01545.

Michael D. Rabin (173), International Flavors and Fragrances Research and Development, 1515 Highway 36, Union Beach, New Jersey, USA 07735.

Amos B. Smith III (1), Department of Chemistry, University of Pennsylvania, Philadelphia, Pennsylvania, USA 19104; Monell Chemical Senses Center, 3500 Market Street, Philadelphia, Pennsylvania, USA 19104.

David V. Smith (149), Department of Otolaryngology and Maxillofacial Surgery, University of Cincinnati College of Medicine, 231 Bethesda Avenue, Mail Location 528, Cincinnati, Ohio, USA 45267.

David A. Stevens (297), Department of Psychology, Clark University, Worcester, Massachusetts, USA 01610.

Henry G. Trapido-Rosenthal (27), C.V.Whitney Laboratory and Department of Zoology, University of Florida, 9505 AIA South, St Augustine, Florida, USA 32086.

Rainer Voigt (83), Boston University Marine Program, Marine Biological Laboratory, Woods Hole, Massachusetts, USA 02543.

Preface

The air we breathe, the foods we eat, and the beverages we drink are usually complex mixtures of odors or tastes, sometimes both. So too are the worlds of fish, insects and other animals. Many species communicate by complex chemical signals, and many examples exist of secretions that can attract, repel, mark territories and control reproduction. Despite the chemical complexity of these mixtures, the senses of smell and taste can sift out relevant information about the source and acceptability of a food, the presence of a stranger, danger, or the sexual status of a female.

Learning how the senses achieve these feats is a goal of physiologists, psychologists, biochemists, behavioralists and marine biologists. With the advent of modern chemical instruments, scientists have been able, in many instances, to identify the components of mixtures, and to determine which of the many components are important. They have learned about the amazing sensitivity of insect receptor cells to odorants of biological significance (e.g. sex attractants), and that the perception of certain components may be masked or enhanced by others.

The benefits of understanding mixture perception would be considerable. For example, it would allow the flavorist and perfumer direct insight into how components 'interact' — how they mask or enhance one another, and which components in a product are redundant and might therefore be eliminated. The ecologist and biologist, by understanding the language of animal and insect communication, could develop better methods of biological control of pests, or could improve animal husbandry. Marine biologists would know how it is that fish 'home' to a particular stream, what chemical environment is necessary for successful aquaculture, and why fish eat what they eat.

Interaction of the chemical senses in the perception of complex mixtures is also important. As well as stimulating taste and smell, food and drink often stimulate the trigeminal (pain) nerve in the nose and tongue. This nerve senses those 'cool' (e.g. menthol) and 'hot' (e.g. capsaicin) chemicals that can contribute greatly to flavor (as afficionados of Mexican food will testify!); but it also monitors irritants in the air around us, and alerts us to dangerous levels of chemicals in our environment.

So far, no book, conference or workshop has dealt exclusively with the perception of complex odors or tastes. In order to rectify this omission we set two goals. Our first aim was to bring together, at an international workshop, those currently working on mixtures in the chemical senses, so that the latest results and views could be presented, discussed and exchanged. This aim was achieved in April 1987 in Sarasota, Florida. Second, we wished to allow contributors the opportunity to expand their views and research in a book. Of cross-disciplinary character, the book would afford readers from different backgrounds the chance to discover what is happening across a number of fields.

Thus this book contains chapters on the nature and complexity of odor and taste mixtures; the physiology of mixture perception; human perception of mixtures; and interaction of the chemical senses. Each chapter clearly demonstrates that our knowledge and understanding of mixture perception is in its infancy and that progress will require workers to be aware of and apply the findings of others in different disciplines to their specific research problem. Finally, we hope the wide range of approaches described in this book will stimulate others to join the small band of workers presently engaged in this challenging area of research.

PART ONE
COMPLEXITY OF NATURAL MIXTURES AND THEIR PERCEPTION

1

Scent mixtures used as social signals in two primate species: *Saguinus fuscicollis* and *Saguinus o. oedipus*

G. Epple[1,2], A. Belcher[1,2], K. L. Greenfield[2,4],
I. Küderling[1], K. Nordstrom[3] and A. B. Smith III[2,4]

[1] *German Primate Center, Göttingen, Federal Republic of Germany*
[2]*Monell Chemical Senses Center, Philadelphia, Pennsylvania, USA*
[3] *Department of Dermatology, University of Pennsylvania, Philadelphia, Pennsylvania, USA*
[4] *Department of Chemistry, University of Pennsylvania, Philadelphia, Pennsylvania, USA*

I. Introduction

Many mammalian species make extensive use of chemical signals in socio-sexual communication and in priming of reproductive functions. Chemical signals are contained in urine, feces, genital discharge, saliva and the secretion of specialized skin glands, of which several types may be present on an individual. Often, ingredients from several sources are combined to form complex mixtures. In addition, bacteria may act on substrates such as vaginal discharge or skin secretions to develop volatiles which carry signal function (Albone, 1984; Brown & Macdonald, 1985; Duvall, Müller-Schwarze & Silverstein, 1986; Halpin, 1986; Marchlewska-Koj, 1983; Müller-Schwarze & Silverstein, 1980; 1983; Stoddart, 1980; Vandenbergh, 1983).

During recent years our knowledge of the chemical composition of scent material employed in communication has increased considerably (Albone, 1984). However, decoding chemical signals to the point where chemical structure can be related to biological function continues to be

Copyright © 1989 by Academic Press Australia.
All rights of reproduction in any form reserved.

difficult. In a few mammals, for example the golden hamster (Singer, Clancy, Macrides & Agosta, 1984; Singer, Macrides, Clancy & Agosta, 1986) and the domestic pig (Melrose, Read & Patterson, 1971), specific biological activities rest with a single compound or a small number of compounds. In many cases, however, the biological activity resides in chemically complex mixtures, containing a number of classes of compounds (e.g. Belcher, Smith, Jurs, Lavine & Epple, 1986; Crump, Swigar, West, Silverstein, Müller-Schwarze & Altieri, 1984; Müller-Schwarze, Morehouse, Corradi, Cheng-hua & Silverstein, 1986; Raymer, Wiesler, Novotny, Asa, Seal & Mech, 1985, 1986). These mixtures may contain compounds of additive or redundant bioactivity, compounds that work in synergy and compounds that inhibit the biological response (Albone, 1984; Müller-Schwarze *et al.* 1986; Novotny, Harvey, Jemiolo & Alberts, 1985). Indeed, Albone (1984) and Müller-Schwarze *et al.* (1986) have pointed out that the complexity of many mammalian chemosignals is best described in terms of an 'odor image' or a scent 'Gestalt'.

The 'Gestalt' concept appears to be well suited to describe chemical signals employed by some primates (Crewe, Burger, Le Roux & Katsir, 1979; Keverne, 1978; Schilling, unpublished doctoral dissertation, Pierre & Marie Curie University, Paris, 1980; Schilling & Perret, 1987; Wheeler, Blum & Clark, 1977). Among primates, the prosimians and the South American monkeys make extensive use of complex scent mixtures for the purpose of social communication (Epple, 1986; Schilling, 1979). One family, the South American Callitrichidae, has been the focus of our research for several years. These small arboreal primates produce scent marks by mixing the secretions of specialized skin glands with urine. This paper discusses our recent comparative studies on the production, communicatory function and chemical composition of scent mixtures in two closely related callitrichid species, the cotton-top tamarin (*Saguinus o. oedipus*, Fig. 1.1) and the saddle-back tamarin (*Saguinus fuscicollis*, Fig. 1.2). The long-term goal of these comparative studies is to obtain insight into the way in which communicatory signals are structured within a group of closely related species.

II. Scent glands and scent-marking behaviors

Circumgenital scent glands are present in all species of the Callitrichidae (Epple, Belcher & Smith, 1986). However, the morphology of these glands varies considerably among tamarins and marmosets. Adult male and female saddle-back tamarins possess a large scent gland in the circumgenital area. The scent gland involves the *perineum*, the deeply

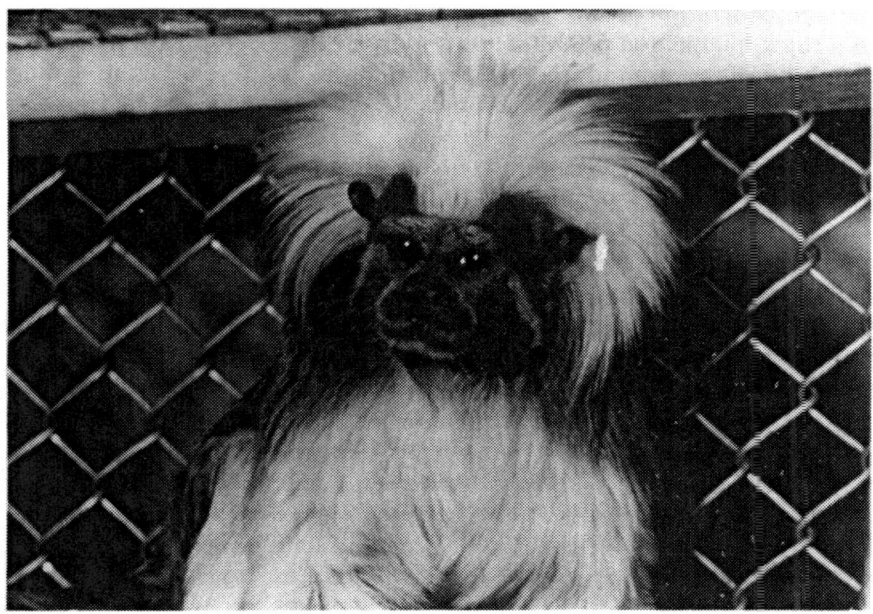

Fig. 1.1. The cotton-top tamarin, *Saguinus o. oedipus.*

Fig. 1.2. Suprapubic marking of a branch by the saddle-back tamarin, *Saguinus fuscicollis.*

pigmented *scrotum* or *labia majora*, and extends into the suprapubic area as a thick, pigmented pad which is only sparsely covered with hair. The gland develops around the time of puberty, and reaches its adult morphological appearance in animals of 1.5 to 2 years of age, or even older.

A very complex glandular organ, composed of holocrine and apocrine glands, is located beneath the epidermis of the circumgenital skin (Perkins, 1966; Zeller, Epple, Küderling & Kuhn, 1988). Our recent studies show that in males, specialized holocrine glands, associated with hair follicles, predominate. They have a complex alveolar structure and possess numerous branched excretory ducts. Each group of glands empties into a common duct which enters the hair follicle. The apocrine glands are located predominantly at the periphery of the glandular pad and between the scrotal and perineal areas. The excreting ducts of most apocrine glands empty onto the skin surface in close spatial association with hair follicles. In females, the specialized holocrine glands resemble those of males but are more frequently interspersed with apocrine glands. The apocrine glands are larger and much more numerous than in males, especially in the region of the *labia majora*. Based on their size and complexity, the holocrine glands of saddle-backs are among the most highly specialized among primates. In males they form a continuous layer and almost entirely replace the dermis. The apocrine glands of females are exceptionally large and are among the most highly specialized apocrine glands in mammals (cf. Schaffer, 1940).

Female *Saguinus o. oedipus* possess a circumgenital glandular organ analogous to that of female *Saguinus fuscicollis* (Perkins, 1969; Wislocki, 1930). The glandular organ occupies the suprapubic, labial and perineal areas and is composed of large holocrine and apocrine glands. In males of this species, the scent glands of the circumgenital area are much smaller than in females and are nearly absent in the scrotal skin (Perkins, 1969; Wislocki, 1930; Zeller & Epple, unpublished data).

Saddle-back and cotton-top tamarins employ their scent glands in scent-marking behavior (Fig. 1.2). Scent-marking involves two basic motor patterns in both species; anogenital marking and suprapubic marking. During anogenital marking, the animals rub the genital and perineal areas on the substrate in a sitting position, depositing secretions from these parts of the scent gland. At the same time, a small amount of urine is discharged. In females discharge from the genital tract also appears to be mixed into the scent mark (Epple *et al.*, 1986; French & Snowdon, 1981; French & Cleveland, 1984). Behavioral observations suggest that the animals do not always add urine and genital discharge to the marks. However, we do not know which factors cause them to vary the amounts they add. Fecal matter, adhering to the anogenital area, may also be deposited during marking, but defecation is not part of the marking behavior.

Suprapubic marking is accomplished by rubbing the suprapubic part of the glandular pad against the substrate while assuming a sprawling position. In saddle-backs, but not in cotton-tops, urine is added to suprapubic marks in a variable manner (Epple *et al.*, 1986; French & Cleveland, 1984). We have no information on the addition of genital discharge during this type of marking.

Although the two basic scent-marking patterns are similar in both species, there are considerable species differences in the manner in which they are carried out. One of the differences relevant to the work reported here is the sexual dimorphism in the frequencies with which adult animals of comparable social status display scent-marking. In saddle-back tamarins, both sexes scent-mark frequently, but females show somewhat more marking than males. Although this difference is statistically significant, it is not dramatic (Epple, 1980). Cotton-top females on the other hand display much higher frequencies of scent-marking than cotton-top males (French & Snowdon, 1981; French & Cleveland, 1984; Epple, Küderling & Belcher, 1988).

The two species also differ in the degree of variability in the motor patterns of anogenital and suprapubic marking and in the behavioral contexts in which both types of marking are performed. In male and female saddle-back tamarins, scent-marking is quite variable. The animals show the typical anogenital and suprapublic marking patterns, but may also display various combinations of both. On rare occasions, the use of another scent gland located above the sternum is combined with that of the circumgenital and suprapubic glands as the animal rubs the whole ventral surface against the substrate. At low intensities of arousal, anogenital marking predominates in male and female saddle-backs. At high intensities, suprapubic marking is added and combined with anogenital marking in various patterns. There is no evidence that in this species anogenital and suprapubic marking are associated with different behavioral contexts or fulfill different biological roles. For practical purposes, we therefore refer to the complex of both patterns as 'circumgenital marking'. The scent marks collected from saddle-backs for the purpose of the behavioral and chemical studies reported here were anogenital marks, suprapubic marks and combinations of both.

In cotton-top tamarins, the typical marking pattern shown by undisturbed individuals at low arousal is anogenital marking (French & Snowdon, 1981). For this type of marking, these tamarins prefer small protruberances such as short twigs. Unlike in saddle-backs, suprapubic marking does not seem to be combined with anogenital marking (French & Snowdon, 1981; French & Cleveland, 1984). It is much less frequently performed and in the cotton-top appears to be a pattern that is functionally distinct from anogenital marking. Suprapubic marking is displayed predominantly by females of breeding status during agonistic encounters with conspecifics (French & Snowdon, 1981). All scent marks collected from cotton-top donors for our behavioral and chemical studies were anogenital marks.

In summarizing the information presented above, it becomes clear that both species produce highly complex scent mixtures, consisting mainly of the secretions of specialized holocrine and apocrine skin glands. Variations in the amount of urine and vaginal discharge added to the mixture may result in variation in the sensory qualities of the mixture. In addition, manual stimulation of the gland pad by means of scratching and kneading, which is frequent in both species (personal observation), may result in selective secretion by part of the gland. This may further influence the composition of the scent and its sensory qualities.

Behavioral studies on scent-marking in a number of callitrichid species have suggested that both scent-marking itself and the chemical signals thus deposited are important in a number of sexual and social interactions, such as pair bonding, socialization, dominance interactions and territorial defense. Their functions within these different interactional contexts, however, is beyond the scope of this paper and the reader is referred to a recent review by Epple *et al.* (1986). The present paper is concerned with the informational content of the scent marks, their chemical composition, and efforts to correlate chemical composition with communicatory messages.

III. Communicatory content of scent material from saddle-back and cotton-top tamarins

Complex scent mixtures, such as the ones produced by the two tamarin species, have the potential to encode a large variety of communicatory messages by means of both qualitative and quantitative composition patterns. It is important, therefore, to determine which messages are contained in the scent marks from each species. Toward this end, a series of behavioral studies were conducted on saddle-back and cotton-top tamarins in order to determine the ability of the tamarins to distinguish between scents from various types of donors. Methodological details varied somewhat from experiment to experiment and are specified in the original publications. In general, all behavioral tests consisted of the presentation of either a single scented stimulus object, or of two identical stimulus objects, each carrying a different scent. The stimulus objects were introduced into the home cage of a subject monkey for a prescribed period of time. The subject was allowed to investigate the sample(s) freely for the entire test period. All tests conducted with saddle-back tamarins lasted five minutes. Because cotton-top tamarins required more time to approach and investigate novel objects introduced into their cages, tests with this species lasted 10 or 15 minutes.

The stimulus objects used for saddle-backs were wooden perches, aluminum plates or frosted glass plates, all measuring 61 × 5 × 0.6 cm (approx. 24 × 2 × ¼ ins). They were placed on the floor of the home cage or into holders suspended about 61 cm (2 ft) above the cage floor. Cotton-tops prefer to scent-mark small protruberances rather than level surfaces. Because of this, the stimulus objects used for this species were frosted glass rods. These rods were made by closing one end of a frosted glass tube (10 × 1.5 cm or approx. 4 × ½ ins) to form a round, smooth top. The glass rods could be inserted into holes drilled into heavy wooden shelves, and in this way mimicked small protruding twigs. The shelves, suspended about 61 cm (2 ft) above the cage floor, were part of the permanent equipment of the home cages of the cotton-tops.

All test periods were divided into intervals of five seconds each. Three categories of behavioral responses were recorded and the subjects received a score of 1 per stimulus in each response category for every interval in which this response was shown. The responses recorded in tests on saddle-backs were contacting each stimulus object, sniffing each stimulus object and scent-marking each stimulus object. Scent-marking included circumgenital marking, suprapubic marking and combinations of both patterns. The responses recorded for cotton-tops were contacting each wooden shelf containing the stimulus object, sniffing each stimulus object and anogenital marking each stimulus object. Suprapubic marking was very rare and was not evaluated.

For each behavioral category, mean response scores were computed for each subject in every experiment. For choice tests, the level of responses directed at one of the two samples was compared with that directed at the second sample by means of statistical tests for related samples. For single-sample tests, the level of responses directed at one type of sample was compared to that of responses directed at other types of samples presented under identical conditions by means of an analysis of variance (several stimuli), or a t-test (two stimuli). Significant differences in the level of responses directed toward different stimuli were interpreted as evidence of the tamarins' ability to distinguish the stimuli from one another.

At the beginning of our studies, a number of control tests were performed on saddle-back tamarins in order to document that the levels of contacting, sniffing and scent-marking shown in response to scent marks from conspecifics were higher than those shown in response to control stimuli. The results of these tests show that unscented stimulus objects elicit those responses at low levels. Objects scented with synthetic exaltolide or with synthetic mixtures of butyric acid esters, which are among the volatile components of the scent marks of this species (see below), do not elicit higher levels of contacting, sniffing and marking than unscented plates (Epple, Alveario, Golob & Smith, 1980). Urine, which is one of the ingredients of the scent marks and appears to contain cues for species and gender recognition, is also investigated by the tamarins. However, it is a much less attractive stimulus than scent marks from the same donor animal (Epple, 1978a). Fresh scent marks from males elicit higher levels of response in male and female test subjects than any of the control odors (Epple et al., 1980). In choice tests, scent marks from males elicit higher responses than those from females (see below). However, stimulus objects scent-marked by females elicit higher responses than blank stimulus objects, showing that scent from females is also attractive.

These control studies show that complex scent marks from conspecifics elicit specifically high levels of contacting, sniffing and scent marking. The following series of behavioral tests were conducted to determine among which categories of donors the tamarins can discriminate on the basis of various scent materials (scent marks, urine samples etc). In one series of tests, the ability of both tamarin species to discriminate between scents from conspecifics and from other callitrichids was investigated (Epple, Golob & Smith, 1979; Epple et al., 1988; Epple, Alveario, Belcher & Smith, 1987). It was found that saddle-backs and cotton-tops investigate scents from conspecifics more frequently than scents from closely related callitrichid species. This discrimination is made when the stimuli are complex, natural scent marks. However, urine samples from saddle-backs and gland secretion from cotton-tops (removed from the surface of the glands of trained animals by gentle wiping with a frosted glass rod) also appear to contain cues on which discrimination between scent from conspecifics and related species can be based.

S. fuscicollis is distributed in numerous subspecies throughout upper Amazonia (Hershkovitz, 1978). The scent marks from these closely related subspecies appear to offer cues for subspecies discrimination. Choice tests have shown that *S. f. fuscicollis* and *S. f. illigeri*, the two subspecies maintained in our colony, discriminate between scent marks in choice tests (Epple et al., 1979; 1987).

Saddle-back tamarins discriminate between scent from male and female conspecifics under a variety of experimental conditions. Scent marks and urine samples from males are more frequently contacted, sniffed and marked than corresponding material from females (Epple, 1974; 1978a). Even mixtures of scent marks from males and females pooled together in methanol appear to be discriminated from equal amounts of marks from females alone (Epple, 1978a). Cotton-top tamarins, in contrast to saddle-backs, do not discriminate between urine samples from males and from females (Epple, unpublished data). Discrimination between scent marks from males and from females was not investigated in cotton-tops because the infrequent marking of males makes collection of stimulus material difficult.

Gonadal hormones appear to influence the stimulus characteristics of the scent marks of male saddle-backs. However, in contrast to some other mammalian species (Ebling, 1977), gonadectomy has no permanent effect on scent-marking behavior (Epple, 1978 b, 1981, 1982). Scent marks from castrated saddle-back males are discriminated from those of intact males but not from those of females (Epple, 1979).

Discrimination between scents from two individuals of the same sex is made by both tamarin species. In saddle-backs, this was tested by offering the subjects choices between scent marks from two familiar males or females, one of which had had a recent aggressive encounter with the subjects. Under these conditions, scent marks, but not urine, from the recent opponent elicit higher responses than corresponding samples from the neutral familiar donor. Moreover, threat behaviors are shown frequently by subjects investigating scent marks from an opponent, suggesting individual recognition of the marks rather than merely discrimination between two different stimuli (Epple, 1974).

In cotton-tops, discrimination between individuals was tested using glandular secretion collected on glass rods from trained animals, as described above. These tests incorporated a habituation paradigm (Epple et al., 1988). It is based on the assumption that the animals will habituate to a scent stimulus which is presented several times in close succession, showing a decrease in investigatory responses. Investigatory responses should increase again when a novel stimulus is presented after the animals have habituated to the original one. The animals were habituated by offering them two successive samples of secretion from one donor female. When secretion from a novel female was offered as a third sample, the subjects displayed an increase in sniffing and contacting activities (Epple et al., 1988).

The scent marks of saddle-backs appear to contain information relating to the social status of the donor. When given a choice between stimulus objects scent-marked by unfamiliar, socially dominant males and objects scent-marked for the same period of time by unfamiliar, subdominant males, saddle-backs direct more attention toward the marks from dominant males (Epple, 1974). Dominant males normally show more scent-marking than subdominant ones. Therefore, the stimulus object marked by the dominant male probably carried more scent and this discrimination between the two samples might have been based on quantitative cues.

Studies on saddle-backs have suggested that the scent marks also contain cues relating to the age of the material (Epple et al., 1980). Freshly deposited scent marks are discriminated from scent marks deposited by the same donor animal 24 hours or more prior to testing. Freshly deposited scent marks and marks deposited 24 hours earlier elicit higher levels of investigation than do blank stimulus objects. Older scent marks, however, are not investigated more intensively than blank stimulus objects. Scent marks from males are discriminated from those from females when both stimuli are fresh, 24 and 48 hours old. However, no discrimination is made on the basis of marks older than 48 hours (Epple et al., 1980).

The failure of the saddle-backs to discriminate aged scent marks from blank stimulus objects and to discriminate between scents from males and females when the material is older than 48 hours does not necessarily mean that the animals are no longer perceiving information relevant to species or gender recognition. Some of our analytical studies have shown that aged scent marks contain higher concentrations of volatile compounds than fresh material (Belcher, Epple, Kostelc & Smith, 1982). Therefore, it is possible that during the aging process scent marks develop additional cues on which recognition of the age of the material could be based. Such information may be important in territorial species such as these (Terborgh, 1983), in which groups may keep track of the movements of neighboring groups by means of scent.

IV. Chemical composition of scent marks

Saddle-back tamarins discriminate between scent marks from males and females when the marks are offered underneath a screen, so that they can be sniffed but not contacted (Epple, 1978a). Cotton-top tamarins are capable of discriminating between scent marks from conspecifics and from saddle-backs under similar conditions, although contact with the scent material appears to be necessary for the perception of its full attractiveness (Belcher, Epple, Küderling & Smith, 1988). These results suggest that at least some of the informational content in the scent marks from both species is based on volatile compounds. Therefore, our initial studies on the chemical composition of the scent marks of both tamarins concentrated on the identification of volatile components.

Table 1.1. Compounds present in scent marks from *Saguinus fuscicollis (S.f)* and *Saguinus o. oedipus (S.o.)*

Class of compounds	S.f.	S.o.	Class of compounds		S.f.	S.o.
AROMATICS			BUTYRATE ESTERS			
p-CH₃O- ◎ CHO	+	+	C16:0	A	+	−
p-CH₃O- ◎ CH₂COOH	+	+	C18:1Δ9	B	+	−
			C18:1Δ11	C	+	−
ACETATE ESTERS			C18:0	D	+	−
C16:0	−	+	C20:2Δ11,14	E	+	−
			C20:1Δ11	Z		
AΓIΔT			C20:1Δ13	G	+	−
C8:0			C20:0	H	+	−
C10:0	+	−	C22:2Δ13, 16	I	+	−
C12:0	+	−	C22:1Δ13	J	+	−
C14:0	+	−	C22:1Δ15	K	+	−
C16:0	+	+	C22:0	L	+	−
C18:0	+	−	C24:2Δ15,18	O	+	−
			C24:1Δ15	P	+	−
STEROIDS			C24:1Δ17	Q	+	−
Cholesterol	+	+				
Estrone	+	+	HYDROCARBONS			
Estradiol	+	+	Squalene	N	+	+
Testosterone	+	?				
			TRIGLYCERIDES		+	+
			PROTEINS			
			MW range 6,000–200,000		+	+
STERYL ESTERS	+	+				

Chemical terminology: C24:2Δ15,18 signifies a 24 carbon moiety with two double bonds; one between carbons 15 and 16, the second between carbons 18 and 19.

Scent material for analytical studies was collected by allowing donors to scent-mark frosted glass plates (saddle-backs) or glass rods (cotton-tops). Male and female saddle-backs donated circumgenital and suprapubic marks, and combinations of both. Only female cotton-tops donated marks, providing anogenital marks only. The marked plates and rods were washed with a mixture of nanograde methanol and methylene chloride (1:3). The extracts were concentrated in a rotary evaporator under reduced pressure and dissolved in methanol or hexane. Gas chromatography (GC) and gas chromatography/mass spectrometry (GC/MS) were employed as the major tools for the analysis of the organic soluble material (Belcher *et al.*, 1986; Smith, Belcher, Epple, Jurs & Lavine, 1985; Yarger, Smith, Preti & Epple, 1977).

Table 1.1 summarizes our analytical findings of compounds currently known to be present in the scent marks from both tamarin species.

This includes the volatile constituents analyzed by means of GC/MS and other components whose presence in the marks was demonstrated recently by methods outlined below.

Our analytical studies have shown that the major volatile constituents of the scent marks from saddle-back tamarins are squalene, cholesterol and 15 esters of n-butyric acid (Fig. 1.3, Table 1.1). In addition to these compounds, which constitute approximately 90% of the volatile material of the marks of this species, a number of more highly volatile compounds, including fatty acids, are present in relatively low concentrations (Table 1.1). Gas chromatographic analyses performed over many years have documented the constant presence of squalene and the butyrates in scent marks from males and females of two subspecies of saddle-backs (*S. f. fuscicollis* and *S. f. illigeri*) and from hybrids between these subspecies. However, the presence of the more highly volatile

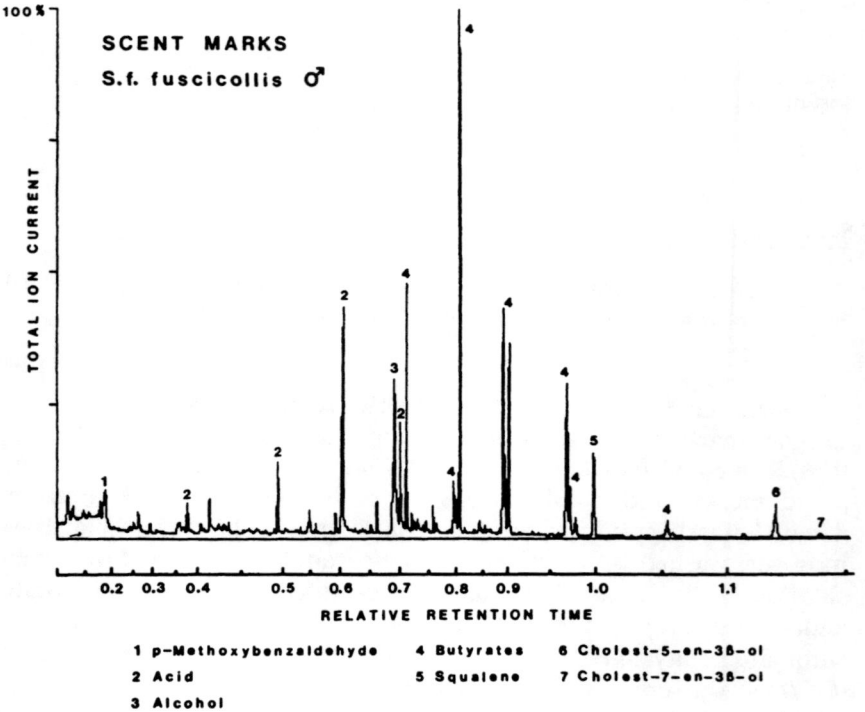

Fig. 1.3. Gas chromatogram of the major volatile constituents in scent marks from saddle-back tamarins. (From Belcher *et al.*, 1988)

components is more variable. To date, we have not been able to find qualitative differences in the chemical composition of the volatile material between scent marks from two subspecies and between scent marks from males and females (Yarger *et al.*, 1977; Belcher *et al.*,1986).

The anogenital scent marks from cotton-tops contain much lower concentrations of volatile compounds than the scent marks from saddle-backs. Estimations obtained by comparison of the squalene content of the scent marks from both species using a known concentration of squalene as an external standard indicate that the concentration of volatile material in the scent marks of saddle-backs is about four times higher than that in scent marks of cotton-tops (Belcher *et al.*, 1988).

Chemical analysis of the volatile components of scent marks from cotton-top tamarins reveals a much different profile from that found in saddle-backs (Fig. 1.4, Table 1.1). The relative concentrations of the

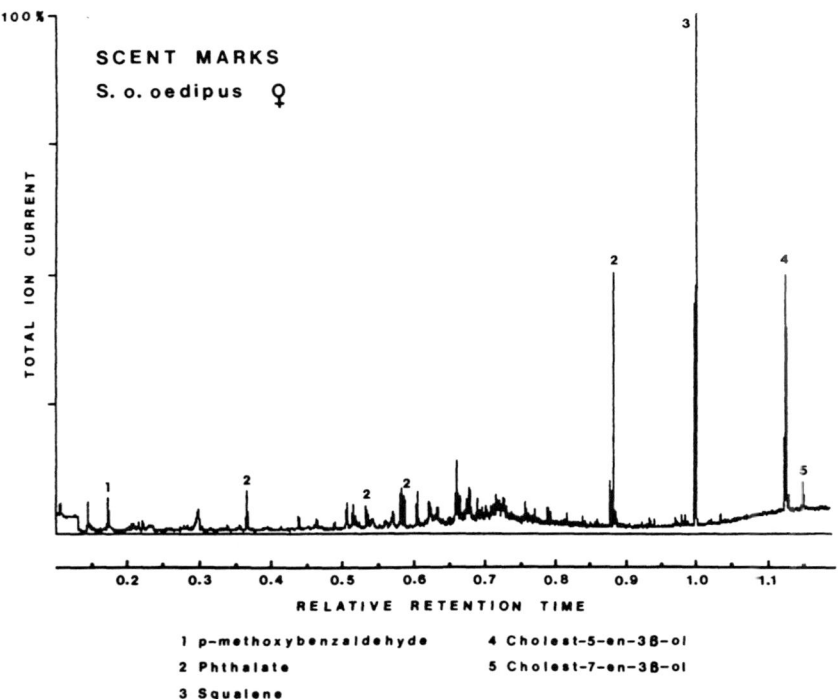

Fig. 1.4. Gas chromatogram of the major volatile constituents in scent marks from cotton-top tamarins. (From Belcher *et al.*, 1988)

volatile components in scent marks from cotton-top tamarins do not show a consistent pattern. Only the major components, squalene, cholesterol and p-methoxybenzaldehyde, were found in all scent mark samples that were investigated. All three are also present in the scent marks of the saddle-back tamarin (Belcher *et al.*, 1988). Squalene, cholesterol and p-methoxybenzaldehyde account for approximately 36% of the total volatile material in the scent marks of female cotton-tops. In addition, 10 other peaks were found in very low concentrations, their presence being quite variable.

Behavioral studies on both tamarin species suggest that volatile cues alone, although important, do not convey the full information content of the scent marks. Cotton-top tamarins reliably prefer scent marks from cotton-tops over scent marks from saddle-backs only when they can contact the stimuli. No discrimination is made when they are allowed to investigate scent marks presented under a screen (Belcher *et al.*, 1988). This result was obtained with a group of 12 subjects. Four of these subjects appeared to be more interested in the screened marks than the other eight animals. Subsequently, more extensive testing of these four tamarins showed that each individual was able to discriminate between the scent marks from the two donor species when tested with screened marks (Belcher *et al.*, 1988). These findings suggest that the cotton-tops can derive some information on species identity from volatile cues alone, but that additional compounds in the scent are necessary to elicit behavioral responses similar to those shown when the scent material is fully accessible to the monkeys (Belcher *et al.*, 1988). Unpublished studies on male-female discrimination in saddle-back tamarins suggest that in this species as well, volatile cues alone do not communicate the full informational content of the scent marks (see V. Structure-activity relationships).

These bioassay results clearly indicate that in both species, compounds of lower volatility are important for the full informational content and/or attractiveness of the scent marks. Therefore, we have begun to direct our attention to those constituents of the scent mark that are of relatively low volatility. The results will be reported in due course. Some preliminary analytical results are reported below.

Electrophoretic and high pressure liquid chromatography (HPLC) studies have shown that the scent marks from both species contain a number of proteins. These proteins range in molecular weight between 6000 and 200 000 daltons. In addition, thin layer chromatography of scent marks from both tamarin species has suggested the presence of steryl esters and triglycerides. In human skin lipid, free fatty acids are derived from triglycerides by the action of skin microorganisms, while many other species of mammals produce other classes of esters which are less readily hydrolyzed by bacteria (Albone, 1984). Although we know nothing about the biological role of the triglycerides in the scent marks of our tamarins, it is possible that they are the sources of the fatty acids found in the material, and that bacteria present on the surface of the scent glands are involved in their production. This notion is supported by the fact that in saddle-back tamarins, the scent-gland region supports a resident population of bacteria (Nordstrom, Belcher, Epple, Greenfield, & Smith, in press).

Additional compounds present in the scent marks of saddle-backs and cotton-tops are metabolites of the gonadal hormones estrone, estradiol and testosterone (Table 1.1). These metabolites have been documented in the urine of males and females of both species by means of radioimmunoassay (Epple & Katz, 1984; French, Abbott & Snowdon, 1984; Küderling & Epple, unpublished data). Estrone and estradiol were also detected in the scent marks of saddle-back females (Katz, unpublished data). The steroid metabolites probably are urinary constituents rather than products of the scent glands.

V. Structure-activity relationships

Our attempts to correlate the chemical constituents of the scent marks with their communicatory content have extended along two lines: fractionation studies and pattern recognition studies. Our fractionation studies are still in progress. Preliminary results are summarized below. The goal of fractionation studies on scent marks of saddle-backs is the isolation of compounds involved in gender identification; i.e., the attempt to obtain fractions of scent marks from males and females that can be discriminated from each other in a bioassay choice test. If during these choice tests the subjects contact, sniff and scent-mark fractions of material from males more frequently than fractions of material from females, we assume that gender identity is retained in the fraction. When a fraction is found to be active in terms of gender discrimination, its attractiveness is compared to the attractiveness of the natural scent marks of the same donor individual by means of choice tests.

For the purpose of fractionation, natural scent marks are collected on frosted glass plates. The marked plates are rinsed with methylene chloride: methanol (3:1). After rinsing, a thin film, visible on the plates, indicates the presence of residual material.

The methylene chloride-methanol fraction appears to contain much of the biological activity. After evaporation of this solvent, the material is dissolved in hexane or in methanol. Bioassay choice tests were conducted both with scent material dissolved in hexane and with material dissolved in methanol. In both cases, fractions of material from males are preferred over fractions of material from females, a response similar to that given to natural scent marks. However, when fractions of the marks from individual donors are tested against the chemically untreated marks from the same individual, natural scent marks are preferred over hexane fractions. Methanol appears to be a more suitable solvent than hexane in terms of retaining activity. No significant discriminatory responses are shown when plates carrying such fractions are tested against plates scent-marked by the same individual, although a strong trend toward a preference for natural marks is obvious, suggesting that even methanol fractions do not contain the full biological activity.

These findings show that even the initial step of dissolving natural scent material in organic solvent reduces its attractiveness. A recent study has shown that some of the material on which this attractiveness is based is contained in the residue remaining on the marked glass plate after this has been rinsed with methanol-methylene chloride (Epple, Belcher, Greenfield & Smith, unpublished data). For bioassay, the residue was combined with the methanol-methylene chloride soluble fraction of the scent mark and tested against the fraction alone. The combination of fraction and residue is more attractive to the tamarins than the fraction alone, eliciting higher levels of investigative behavior. However, the combination of residue and fraction is investigated more frequently only in those bioassay choice tests in which the subjects can contact the material directly. When both samples are presented under a screen so that they can be sniffed but not contacted, no discrimination is made. This result suggests that organic insoluble compounds of high molecular weight, which may not be perceived via olfaction, contribute to the full bioactivity of the scent marks.

Bioassays of gas chromatographic fractions of the hexane-soluble scent material were also performed. For this preparative work, the scent material was chromatographed through packed columns (SE-30, Carbowax or Dexsil on solid supports) and trapped via a Brownlee-Silverstein thermal gradient collector (Brownlee, Silverstein, Müller-Schwarze & Singer, 1969). When the total eluate is collected as one fraction, material from males elicits higher scent-marking responses than material from females. However, it is not preferred over material from females in terms of contacting and sniffing. Gas chromatographic fractionation of scent marks from males and from females into fractions containing the more highly volatile compounds fixed in squalene, and into fractions containing only the butyrates and squalene, results in a loss of biological activity in both fractions (Belcher *et al.*, 1986). These results indicate that gas chromtography further reduces the activity of the material and show that neither the highly volatile compounds by themselves, nor the butyrates and squalene alone, contain information on gender.

The studies reviewed above indicate that the organic soluble portion of the scent marks of saddle-back tamarins retains information on gender. However, even the initial fractionation by means of methylene chloride–methanol results in the loss of some of the biological activity of the scent. The methanol–methylene chloride extracts of the scent marks contain none of the proteins present in unfractionated marks. Electrophoretic studies show that the proteins remain in the filmy residue on the marked plate (Greenfield, unpublished data). The fact that this residue contributes to the attractiveness of the material suggests that the proteins might be responsible for the attractiveness of this fraction.

Because of the possibility that proteins contained in the residue play a role in determining the biological activity of scent marks, we are currently working with aqueous suspensions of scent material rather than with organic solvent fractions. Studies on aqueous suspensions have included both species of tamarins. Glass plates scent-marked by saddle-backs and glass rods scent-marked by cotton-tops are rinsed with deionized water, which has been saturated with nitrogen. Rinsing results in a cloudy suspension of scent material, leaving no residue on the glass surface. The suspensions are centrifuged for 15 minutes at 4000 rpm. The pellet, containing bacteria, cells from fruit consumed by the animals and other contaminants, is discarded. The supernatant is either tested as an aqueous suspension or freeze-dried, reconstituted with water and then bioassayed.

These bioassays show that saddle-back tamarins discriminate between aqueous suspensions of scent marks from males and from females. Cotton-top tamarins discriminate between aqueous suspensions of scent marks from conspecifics and from saddle-back tamarins. These discriminations are made with fresh pools as well as material that has been freeze-dried and reconstituted with water.

Parallel to our studies with fractionated scent material from both species, 'image guided' studies (Albone, 1984) have been conducted on scent material from saddle-back tamarins . As pointed out above, no qualitative differences between the volatile constituents of the scent marks from males and females and from our two subspecies have been detected. Therefore, we conducted quantitative studies to determine whether the scent marks from the two subspecies in our colony and those from males and females are characterized by different 'scent image' – like patterns of their major volatile constituents.

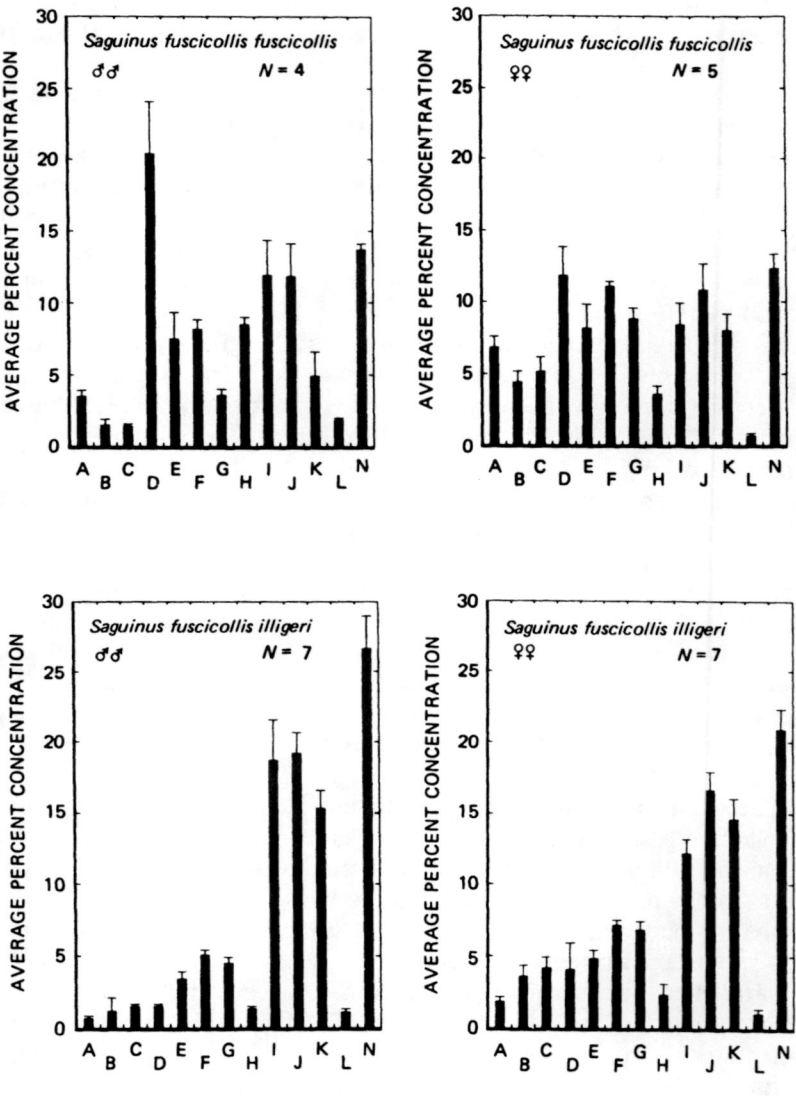

Fig. 1.5. Specific concentration profiles of squalene (N) and the butyrates (A-L) (see Table 1.1) in scent marks from males and females of two subspecies of *Saguinus fuscicollis*. (From Smith *et al.*, 1985, © American Association for the Advancement of Science.)

In an attempt to examine differences between *S. f. fuscicollis* and *S. f. illigeri*, as well as between males and females, computerized pattern recognition analyses were used to classify the concentration profiles of the major volatiles in the scent marks from individuals belonging to these four categories (Smith *et al.*, 1985). In general, pattern recognition methods are employed to classify patterns according to a particular set of measurements. The sample or pattern is represented as a point in *n*-dimensional space where the axes of this space are defined by the set of measurements used to describe the pattern.

In our study, pattern recognition techniques were employed to classify the individual scent prints according to the relative concentrations of 13 major volatile components of the scent marks. Means and standard errors for the relative concentrations of 12 butyrate esters (A–L) and squalene (N) were calculated for the scent print of each individual and these data were submitted to pattern recognition analysis. The results of the pattern recognition studies are shown in Figure 1.5. They document significant differences in the concentration profiles between males and females and between the two subspecies. In addition the scent prints of hybrids were different from those of each pure-bred subspecies (Epple *et al.*, 1979). These analyses demonstrated that the concentration patterns of the major volatile consitituents of the scent marks can be used mathematically to classify individuals as to their subspecies and gender (Smith *et al.*, 1985; Belcher *et al.*, 1986).

During the course of the pattern recognition studies, it became apparent that the relative concentrations of squalene and the butyrates in the scent marks of individual tamarins remained remarkably constant when monitored over extended periods of time. Figure 1.6, showing the 'scent print' of a hybrid male over a period of four years, documents this finding.

Pattern recognition methods are indeed suitable tools for the detection of differences between donor types in complex chemical signal mixtures. The existence of quantitative patterns of volatile compounds that are characteristic of subspecies and gender in these tamarins suggests that these compounds may be involved in structuring communicatory messages in the marks. However, the demonstration of a mathematical classification of the butyrates and squalene concentrations into subspecies and gender categories does not necessarily imply that this information is utilized by the animals for biological purposes.

We have begun to evaluate the behavioral significance of these butyrate-squalene concentrations. A number of studies were carried out in which we altered natural marks by adding synthetic butyrates in concentrations sufficient to change the total butyrate-squalene profiles,

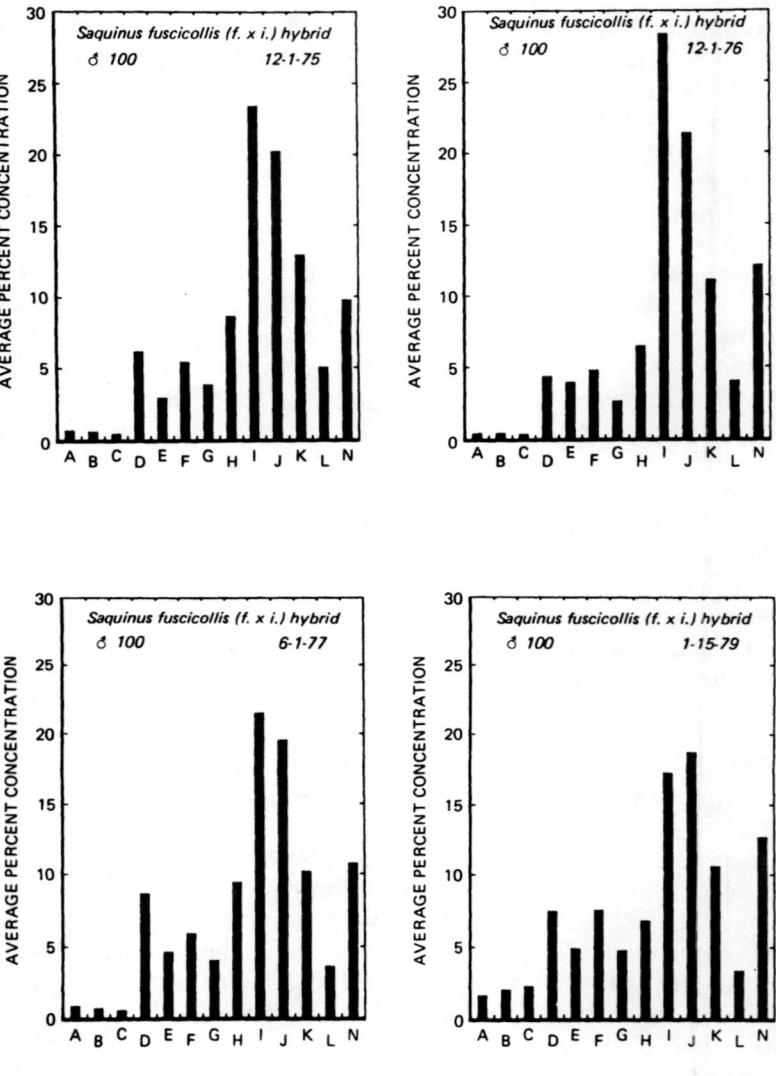

Fig. 1.6. Long-term stability in the concentration profiles of squalene (N) and the butyrates (A-L) (see Table 1.1) in scent marks from an adult male *Saquinus fuscicollis* hybrid. (From Smith *et al.*, 1985, © American Association for the Advancement of Science.)

or in which we employed entirely synthetic formulations of squalene and the butyrates which were synthesized in our laboratory (Golob, Yarger & Smith, 1979). These studies have recently been summarized by Belcher *et al.*, (1986). Their results provide no evidence that changes in the relative concentrations of squalene and the butyrates alone can change the stimulus characteristics of scent marks from one subspecies into those of scent marks from the other, or change the stimulus characteristics of scent marks from males into those from females. Moreover synthetic formulations, mimicking the butyrate-squalene concentration profiles of two individuals, were not discriminated from each other in bioassays using habituation techniques (Epple, Yarger & Smith, unpublished data). In an additional study, synthetic mixtures of squalene and all butyrates were formulated so that their GC profiles mimicked the profile of an individual male of each subspecies. In one bioassay, the subjects showed significantly more scent-marking in response to synthetic *S. f. fuscicollis* formulations, while sniffing and contacting both formulations with equal intensity (Epple *et al.*, 1979). However, the differences in scent-marking behavior, although significant, were small, and a replication of this experiment did not yield significant differences in response to the two formulations.

VI. Conclusions

Our behavioral studies indicate that scent marks from saddle-backs and probably also from cotton-tops communicate a large body of information about the individual that produces them. This informational complexity is paralleled by chemical complexity. Much of our chemical and behavioral work involving fractionations, work with synthetic and semi-synthetic mixtures and pattern recognition studies has shown that the signal content of the scent material is very likely based on overlapping qualitative and quantitative patterns of a variety of components.

In both species, the volatile constituents of the scent mark communicate some of its behaviorally significant information. Furthermore, in saddle-backs, gender and subspecies categories are mathematically characterized by specific concentration patterns of the major volatile constituents of the marks. However, these patterns alone do not communicate the identity of subspecies, gender or indivudual. We have evidence that non-volatile as well as volatile components are necessary to fully constitute the chemical messages. These components may derive from several sources, including glandular secretion, urine, genital discharge and possibly products of bacterial action. It appears that all of these components together present the total 'Gestalt' or 'image' which contains encoded messages of a wide variety. This total image reflects a quite complex system which was evolved to contain many communicatory messages.

Our future studies will concentrate on obtaining a better understanding of the chemical complexity of the material from both species. This knowledge will be necessary before we can target additional compounds or patterns of compounds of biological significance.

Acknowledgments

The long-term support of our research by the US National Science Foundation is gratefully acknowledged. Our more recent studies were supported by the US National Institutes of Health (RO1 NS-21790) and the Deutsche Forschungsgemeinschaft.

References

Albone, E. S. (1984). *Mammalian semiochemistry*. John Wiley, New York.

Belcher, A. M., Epple, G., Kostelc, J. G., & Smith III, A. B. (1982). Changes in chemical composition due to aging in the scent marks of a primate, *Saguinus fuscicollis* (abstract). Paper presented at the 4th Annual Meeting, Association for Chemoreception Sciences, Sarasota, Florida.

Belcher, A. M., Smith III, A. B., Jurs, P. C., Lavine, B., & Epple, G. (1986). Analysis of chemical signals in a primate species (*Saguinus fuscicollis*): use of behavioral, chemical and pattern recognition methods. *Journal of Chemical Ecology* 12, 513–531.

Belcher, A. M., Epple, G., Küderling, I., & Smith III, A. B. (1988). Volatile components of scent material from the cotton-top tamarin (*Saguinus o. oedipus*): a chemical and behavioral study. *Journal of Chemical Ecology* 19, 1367.

Brown, R. E., & Macdonald, D. W. (eds) (1985). *Social odours in mammals*. Oxford University Press, Oxford.

Brownlee, R. G., Silverstein, R. M., Müller-Schwarze, D., & Singer, A. G. (1969). Isolation, identification and function of the chief component of the male tarsal scent in black-tailed deer. *Nature* 221, 284–285.

Crewe, R. M., Burger, B. V., Le Roux, M., & Katsir, Z. (1979). Chemical constituents of the chest gland secretion of the thick-tailed galago (*Galago crassicaudatus*). *Journal of Chemical Ecology* 5, 861–868.

Crump, D. R., Swigar, A. A., West, J. R., Silverstein, R. M., Müller-Schwarze, D., & Altieri, R. (1984). Urine fractions that release flehmen in black-tailed deer, *Odocoileus hemionus columbianus*. *Journal of Chemical Ecology* 10, 203–215.

Duvall, D., Müller-Schwarze, D., & Silverstein, R. M., eds (1986). *Chemical signals in vertebrates — ecology, evolution and comparative biology*, vol. 4. Plenum Press, New York.

Ebling, J. F. (1977). Hormonal control of mammalian skin glands. In D. Müller-Schwarze & M. M. Mozell, eds, *Chemical signals in vertebrates*, vol. 1, pp. 17–38. Plenum Press, New York.

Epple, G. (1974). Pheromones in primate reproduction and social behavior. In W. Montagna & W. A. Sadler, eds, *Reproductive behavior*, pp. 131–155. Plenum Press, New York.

Epple, G. (1978*a*). Studies on the nature of chemical signals in scent marks and urine of *Saguinus fuscicollis* (Callitrichidae, Primates). *Journal of Chemical Ecology* **4**, 383–394.

Epple, G. (1978*b*). Lack of effects of castration on scent marking, displays and aggression in a South American Primate (*Saguinus fuscicollis*). *Hormones & Behavior* **11**, 139–150.

Epple, G. (1979). Gonadal control of male scent in the tamarin, *Saguinus fuscicollis* (Callitrichidae, Primates). *Chemical Senses & Flavour* **4**, 15–20.

Epple, G. (1980). Relationships between aggression, scent marking and gonadal state in a primate, the tamarin *Saguinus fuscicollis*. In D. Müller-Schwarze & R. M. Silverstein, eds, *Chemical signals — vertebrates and aquatic invertebrates*, vol. 2, pp. 87–105. Plenum Press, New York.

Epple, G. (1981). Effects of prepubertal castration on the development of the scent glands, scent marking and aggression in the saddle-back tamarin (*Saguinus fuscicollis*, Callitrichidae, Primates). *Hormones & Behavior* **15**, 54–67.

Epple, G. (1982). Effects of prepubertal ovariectomy on the development of scent glands, scent marking and aggressive behaviors of female tamarin monkeys (*Saguinus fuscicollis*). *Hormones & Behavior* **16**, 330–342.

Epple, G. (1986). Communication by chemical senses. In G. Mitchell & J. Erwin, eds, *Comparative primate biology — behavior, conservation and ecology*, vol. 2, pp. 531–580. Alan Liss, New York.

Epple, G., & Katz, Y. (1984). Social influences on estrogen excretion and ovarian cyclicity in saddle-back tamarins (*Saguinus fuscicollis*). *American Journal of Primatology* **6**, 215–227.

Epple, G., Golob, N. F., & Smith III, A. B. (1979). Odor communication in the tamarin *Saguinus fuscicollis*. In F. J. Ritter, ed., *Odour communication in animals*, pp. 117–130. Elsevier, North Holland, Amsterdam.

Epple, G., Alveario, M. C., Golob, N. F., & Smith III A. B. (1980). Stability and attractiveness related to age of scent marks of saddle-back tamarins (*Saguinus fuscicollis*). *Journal of Chemical Ecology* **6**, 735–748.

Epple, G., Belcher, A. M., & Smith III, A. B. (1986). Chemical signals in callitrichid monkeys — a comparative review. In D. Duvall, D. Müller-Schwarze & R. M. Silverstein, eds, *Chemical signals in vertebrates — ecology, evolution and comparative biology*, vol. 4, pp. 653–672. Plenum Press, New York.

Epple, G., Alveario, M. C., Belcher, A. M., & Smith III, A. B. (1987). Species and subspecies specificity in urine and scent marks of saddle-back tamarins (*Saguinus fuscicollis*). *International Journal of Primatology* **8**, 663–680.

Epple, G., Küderling, I., & Belcher, A. M. (1988). Some communicatory functions of scent marking in the cotton-top tamarin (*Saguinus oedipus oedipus*). *Journal of Chemical Ecology* **14**, 503–515.

French, J.A. , & Cleveland, J. (1984). Scent marking in the tamarin, *Saguinus oedipus*: sex differences and ontogeny. *Animal Behavior* **32**, 615–623.

French, J.A., & Snowdon, C. T. (1981). Sexual dimorphism in intergroup spacing behavior in the tamarin, *Saguinus oedipus*. *Animal Behavior* **29**, 822–829.

French, J. A., Abbott, D. H., & Snowdon, C. T. (1984). The effect of social environment of estrogen excretion, scent marking and socio-sexual behavior in tamarins (*Saguinus oedipus*). *American Journal of Primatology* **6**, 155–167.

Golob, N. F., Yarger, R. G., & Smith III, A. B. (1979). Primate chemical communication. Part III: synthesis of the major volatile constituents of the marmoset (*Saguinus fuscicollis*) scent mark. *Journal of Chemical Ecology* **5**, 543–555.

Halpin, Z. T. (1986). Individual odors among mammals: origins and functions. *Advances in the Study of Behavior* **16**, 39–70.

Hershkovitz, P. (1978). *Living new world monkeys (Platyrrhini)*. The University of Chicago Press, Chicago.

Keverne, E. B. (1978). Olfactory cues in mammalian behaviour. In J. B. Hutchison, ed., *Biological determinants of sexual behavior*, pp. 727–763. Wiley, New York.

Marchlewska-Koj, A. (1984). Pheromones and mammalian reproduction. *Oxford Review of Reproductive Biology* **6**, 266–302.

Melrose, D. R., Reed, H. C. B., & Patterson, R. L. S. (1971). Androgen steroids associated with boar odour as an aid to the detection of oestrus in pig artificial insemination. *British Veterinary Journal* **127**, 497–502.

Müller-Schwarze, D., & Silverstein, R. M., eds, (1980). *Chemical signals — vertebrates and aquatic invertebrates*, vol. 2. Plenum Press, New York.

Müller-Schwarze, D., & Silverstein, R. M., eds, (1983). *Chemical signals in vertebrates*, vol. 3. Plenum Press, New York.

Müller-Schwarze, D., Morehouse, L., Corradi, R., Cheng-hua, Z., & Silverstein, R. M. (1986). Odor images: responses of beaver to castoreum fractions. In D. Duvall, D. Müller-Schwarze & R. M. Silverstein, eds, *Chemical signals in vertebrates — ecology, evolution and comparative biology*, vol. 4, pp. 561–570. Plenum Press, New York.

Nordstrom, K. M., Belcher, A. M., Epple, G., Greenfield, K. L., & Smith III, A. B. (in press). Microbial flora of the skin of the saddle-back tamarin monkey, *Saguinus fuscicollis*. *Journal of Chemical Ecology*.

Novotny, M., Harvey, S., Jemiolo, B., & Alberts, A. (1985). Synthetic pheromones that promote inter-male aggression in mice. *Proceedings of the National Academy of Science (USA)* **82**, 2059–2061.

Perkins, E. M. (1966). The skin of the black-collared tamarin (*Tamarinus nigricollis*). *American Journal of Physical Anthropology* **25**, 41–69.

Perkins, E. M. (1969). The skin of the cotton-top pinche *Saguinus* (=*Oedipomidas*) *oedipus. American Journal of Physical Anthropology* **30**, 13–27.

Raymer, J., Wiesler, D., Novotny, M., Asa, C., Seal, U. S., & Mech, L. D. (1985). Chemical investigations of wolf (*Canis lupus*) anal-sac secretion in relation to breeding season. *Journal of Chemical Ecology* **11**, 593–608.

Raymer, J., Wiesler, D., Novotny, M., Asa, C., Seal, U. S., & Mech, L. D. (1986). Chemical scent constituents in urine of wolf (*Canis lupus*) and their dependence on reproductive hormones. *Journal of Chemical Ecology* **12**, 291–314.

Schaffer, J. (1940). *Die Hautdrüsenorgane der Saügetiere*, p. 464. Urban und Schwarzenberg, Berlin.

Schilling, A. (1979). Olfactory communication in prosimians. In G. A. Doyle, & R. D. Martin, eds, *The study of prosimian behavior*, pp. 542–561. Academic Press, New York.

Schilling, A., & Perret, M. (1987). Chemical signals and reproductive capacity in male prosimian primates (*Microcebus murinus*). *Chemical Senses* **12**, 143–158.

Singer, A. G., Clancy, A. N., Macrides, F., & Agosta, W. C. (1984). Chemical studies of hamster vaginal discharge: male behavioral responses to a high molecular weight fraction require physical contact. *Physiology & Behavior* **33**, 645–651.

Singer, A. G., Macrides, F., Clancy, A. N., & Agosta, W. C. (1986). Purification and analysis of a proteinaceous aphrodisiac pheromone from hamster vaginal discharge. *Journal of Biological Chemistry* **261**, 13 323–13 326.

Smith III, A. B., Belcher, A. M., Epple, G., Jurs, P. C., & Lavine, B. (1985). Computerized pattern recognition: a new technique for the analysis of chemical communication. *Science* **228**, 175–177.

Stoddart, D. M. (1980). *The ecology of vertebrate olfaction.* Chapman and Hall, London.

Terborgh, J. (1983). *The behavioral ecology of five new world primates.* Princeton University Press, Princeton.

Vandenbergh, J. G., ed. (1983). *Pheromones and reproduction in mammals.* Academic Press, New York.

Wheeler, J. W., Blum, M. S., & Clark, A. (1977). β-(p-hydroxyphenyl) ethanol in the chest gland secretion of a galago (*Galago crassicaudatus*). *Experientia* **33**, 988–989.

Wislocki, G. B. (1930). A study of scent glands in the marmosets, especially *Oedipomidas geoffroyi. Journal of Mammalogy* **11**, 475–483.

Yarger, R. G., Smith III, A. B., Preti, G., & Epple, G. (1977). The major volatile constituents of the scent mark of a South American primate, *Saguinus fuscicollis,* Callitrichidae. *Journal of Chemical Ecology* **3**, 45–56.

Zeller, U., Epple, G., Küderling, I., & Kuhn, H. J. (1988). The anatomy of the circumgenital scent gland of *Saguinus fuscicollis* (Callitrichidae, Primates). *Journal of Zoology* (London) **214**, 141–156.

2

Stimulants of feeding behavior in marine organisms: receptor and perireceptor events provide insight into mechanisms of mixture interactions

William E. S. Carr, Henry G. Trapido-Rosenthal
and Richard A. Gleeson

*C. V. Whitney Laboratory and Department of Zoology,
University of Florida, St Augustine, Florida USA*

I. Introduction

In marine organisms, it is well established that external chemicals control or influence many behavioral responses, including those associated with the recognition and ingestion of food (Carr & Derby, 1986b), avoidance of predators (Mackie & Grant, 1974; Sleeper, Paul & Fenical, 1980), sexual recognition (Atema & Engstrom, 1971; Gleeson, 1980), discharge of gametes (Müller, Gassmann & Luning, 1979), and selection of a suitable substratum for metamorphosis by larvae (Morse, 1984). For general reviews of chemically stimulated behavior in marine organisms, see Mackie & Grant (1974), Gleeson (1978), Atema (1985), and Carr (1987).

  In this chapter we describe the types of substances that evoke aspects of feeding behavior in marine animals. The behavioral studies demonstrate the importance of nucleotides as feeding stimulants, and show that the stimulatory capacity of food extracts is often due to a mixture of substances. It is also shown that behavioral responses to defined mixtures are indicative of synergistic mixture interactions.

  A nucleotide-sensitive chemosensory system in the olfactory organ of the spiny lobster is described to demonstrate that in addition to receptor sites activated by nucleotides, the olfactory organ also contains perireceptor components including enzymes that dephosphorylate nucleotides and an uptake system that internalizes the dephosphorylated product,

*Copyright © 1989 by Academic Press Australia.
All rights of reproduction in any form reserved.*

Table 2.1. Studies of chemically stimulated feeding behavior in marine animals in which a substantial part of the activity of an extract was mimicked with artificial solutions

Animal studied[a]	Substances required to mimic activity of extract	Source of extract	Behavioral response
A. FISH			
	Single substances		
1. Dover sole	Betaine or betaine + glycine	Mussel	Ingest pellet
2. Pigfish	Betaine	Oyster	Biting response
3. Turbot	Inosine	Squid	Ingest pellet
	Mixture of substances		
4. Rainbow trout	Amino acids	Squid	Ingest pellet
5. Japanese eel	Amino acids	Clam	Attraction
6. European eel	Amino acids	Squid	Ingest ration
7. Red sea bream	Amino acids	Worm	Ingest pellet
8. Cod	Amino acids	Shrimp	Bottom food search
9. Pigfish	Amino acids and betaine	Crab, shrimp	Biting response
10. Pinfish	Amino acids and betaine	Shrimp	Biting response
11. Puffer	Amino acids and betaine	Clam	Ingest pellet
12. Plaice	Amino acids, quaternary amines, nucleotides, nucleosides and lactate	Squid	Ingest pellet
B. CRUSTACEANS			
13. Lobster	Amino acids, quaternary amines, nucleotides, nucleosides and lactate	Squid	Attraction
14. Shrimp	Amino acids, quaternary amines, nucleotides, nucleosides and lactate	Crab, shrimp	Grasp delivery device

Modified from Carr & Derby, 1986b.

[a] References for animal number cited: 1. Mackie, Adron & Grant, 1980; Mackie & Mitchell, 1982. 2. Carr, Blumenthal & Netherton, 1977. 3. Mackie & Adron, 1978. 4. Mackie, 1982. 5. Hashimoto, Konosu, Fusetani & Nose, 1968. 6. Mackie & Mitchell, 1983. 7. Fuke, Konosu & Ina, 1981. 8. Ellingsen & Døving, 1986. 9. Carr, 1976; Carr *et al.*, 1977. 10. Carr & Chaney, 1976. 11. Ohsugi, Hidaka & Ikeda, 1978; Hidaka, 1982. 12. Mackie, 1982. 13. Mackie, 1973. 14. Carr & Derby, 1986a.

adenosine. Potential contributions of these perireceptor components to the chemosensory process are described. Knowledge of the biochemical systems whereby chemostimulants are altered and internalized by a chemosensory organ could make an important contribution to our eventual understanding of the mechanisms underlying certain types of mixture interactions.

II. Substances in foods that stimulate feeding behavior

For marine animals, potent stimulants of feeding behavior are often obtained by preparing aqueous extracts of acceptable food organisms. Table 2.1 summarizes the results of behavioral studies in marine animals in which a significant portion of the stimulatory capacity of an aqueous extract was mimicked with artificial solutions containing substances identified in the extract. Note that in each example the principal stimulants of feeding behavior are common low molecular weight substances including amino acids, quaternary ammonium compounds, nucleotides and organic acids. Moreover, in 11 of the 14 examples cited in the table, the stimulatory capacity of the extract was due to a mixture of substances rather than to a single dominant substance in the extract.

III. Responses to single and multiple food components

Quantitative behavioral procedures were used to identify food components serving as chemoattractants of the glass shrimp, *Palaemonetes pugio* (for procedural details, see Carr & Derby, 1986a). Analyses of substances present in food extracts served as 'recipes' for identifying individual chemostimulants and for measuring the effectiveness of complex artificial mixtures having compositions similar to those of natural tissues (see Table 2.2).

When tested individually, only six of the 31 substances identified in the two extracts shown in Table 2.2 functioned as chemoattractants of the glass shrimp (Fig. 2.1A). The importance of nucleotides as behavioral stimulants is evident since three of the attractants are the nucleotides adenosine 5'-monophosphate (AMP), adenosine 5'-diphosphate (ADP) and inosine 5'-monophosphate (IMP). Further, AMP is the single most potent behavioral stimulant identified in the extracts, being greater than 100-times more potent than the second most potent substance, ADP.

Comparisons of the dose-response (D-R) curves obtained for AMP alone, and for artificial mixtures containing all of the substances identified in crab or shrimp tissue, might suggest that the stimulatory capacity of each mixture is due primarily to the AMP it contains (Fig. 2.1A). However, each mixture has a stimulatory capacity that is 18 to 400 times greater than is predicted on the basis of the small amount of AMP it

Table 2.2.   Concentration of low molecular weight substances in aqueous extracts of crab and shrimp (from Carr & Derby, 1986*a*).

Substance	Concentration (mM)	
	Crab	Shrimp
I. AMINO ACIDS		
Alanine	2.14	3.54
Arginine	2.29	1.47
Asparagine	0.195	0.167
Aspartic acid	0.036	0.138
Cysteine	0.120	
Glutamic acid	0.284	0.275
Glutamine	2.95	1.05
Glycine	11.34	15.62
Histidine	0.138	0.054
Isoleucine	0.107	0.176
Leucine	0.245	0.317
Lysine	0.227	0.081
Methionine	0.392	0.173
Phenylalanine	0.093	0.090
Proline	5.57	1.90
Serine	0.265	0.224
Taurine	1.88	5.50
Threonine	0.404	0.106
Tryptophan	0.131	
Tyrosine	0.101	0.153
Valine	0.318	0.384
Total	29.226	31.418
II. NUCLEOTIDES,-SIDES AND RELATED		
Adenosine 5'-monophosphate	0.025	0.813
Adenosine 5'-diphosphate	0.182	0.238
Adenosine 5'-triphosphate	0.878	0.066
Inosine 5'-monophosphate	0.158	0.333
Hypoxanthine	0.033	0.038
Inosine	0.069	0.010
Total	1.345	1.463
III. QUATERNARY AMMONIUM COMPOUNDS		
Betaine	3.70	8.40
Homarine	0.469	1.15
Trimethylamine oxide	3.38	9.0
Total	7.549	18.55
IV. ORGANIC ACID		
Lactic acid	4.35	4.58
GRAND TOTAL	42.47	56.06

contains (Fig. 2.1, B and C). Hence it is clear that additional substances make major contributions to the stimulatory capacity of each mixture.

The question of whether the complex mixtures shown in Table 2.2 express themselves in an additive or an interactive manner was addressed by Carr and Derby (1986*a*, *b*). In these studies, the D-R curves obtained from the behavioral assays of the mixtures and the individual substances were compared with the D-R curves predicted by applying the additive methods of response summation and stimulus summation. It was proposed that synergistic mixture interactions occurred, since each mixture was far more effective than predicted by either of the additive methods.

As discussed by Carr and Derby (1986*a*, *b*), synergistic interactions provide an organism with a capacity to detect and respond to mixtures at concentrations that are subthreshold for each of the individually stimulatory components in the mixture. A mechanism for broadening an organism's range of sensitivity, by extending that range downward to concentrations below the response thresholds for individual substances, should be very advantageous to animals whose survival is highly dependent upon the detection and response to external chemical signals.

We have not yet identified the specific substances that contribute to the synergistic interactions described above. However, purine nucleotides are likely candidates for interactions, because in addition to being major constituents of the mixtures, they are also known for their synergistic or potentiating effects on the response of the vertebrate gustatory system to glutamate (Sato, Yamashita & Ogawa, 1970; Rifkin & Bartoshuk, 1980; Cagan, Torii & Kare, 1979; Torii & Cagan, 1980) and to other amino acids (Yamaguchi, 1979; Yoshii, Yokouchi & Kurihara, 1986). Moreover, the purine nucleotides AMP and ATP are known to function in internal tissues as co-transmitters or modulators that effect the activity of other putative neurotransmitters (Stone, 1983; Burnstock, 1985; Richardson, Brown, Bailyes & Luzio, 1987).

IV. Properties of nucleotide-sensitive cells in the olfactory organ of the spiny lobster

To define the physiological properties of chemoreceptor cells sensitive to purine nucleotides, the Florida spiny lobster (*Panulirus argus*) was employed because of its large size and exceptional suitability for electrophysiological studies (Ache, 1982). The functional organization of the olfactory system in this animal has been reviewed recently by Ache and Derby (1985). The olfactory organ is composed of a tuft of sensilla, called aesthetasc sensilla, present on the lateral filament of the antennule. Each antennule has about 2000 of these olfactory sensilla, with each sensillum being innervated by the ciliated dendrites of an estimated 320

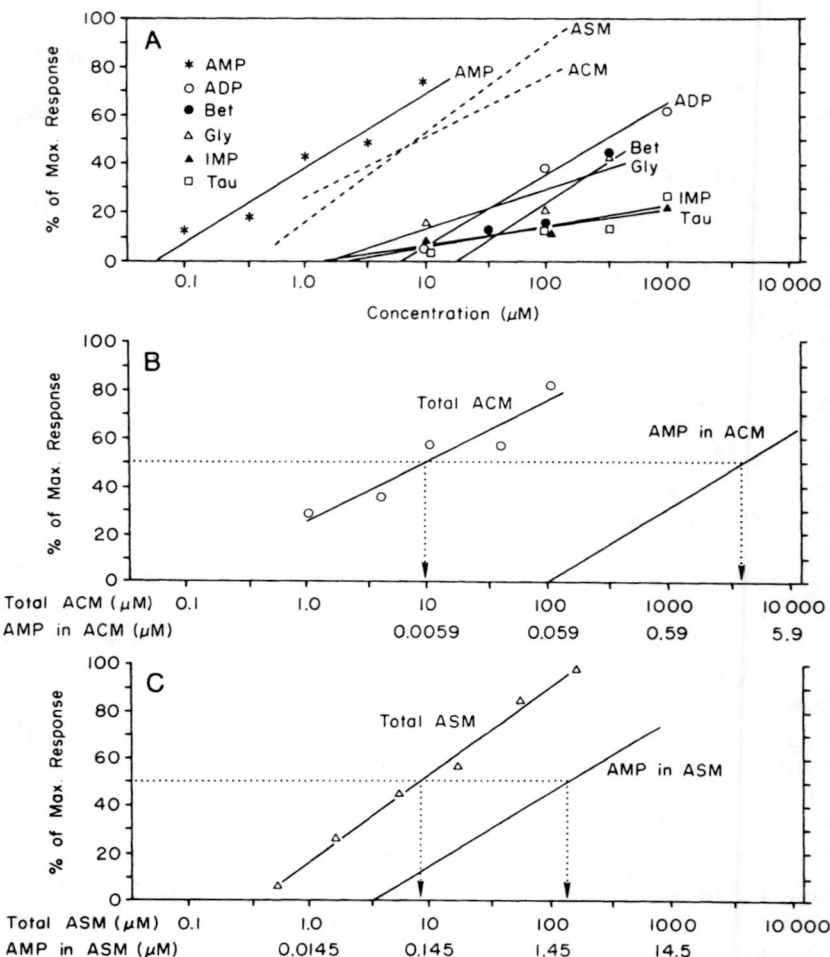

Fig. 2.1. Comparisons of the dose-response functions for individual substances and complex mixtures tested behaviorally with the shrimp, *Palaemonetes pugio*. A response is expressed as a percentage of the maximum response obtained with any stimulus. Individual substances: adenosine 5′-monophosphate (AMP), adenosine 5′-diphosphate (ADP), betaine (Bet), glycine (Gly), inosine 5′-monophosphate (IMP), and taurine (Tau). The mixtures are an artificial shrimp mixture (ASM) and an artificial crab mixture (ACM) formulated according to the analysis of extracts shown in Table 2.2. **A.** AMP is the single most effective substance. On a molar basis, AMP is also more effective than either artificial mixture. **B and C.** Each mixture (ACM or ASM) is more effective than predicted on the basis of the amount of AMP it contains. On the horizontal axes of **B** and **C**, two concentrations are given: total concentration of solute in the mixture (upper); concentration of AMP present in each defined concentration of the mixture (lower). Horizontal dotted lines compare ED_{50} values (= concentration evoking a 50% maximum response) for the mixture and for AMP in the mixture. (Data from Carr & Derby 1986*a*.)

bipolar sensory neurons (Laverack & Ardill, 1965; Grünert & Ache, 1988).

Physiological studies have shown that the antennule of the spiny lobster contains several populations of chemosensory cells with narrow response spectra; some of these cells are stimulated primarily by specific amino acids or specific nucleotides found in food organisms (Fuzessery, Carr & Ache, 1978; Derby & Atema, 1987). Included among these 'specialist-type' olfactory cells are three populations activated primarily by purine nucleotides; these cells are referred to hereafter as purinergic cells.

One population of purinergic cells has 'AMP-best' receptors showing a potency sequence of AMP > ADP > ATP or adenosine (Fig. 2.2A) (Derby, Carr & Ache, 1984). The second population of purinergic cells has 'ATP-best' receptors showing a potency sequence of ATP > ADP > AMP or adenosine (Fig. 2.2B) (Carr, Gleeson, Ache & Milstead, 1986). The third population shows an 'ADP-best' response at concentrations greater than 10 μM, but is about equally stimulated by ADP, AMP and ATP at lower concentrations (Fig. 2.2C). It is important to note that the non-phosphorylated nucleoside, adenosine, is virtually non-stimulatory to all three types of purinergic chemoreceptors (see Fig. 2.2). It is also significant that the removal of the terminal phosphate group from either ATP, ADP or AMP produces in each case a product that is far less effective for the ATP-, ADP- or AMP-best cells, respectively.

V. Biochemical fate of nucleotides in the olfactory organ

Getchell, Margolis & Getchell (1984) introduced the term 'perireceptor event' as a descriptor of an event or process contributing to the movement of chemostimulants into or out of the immediate receptor environment. As shown below, perireceptor events occurring within the olfactory organ of the spiny lobster affect the residence time and structural integrity of chemostimulants.

The nucleotides AMP and ATP, the major excitants for two populations of antennular purinergic receptors, undergo rapid extracellular dephosphorylation by enzymes in the olfactory sensilla. The dephosphorylated product, adenosine, is then internalized by a specific nucleoside uptake system (Trapido-Rosenthal, Carr & Gleeson, 1987). The dephosphorylation of both AMP and ATP occurs prior to the uptake step. This sequence was demonstrated by incubating sensilla with ring-labelled ^{3}H-AMP or ^{3}H-ATP, and with phosphate-labelled ^{32}P-AMP or ^{32}P-ATP. In both cases, ^{3}H was rapidly internalized, whereas ^{32}P was not (Fig. 2.3, A and B). In the case of AMP, we know that the dephosphorylation step occurs very rapidly because ^{3}H in ring-labeled AMP is internalized as rapidly as ^{3}H in adenosine (Fig. 2.3C).

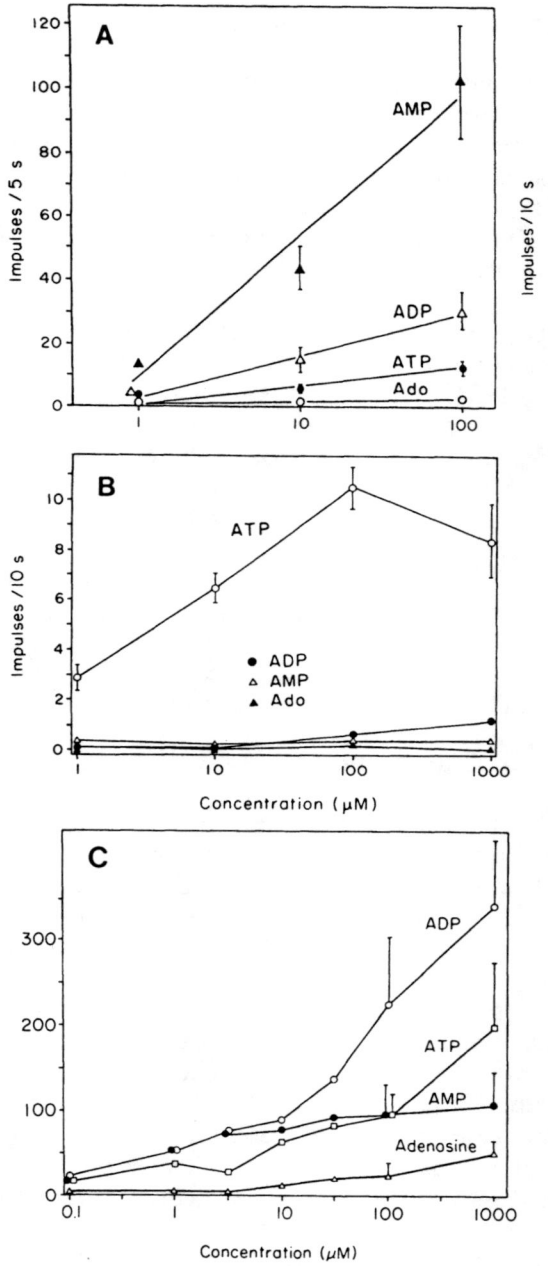

Fig. 2.2.   Physiological dose-response relationships for adenine nucleotides and adenosine in three populations of antennular chemosensory cells of the spiny lobster. **A.** AMP-best cells. (From Derby *et al.* (1984). **B.** ATP-best cells. **C.** ADP-best cells (N = 5). Data points are means (± SEM).

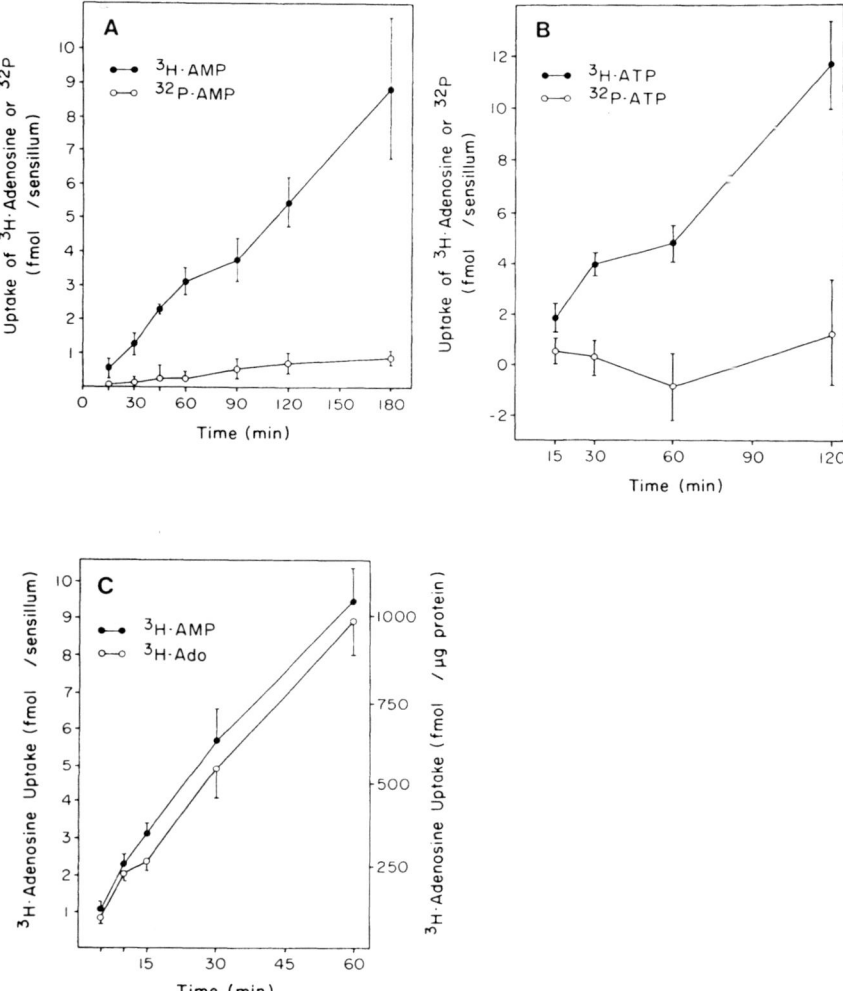

Fig. 2.3. Demonstration that sensilla of the spiny lobster dephosphorylate nucleotides prior to the uptake of adenosine. **A and B.** Sensillar uptake of ^{3}H (•) and ^{32}P (O) from labeled AMP **(A)** and ATP **(B)**. Sensilla are incubated with 0.1 μM AMP or 0.5 μM ATP labeled with tritium in the adenine ring, and with ^{32}P at the α-phosphate of AMP and the γ-phosphate of ATP. With both nucleotides, ^{3}H in the adenine ring is rapidly internalized whereas the ^{32}P-labeled phosphate group is not. **C.** Uptake of ^{3}H by sensilla incubated in 0.5 μM ^{3}H-AMP (•) or 0.5 μM ^{3}H-adenosine (O). Tritium from adenine-labeled AMP is internalized as rapidly as tritium from adenine-labeled adenosine. All points are means (± SEM). (From Trapido-Rosenthal, Carr & Gleeson, 1987.)

Recent results indicate that at AMP concentrations of 1 to 10 μM, less than 500 ms is required to dephosphorylate half of the AMP molecules entering the extracellular lymph of active sensilla (Trapido-Rosenthal, Gleeson & Carr, unpublished data). This rapid rate of dephosphorylation appears to correlate well with the antennular 'flicking' mechanism used by the lobster to obtain intermittent samples of the external chemical environment (e.g. Schmitt & Ache, 1979). Since antennular flicks occur at intervals of 500 to 2000 ms (Schmitt & Ache, 1979), a nucleotide half-life of less than 500 ms indicates that much of the AMP entering the sensillar lymph during an antennular flick will be dephosphorylated prior to the next flick.

VI. Mechanisms whereby perireceptor events may contribute to interactive effects

A. Dephosphorylation of nucleotides

The perireceptor events involved with the dephosphorylation of nucleotides may affect the chemosensory process in several ways. As illustrated in Figure 2.4, the dephosphorylations create major changes in the structure of excitatory molecules after contact is made with the chemosensory organ. As a consequence, substances are produced that are structurally and functionally different from those initially introduced into the receptor environment. As described below, the loss of a single phosphate group from ATP, ADP or AMP may have several immediate, and simultaneous, effects.

1. With ATP, ADP or AMP, removal of a single phosphate group serves as an *inactivation step* that eliminates the major excitant for a particular 'best' receptor type.
2. With ATP or ADP, dephosphorylation also generates *new excitants* (e.g. ADP and AMP) for additional major receptor types.
3. Recent studies also show that the dephosphorylation of either ATP or ADP generates *suppressants* of one of the major receptor types, since both ADP and AMP serve as potent antagonists of the ATP-best cells (Gleeson, Trapido-Rosenthal & Carr, unpublished data).

Adenosine, the completely dephosphorylated product, is virtually non-stimulatory to any of the purinergic chemoreceptors (see Fig. 2.2). However, even adenosine may be more than an inert, 'inactivated product', since adenosine is known to have receptor-mediated effects in internal tissues (Burnstock & Brown, 1981; Richardson *et al.*, 1987) and

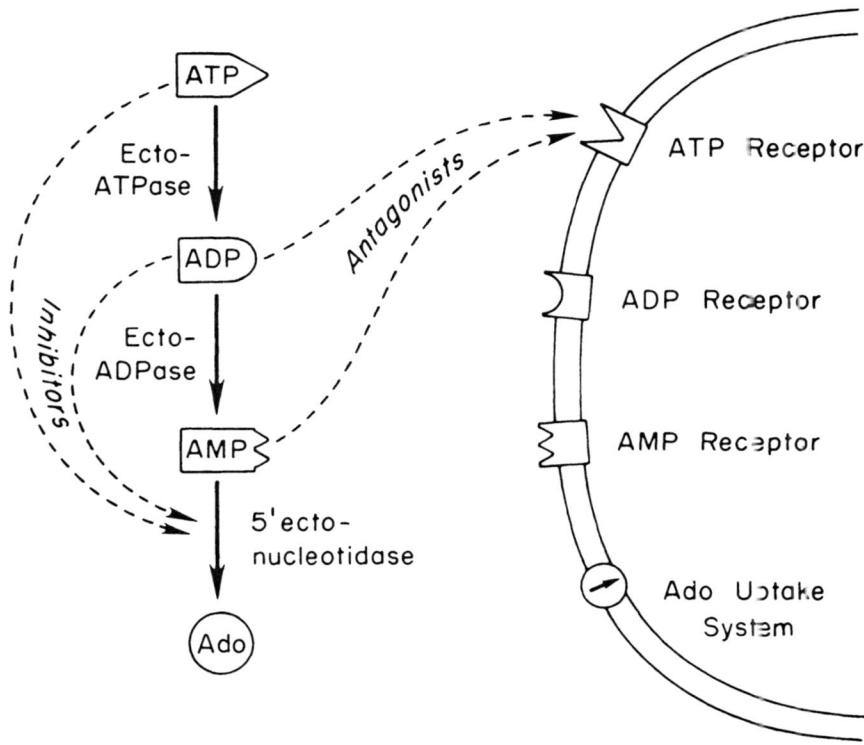

Fig. 2.4. Cartoon depicting perireceptor events that may affect the chemosensory process by changing the structure of excitatory nucleotides. Ectoenzymes in the olfactory sensilla dephosphorylate ATP, ADP and AMP in a stepwise manner to generate products that may be either more or less excitatory to the specific types of purinergic receptors (see Fig. 2.2). Adenosine, the final, completely dephosphorylated product, is virtually non-stimulatory to any of the receptor types. Dotted lines indicate processes for which antagonists or inhibitors have been identified. Additional details are given in the text.

apparently also in human gustation (Schiffman, Diaz & Beeker, 1986). Hence adenosine itself may ultimately be found to contribute to the chemosensory process by inhibiting or enhancing the activity of other types of receptor cells.

The perireceptor events described above suggest mechanisms whereby single nucleotides, or nucleotide mixtures, may contribute a cascade of secondary substances having the potential to affect a variety of receptor types. Moreover, as shown below, at least one of the perireceptor events contributing to the production of these products is itself subject to modulation in a manner that may further regulate the residence time of excitants or suppressants in the receptor environment.

Both ADP and the ADP analog, α, β-methylene ADP (AMPCP), are potent inhibitors of 5'-ectonucleotidase, the sensillar enzyme that dephosphorylates the excitatory AMP molecule and produces the non-excitatory nucleoside, adenosine (see Fig. 2.5A). Moreover, the 5'-ectonucleotidase inhibitor, AMPCP, also enhances the physiological response of some AMP-best cells of AMP (Fig. 2.5B). This response enhancement is thought to be due to an inhibition of the enzymatic inactivation process that removes AMP from the receptor environment. These results suggest a mechanism whereby certain substances can modulate a chemosensory response by affecting the perireceptor event that regulates the residence time of the excitant, or suppressant, in the receptor environment. The type of enhancement of a chemosensory response seen with AMP is quite analogous to the potentiation of the response to acetylcholine that occurs in internal tissues in the presence of a cholinesterase inhibitor such as physostigmine (Brzin, Sketelj & Klinar, 1984).

B. Uptake of excitatory substances

Whenever one mixture component influences the rate of inactivation of a second mixture component, then a mixture interaction may occur. Such an interaction may be mediated by a perireceptor process such as an enzyme-catalyzed dephosphorylation, as described above. Alternatively, as noted earlier by Getchell *et al.* (1984), an uptake process may serve as a perireceptor mechanism for removing an excitatory substance from the receptor environment. Hence substances that influence the rate of an uptake process may also contribute to a mixture interaction by affecting the residence time of an excitant or suppressant in the chemosensory organ. The fact that uptake systems function in internal tissues to terminate the responses of several neuroactive substances (Lodge, 1981) shows that uptake systems do provide viable mechanisms for the rapid removal of certain excitatory molecules. Indeed, it is well known that uptake inhibitors potentiate the activity of transmitter substances such as γ-aminobutyric acid and glutamate by prolonging their residence time in the receptor environment (Brown, Collins & Galvan, 1980; Currie & Kelly, 1981).

We have no evidence that inhibition of adenosine uptake affects the responsiveness of AMP- or ATP-sensitive cells of the lobster. Since adenosine itself is not an excitant of the nucleotide-sensitive cells (see

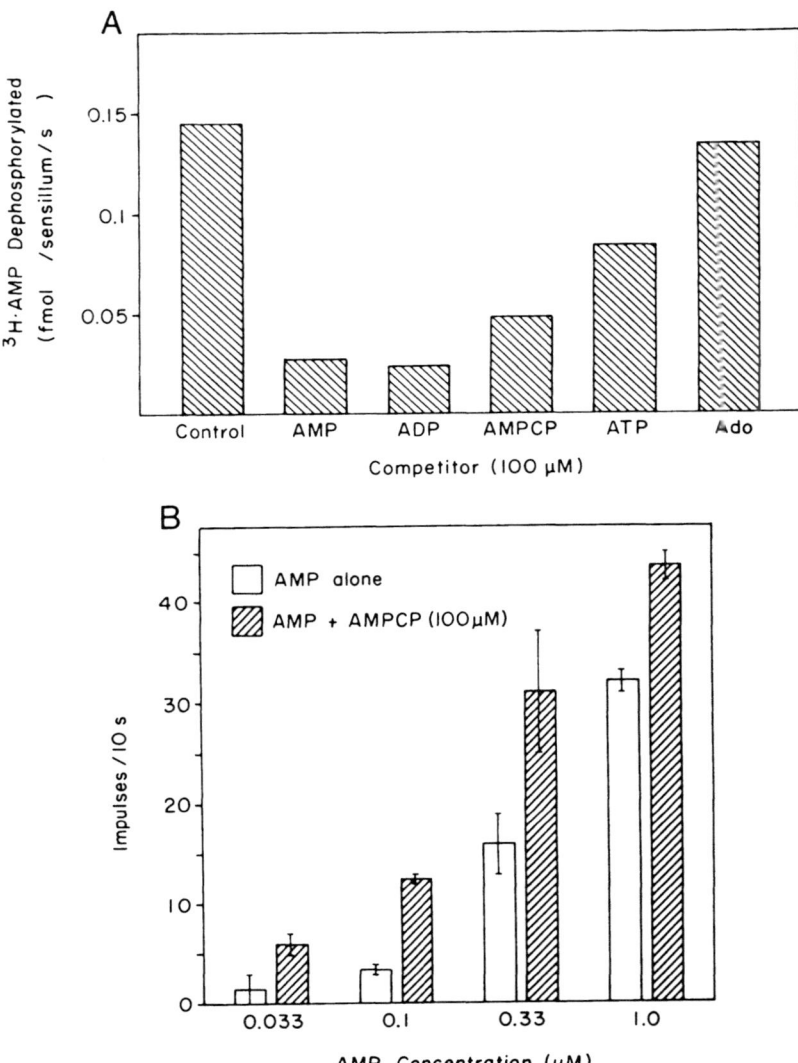

Fig. 2.5. **A.** Dephosphorylation of ^{3}H-AMP by olfactory sensilla of the spiny lobster is inhibited by AMP, ADP, α, β-methylene ADP (AMPCP), and ATP, but not by adenosine. Sensilla are incubated with 1 μM ^{3}H-AMP ± 100 μM of one of the unlabeled substances. **B.** Physiological response of an AMP-best cell to AMP is enhanced by AMPCP, an inhibitor of 5'-ectonucleotidase. The inhibitor (100 μM) and the indicated concentration of AMP are presented simultaneously. AMPCP alone is non-stimulatory. A paired t-test of data derived from nine AMP-best cells indicated significant enhancement ($p < .05$) of the response to AMP when AMPCP was presented simultaneously.

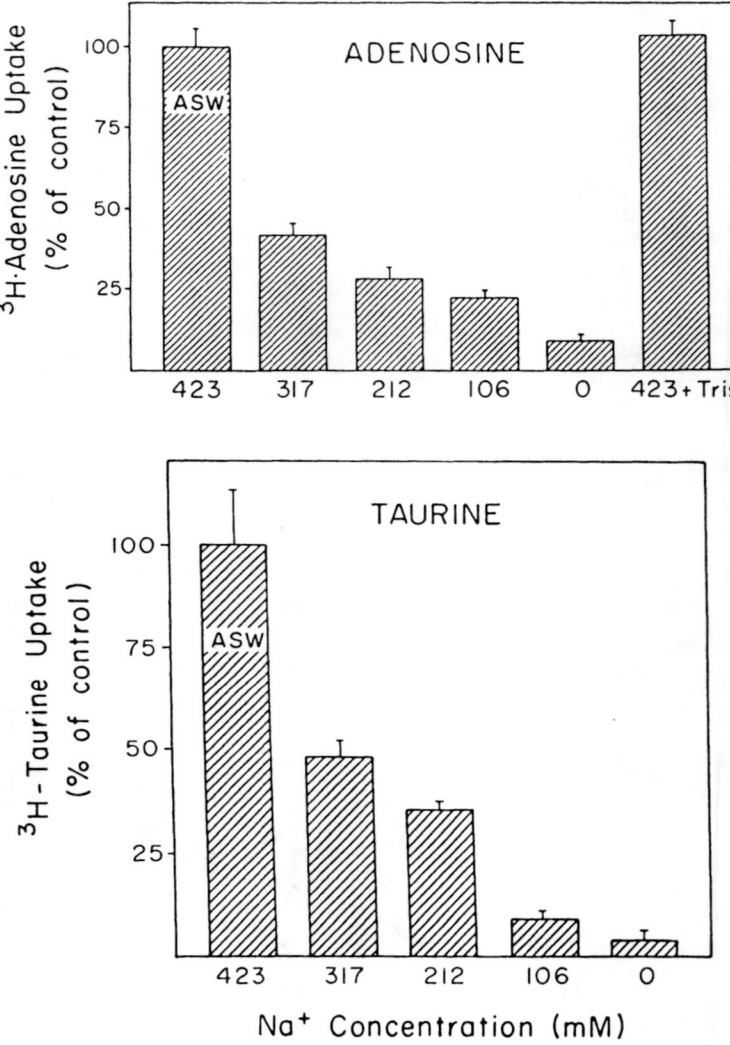

Fig. 2.6. Sodium dependence of the uptake of adenosine (top) and taurine (bottom) by olfactory sensilla of the spiny lobster. Sensilla are incubated with 0.5 μM ^{3}H-adenosine or ^{3}H-taurine in control artificial seawater (ASW; NaCl concentration = 423 mM), and in ASW having the NaCl replaced as indicated with approximately equimolar Tris-Cl. The specific uptake of adenosine or taurine in the control ASW is defined as 100%. Values are means (± SEM). (From Trapido-Rosenthal *et al.*, 1987; and Gleeson, Trapido-Rosenthal & Carr 1987.)

Fig. 2.2), this result is not too surprising. However, the antennule also has a specific uptake system for taurine (Gleeson, Trapido-Rosenthal & Carr 1987), an amino acid that is an excitant of at least two distinct populations of chemosensory cells in the antennule (Fuzessery *et al.*, 1978; Gleeson & Ache, 1985). Studies are currently in progress to determine whether inhibition of taurine uptake affects the physiological response to this excitant. It is also likely that uptake systems exist for other excitatory amino acids such as glutamate and glycine, but we have not yet sought to identify them.

One 'substance' having a dramatic effect upon many uptake processes is the sodium (Na^+) ion. Uptake systems for several types of compounds (e.g. amino acids) are known for their Na^+-dependence (Gunn, 1980; Stevens, Kaunitz & Wright, 1984). This dependence is related to a carrier-mediated co-transport of Na^+ in which a transmembrane Na^+ gradient provides the energy to drive the uptake. The taurine and the adenosine uptake systems in the olfactory organ of the lobster are both markedly Na^+-dependent (Fig. 2.6). This dependence suggests that when such an uptake system serves to remove a chemosensory excitant from the receptor environment, the residence time of the excitatory molecules will be inversely related to the Na^+ content of the milieu. If the residence time of the excitant is indeed regulated by Na^+-dependent uptake, then responses to mixtures of Na^+ plus the excitant may not exhibit simple additivity.

In addition to activating chemosensory receptors in marine animals such as the lobster, amino acids and related substances are known to activate certain gustatory receptors in mammals (Boudreau, Anderson & Oravek, 1975; Boudreau *et al.*, 1985), as well as both gustatory and olfactory receptors in some freshwater fish (Caprio, 1978). If Na^+-dependent inactivation mechanisms (i.e. uptake) play a regulatory role in the responses to amino acids, then at least two interesting questions are raised.

1. Are freshwater organisms able to utilize an Na^+-dependent uptake process to inactivate amino acids, even though their chemosensory surfaces are constantly immersed in a medium with a very low Na^+ content?

2. To what extent do Na^+-dependent inactivation processes contribute to the observations that gustatory responses to certain amino acids are markedly affected by both Na^+ concentration (e.g. Boudreau *et al.*, 1975) and Na^+ transport (e.g. Schiffman, Lockhead & Maes, 1983)?

Acknowledgments

This work was supported by US National Science Foundation grant BNS-8607513. We are grateful for the many discussions and collaborations that occurred with Drs Barry Ache and Charles Derby during several years of interacting at the C. V. Whitney Laboratory. Ms Marsha Lynn Milstead prepared the illustrations.

References

Ache, B. W. (1982). Chemoreception and thermoreception. In H. L. Atwood & D. C. Sandeman, eds, *The biology of Crustacea*, vol. 3, pp. 369–398. Academic Press, New York.

Ache, B. W., & Derby, C. D. (1985). Functional organization of olfaction in crustaceans. *Trends in Neurosciences* **8**, 356–360.

Atema, J. (1985). Chemoreception in the sea: adaptations of chemoreceptors and behavior to aquatic stimulus conditions. *Society of Experimental Biology Symposium* **39**, 387–423.

Atema, J., & Engstrom, D. G. (1971). Sex pheromone in the lobster, *Homarus americanus*. *Nature* **232**, 261–263.

Boudreau, J. C., Anderson, W., & Oravec, J. (1975). Chemical stimulus determinants of cat geniculate ganglion chemoresponsive group II discharge. *Chemical Senses & Flavour* **1**, 495–517.

Boudreau, J. C., Sivakumar, L., Do, L. T., White, T. D., Oravec, J., & Hoang, N. K. (1985). neurophysiology of geniculate ganglion (facial nerve) taste systems: species comparisons. *Chemical Senses* **10**, 89–127.

Brown, D. A., Collins, G. G. S., & Galvan, M. (1980). Influence of cellular transport on the interactions of amino acids with γ-aminobutyric acid (GABA)–receptors in the isolated olfactory cortex of the guinea-pig. *British Journal of Pharmacology* **68**, 251–262.

Brzin, M., Sketelj, J., & Klinar, B. (1984). Cholinesterases. In A. Lajtha, ed., *Handbook of neurochemistry*, vol. 4, 2nd edn, pp. 251–292. Plenum Press, New York.

Burnstock, G. (1985). Purinergic mechanisms broaden their sphere of influence. *Trends in Neurosciences* **8**, 5–6.

Burnstock, G., & Brown, C. M. (1981). An introduction to purinergic receptors. In G. Burnstock, ed., *Purinergic receptors*, pp. 1–45. Chapman & Hall, London.

Cagan, R. H., Torii, K., & Kare, M. R. (1979). Biochemical studies of glutamate taste receptors: the synergistic taste effect of L-glutamate and 5′-ribonucleotides. In L. J. Filer, S. Garattini, M. R. Kare, W. A. Reynolds, & R. J. Wurtman, eds, *Glutamic acid: advances in biochemistry and physiology*, pp. 1–9. Raven Press, New York.

Caprio, J. (1978). Olfaction and taste in the channel catfish: an electrophysiological study of the responses to amino acids and derivatives. *Journal of Comparative Physiology* **123**, 357–371.

Carr, W. E. S. (1976). Chemoreception and feeding behavior in the pigfish, *Orthopristis chrysopterus*: characterization and identification of stimulatory substances in a shrimp extract. *Comparative Biochemistry & Physiology* **155A**, 153–157.

Carr, W. E. S. (1987). The molecular nature of chemical stimuli in the aquatic environment. In J. Atema, R. R. Fay, A. N. Popper, & W. N. Tavolga, eds, *Sensory biology of aquatic animals*, pp. 3–27. Springer-Verlag, New York.

Carr, W. E. S., & Chaney, T. B. (1976). Chemical stimulation of feeding behavior in the pinfish, *Lagodon rhomboides*: characterization and identification of stimulatory substances extracted from shrimp. *Comparative Biochemistry & Physiology* **54A**, 437–441.

Carr, W. E. S., & Derby, C. D. (1986*a*). Behavioral chemoattractants for the shrimp, *Palaemonetes pugio*: identification of active components in food extracts and evidence of synergistic mixture interactions. *Chemical Senses* **11**, 49–64.

Carr, W. E. S., & Derby, C. D. (1986*b*). Chemically stimulated feeding behavior in marine animals: the importance of chemical mixtures and the involvement of mixture interactions. *Journal of Chemical Ecology* **12**, 987–1009.

Carr, W. E. S., Blumenthal, K. M., & Netherton, J. C. (1977). Chemoreception in the pigfish, *Orthopristis chrysopterus*: the contribution of amino acids and betaine to stimulation of feeding behavior by various extracts. *Comparative Biochemistry & Physiology* **58A**, 69–73.

Carr, W. E. S., Netherton, J. C., & Milstead, M. L. (1984). Chemoattractants of the shrimp, *Paleomonetes pugio*: variability in responsiveness and the stimulatory capacity of mixtures containing amino acids, quaternary ammonium compounds, purines and other substances. *Comparative Biochemistry & Physiology* **77A**, 469–474

Carr, W. E. S., Gleeson, R. A., Ache, B. W., & Milstead, M. L. (1986). Olfactory receptors of the spiny lobster: ATP-sensitive cells with similarities to P2-type purinoceptors of vertebrates. *Journal of Comparative Physiology* **158**, 331–338.

Currie, D. N., & Kelly, J. S. (1981). Glial versus neuronal uptake of glutamate. *Journal of Experimental Biology* **95**, 181–193.

Derby, C. D., & Atema, J. (1987). Chemoreceptor cells in aquatic invertebrates: peripheral mechanisms of chemical signal processing in decapod crustaceans. In J. Atema, R. R. Fay, A. N. Popper, & W. N. Tavolga, eds, *Sensory biology of aquatic animals*, pp. 365–385. Springer-Verlag, New York.

Derby, C. D., Carr, W. E. S., & Ache, B. W. (1984). Purinergic olfactory receptors of crustaceans are similar to internal purinergic receptors of vertebrates. *Journal of Comparative Physiology* **155**, 341–349.

Ellingsen, O. F., & Døving, K. B. (1986). Chemical fractionation of shrimp extracts inducing bottom food search behavior in cod (*Gadus morhua* L.). *Journal of Chemical Ecology* **12**, 155–168.

Fuke, S., Konosu, S., & Ina, K. (1981). Identification of feeding stimulants for red sea bream in the extract of the marine worm *Perinereis brevicirrus*. *Bulletin of the Japanese Society of Scientific Fisheries* **47**, 1631–1635.

Fuzessery, Z. M., Carr, W. E. S., & Ache, B. W. (1978). Antennular chemosensitivity in the spiny lobster, *Panulirus argus*: studies of taurine sensitive receptors. *Biological Bulletin* **154**, 226–240.

Getchell, T. V., Margolis, F. L., & Getchell, M. L. (1984). Perireceptor and receptor events in vertebrate olfaction. *Progress in Neurobiology* **23**, 317–345.

Gleeson, R. A. (1978). Functional adaptations in chemosensory systems. In M. A. Ali, ed., *Sensory ecology*, pp. 291–317. Plenum Press, New York.

Gleeson, R. A. (1980). Pheromone communication in the reproductive behavior of the blue crab, *Callinectes sapidus*. *Marine Behavior & Physiology* **7**, 119–134.

Gleeson, R. A., & Ache, B. W. (1985). Amino acid suppression of taurine-sensitive neurons. *Brain Research* **335**, 99–107.

Gleeson, R. A., Trapido-Rosenthal, H. G., & Carr, W. E. S. (1987). A taurine receptor model: taurine-sensitive cells in the lobster. In R. J. Huxtable, F. Franconi, & A. Giotti, eds, *The biology of taurine: methods and mechanisms*, pp. 253–263. Plenum Press, New York.

Grünert, U., & Ache, B. W. (1988). Ultrastructure of the aesthetasc (olfactory) sensilla on the antennules of the spiny lobster *Panulirus argus*. *Cell & Tissue Research* **251**, 95–103.

Gunn, R. B. (1980). Co- and countertransport mechanisms in cell membranes. *Annual Review of Physiology* **42**, 249–259.

Hashimoto, Y., Konosu, S., Fusetani, N., & Nose, T. (1968). Attractants for eels in the extracts of short-necked clams. I. Survey of constituents eliciting feeding behavior by the omission test. *Bulletin of the Japanese Society of Scientific Fisheries* **34**, 78–83.

Hidaka, I. (1982). Taste receptor stimulation and feeding behavior in the puffer. In T. J. Hara, ed., *Chemoreception in fishes*, pp. 243–257. Elsevier Scientific Publishing Company, Amsterdam.

Laverack, M. S., & Ardill, D. J. (1965). The innervation of the aesthetasc hairs of *Panulirus argus*. *Quarterly Journal of Microscopical Science* **106**, 45–60.

Lodge, D. (1981). Uptake inhibitors, amino acids, and spinal neurons. In F. V. DeFeudis & P. Mandel, eds, *Amino acid neurotransmitters*, pp. 327–332. Raven Press, New York.

Mackie, A. M. (1973). The chemical basis of food detection in the lobster *Homarus gammarus*. *Marine Biology* **21**, 103–108.

Mackie, A. M. (1982). Identification of the gustatory feeding stimulants. In T. J. Hara, ed., *Chemoreception in fishes*, pp. 275–291. Elsevier Scientific Publishing Company, Amsterdam.

Mackie, A. M., & Adron, J. W. (1978). Identification of inosine and inosine 5'-monophosphate as the gustatory feeding stimulants for the turbot, *Scophthalmus maximus*. *Comparative Biochemistry & Physiology* **60A**, 79–83.

Mackie, A. M., & Grant, P. T. (1974). Interspecies and intraspecies communication by marine invertebrates. In P. T. Grant & A. M. Mackie, eds, *Chemoreception in marine organisms*, pp. 105–141. Academic Press, London.

Mackie, A. M., & Mitchell, A. I. (1982). Further studies on the chemical stimulation of feeding behavior in the Dover sole, *Solea solea*. *Comparative Biochemistry & Physiology* **73A**, 89–93.

Mackie, A. M., & Mitchell, A. I. (1983). Studies on the chemical nature of feeding stimulants for the juvenile European eel, *Anquilla anquilla*. *Journal of Fish Biology* **22**, 425–430.

Mackie, A. M., Adron, J. W., & Grant, P. T. (1980). Chemical nature of feeding stimulants for the juvenile Dover sole, *Solea solea* (L.). *Journal of Fish Biology* **16**, 701–708.

Morse, D. E. (1984). Biochemical control of larval recruitment and fouling. In J. D. Costlow & R. C. E. Tipper, eds, *Biodeterioration: an interdisciplinary study*, pp. 134–140. Naval Institute Press, Annapolis.

Müller, D. G., Gassmann, G., & Luning, K. (1979). Isolation of a spermatozoid-releasing substance from female gametophytes of *Laminaria digitata*. *Nature* **279**, 430–431.

Ohsugi, T., Hidaka, I., & Ikeda, M. (1978). Taste receptor stimulation and feeding behavior in the puffer, *Fugu pardalis*. II. Effects of mixtures of constituents of clam extracts. *Chemical Senses & Flavour* **3**, 335–368.

Richardson, P. J., Brown, S. J., Bailyes, E. M., & Luzio, J. P. (1987). Ectoenzymes control adenosine modulation of immunoisolated cholinergic synapses. *Nature* **327**, 232–234.

Rifkin, B., & Bartoshuk, L. M. (1980). Taste synergism between monosodium glutamate and disodium 5'-guanylate. *Physiology & Behavior* **24**, 1169–1172.

Sato, M., Yamashita, S., & Ogawa, H. (1970). Potentiation of gustatory response to monosodium glutamate in rat chorda tympani fibers by addition of 5'-ribonucleotides, *Japanese Journal of Physiology* **20**, 444–464.

Schiffman, S. S., Lockhead, E., & Maes, F. W. (1983). Amiloride reduces the taste intensity of Na^+ and Li^+ salts and sweeteners. *Proceedings of the National Academy of Sciences, USA* **80**, 6136–6140.

Schiffman, S. S., Diaz, C., & Beeker, T. G. (1986). Caffeine intensified taste of certain sweeteners: role of adenosine receptor. *Pharmacological Biochemistry & Behavior* **22**, 429–432.

Schmitt, B. C., & Ache, B. W. (1979). Olfaction: responses of a decapod crustacean are enhanced by flicking. *Science* **205**, 204–206.

Sleeper, H. L., Paul, V. J., & Fenical, W. (1980). Alarm pheromones from the marine opisthobranch *Navanax inermis*. *Journal of Chemical Ecology* **6**, 57–70.

Stevens, B. R., Kaunitz, J. D., & Wright, E. M. (1984). Intestinal transport of amino acids and sugars: advances using membrane vesicles. *Annual Review of Physiology* **46**, 417–433.

Stone, T. W. (1983). Interactions of adenosine with other agents. In R. M. Berne, T. W. Rall & R. Rubio, eds, *Regulatory functions of adenosine*, pp. 467–477. Martinus Nijhoff, The Hague.

Torii, K., & Cagan, R. H. (1980). Biochemical studies of taste sensation. IX. Enhancement of L-[^{3}H]glutamate binding to bovine taste papillae by 5'-ribonucleotides. *Biochimica et Biophysica Acta* **627**, 313–323.

Trapido-Rosenthal, H. G., Carr, W. E. S., & Gleeson, R. A. (1987). Biochemistry of an olfactory purinergic system: dephosphorylation of excitatory nucleotides and uptake of adenosine. *Journal of Neurochemistry* **49**, 1174–1182.

Yamaguchi, S. (1979). The umami taste. In J. C. Boudreau, ed., *Food taste chemistry*, pp. 33–51. American Chemical Society, Washington.

Yoshii, K., Yokouchi, C., & Kurihara, K. (1986). Synergistic effects of 5'-nucleotides on rat taste responses to various amino acids. *Brain Research* **367**, 45–51.

PART TWO
PHYSIOLOGY OF MIXTURE PERCEPTION

3

Responses of insect olfactory receptor neurons to biologically relevant mixtures

Robert J. O'Connell and R. Patrick Akers

The Worcester Foundation for Experimental Biology,
Shrewsbury, Massachusetts, USA

I. Introduction

The neural mechanisms responsible for the detection, encoding, transmission and processing of behaviorally relevant odor signals have long been a subject of much theoretical and experimental interest. With the advent of modern microanalytical techniques, it rapidly became clear that in many cases the odor sources responsible for eliciting or modulating a particular behavior were not simple secretions containing a single active compound but were, in fact, complex mixtures of different chemical substances. In some cases individual components of these mixtures were associated with particular components of a behavior. In others, several components of a mixture seemed collectively responsible for modulating a particular behavior. The supposition that mixtures are differentially processed in the olfactory nervous system commonly arises when biologically relevant materials are evaluated, either singly or in multicomponent blends, with modern behavioral assay techniques. It is becoming abundantly clear that this increase in the chemical and behavioral complexity of a particular communication system must be paralleled by an increase in the efficiency of the physiological mechanisms employed for the neural encoding of behaviorally relevant odor compounds and blends.

Here we describe two studies that have examined the electrical activity elicited in primary olfactory receptor neurons when they are each stimulated with single and multicomponent odor mixtures. Particular attention was given to the responses elicited in a subset of the individual

Copyright © 1989 by Academic Press Australia.
All rights of reproduction in any form reserved.

pheromone-sensitive sensilla on the antennae of both the male cabbage looper moth (*Trichoplusia ni*) and the male redbanded leafroller moth (*Argyrotaenia velutinana*). Electrophysiological responses to single and multiple component stimuli, each drawn from among the known behaviorally active compounds for these insects, were obtained at several different stimulus intensities. In the cabbage looper some of the relevant blends elicited electrical responses from primary olfactory receptor neurons which were not readily predicted from a knowledge of the receptor neurons responses to individual components. In the redbanded leafroller, the relevant blends elicited electrical responses from primary olfactory receptor neurons which were indistinguishable from the responses observed with the individual components of the blend.

Complex mixtures of volatile organic compounds play important roles in the chemical communication systems of many animals. Many of these mixtures occur passively as a consequence of organic life; others are especially formulated, synthesized in specialized glands and deposited in the environment so that the organism in question may benefit from their unique communicative capabilities. Much of the challenge in modern odor biology has been associated with attempts to understand the methods devised by nature for discriminating meaningful chemical signals from the enormous, and largely irrelevant, odor background that surrounds living organisms. Within this melange of potential chemical signals one usually finds, for any one species, a set of biologically relevant odorants, a fraction of which are products of the animal's own communication system. These latter substances, when considered together, may constitute a single behaviorally important multicomponent chemical signal. In these cases the relative amounts of the individual components and their absolute concentrations carry important signal value for the receiving animal. Many of the other chemical compounds that remain in an animal's surround may function, either singly or collectively, as chemical signals in some other context or may be of interest to other species. Thus, it is almost always true that the multicomponent pheromonal signals that are of interest to an organism must be discriminated against a chemical background containing very large amounts of many other volatile substances. This becomes an especially challenging problem in those systems that employ multicomponent chemical signals to modulate long distance attraction behaviors, largely because of the downwind dilution factors inherent in this mode of communication. In many cases, we know that these compounds signal vital information to the organism, upon which various life and death decisions are ultimately to be made.

The task of detecting, encoding, transmitting, filtering and interpreting these complex messages, in the face of such a vast volatile background, is the special province of the olfactory system. When viewed from the perspective of a set of individual olfactory receptor neurons, there are two related tasks that must be undertaken. First, the relevant compounds must be detected and discriminated from among the background odors and, second, the relative amounts of each of the target compounds must be determined. One relatively straightforward mechanism to ensure that the resolving power required of the chemical communication system will be adequate for the initial discrimination is to build into the system a very high degree of chemical specificity. That is, the majority of the irrelevant background stimuli can be effectively eliminated by ensuring that the primary chemical detectors in the initial stages of the system are responsive only to behaviorally relevant compounds. In those cases where several different compounds are integral components of an important biological signal one may provide different classes of highly specific receptor neurons, each exclusively responsive to one of the components in the signal, in order to detect the appropriate message. If the message is also involved with long-distance communication and must be processed at considerable distances from a calling insect, then the relevant receptor neurons must also be equipped with a high degree of sensitivity. With this general approach to the problem there is little need for much 'central' processing, as most of the important stages of pheromonal discrimination are handled in the periphery by the design of the response capabilities of the primary receptor neurons. It should also be obvious that the parallel processing of the individual components of a multicomponent blend could be abandoned altogether by designing a type of receptor neuron that is activated only by the appropriate mixture of behaviorally relevant components, and that is insensitive to the remaining individual chemicals found in the blend. In a sense, this is simply a 'higher order' of chemical specificity, which requires a specific mixture of odorants for activation rather than an individual compound.

There are several obvious disadvantages to a coding system based entirely on highly specialized olfactory receptor neurons. In the first case, it requires a specialized receptor neuron class for each of the discriminable stimuli in the mixture. As the number of discriminable odors in a multicomponent blend and, for that matter, as the number of multicomponent blends increases, so too does the number of specialized receptor neuron classes that are required. Coupled to this is the concomitant limitation on the discriminative range available to the organism. That is, only those odor compounds that have been designed into the organism's sensory capabilities are available for the signaling of additional information. Thus the chemical cues emanating from a new predator or a new competitor go undetected if they happen to fall outside the range of sensory capabilities available to the organism. In other words, odor blindness is a relatively rigid strategy and, as such, is relatively nonadaptive. As has been noted many times, this deficiency is easily corrected by simply increasing the qualitative bandwidth of the individual receptor neurons in the sensory system. Efficient detection and encoding of a very large number of different odor compounds can be accomplished by a relatively small number of different olfactory receptor neuron types if each group is uniquely responsive to a range of different odors. The identity of a particular stimulus may then be unambiguously decoded by the central nervous system if it evaluates, in parallel, the relative amounts of neural activity elicited across the whole population of receptor neuron types. It is important to note that the relative breadth of tuning among the different classes of olfactory receptor neurons need not be constant in this type of encoding mechanism. Thus, some of the available input channels could be more narrowly tuned than others and some could be uniquely responsive to particular mixtures of chemical stimuli. This leads us directly to the questions under consideration here. Are the responses elicited in primary olfactory receptor neurons by behaviorally relevant mixtures of chemical stimuli fundamentally different from their responses to pure stimuli? If they are, is this capability a general feature of all olfactory systems?

We chose to explore these issues in two species of Lepidoptera: the cabbage looper moth (CL), *Trichoplusia ni* (Hubner) (Noctuidae), and the redbanded leafroller moth (RBLR), *Argyrotaenia velutinana* (Walker) (Tortricidae). Both have multicomponent pheromones (Bjostad, Linn, Du & Roelofs, 1984a; Bjostad, Linn & Roelofs, 1984b) and have been the subject of a range of field and laboratory behavioral studies (Bjostad *et al.*, 1984b; Linn, Bjostad, Du & Roelofs, 1984; Linn, Campbell & Roelofs, 1986). These studies make it possible to describe at least the broader aspects of each component's significance in the odor-guided behavior of these moths. The stimulus compounds used here, their absolute purity and the relative amount of those identified from calling females are listed in Table 3.1 (Bjostad *et al.*, 1984a, 1984b). We also included in this study (Z)-7-dodecenol (Z-7, 12:OH) which, although not produced in the pheromone gland of either female, is a potent inhibitor of male CL behavior.

The flight responses of both species have been observed in laboratory wind tunnel experiments (Bjostad *et al.*, 1984b; Linn *et al.*, 1984, 1986).

Table 3.1. Stimulus compounds, their absolute purity and the percentage of each measured in volatile collections from female pheromone glands

Stimulus compound	Measured purity	% Composition[a] found in gland effluvia from	
		Trichoplusia ni	*Argyrotaenia velutinana*
Z-7, 12 : AC	99.2	82.8	—
Z-5, 12 : AC	99.1	7.7	—
12 : AC	99.8	6.0	4.8
11, 12 : AC	97.8	2.3	2.4
Z-7, 14 : AC	98.7	0.9	—
Z-9, 14 : AC	98.8	0.4	—
Z-7, 12 : OH	94.4	—	—
Z-11, 14 : AC	99.0	—	79.4
E-11, 14 : AC	99.0	—	7.1
14 : AC	99.0	—	3.9
E-9, 12 : AC	99.0	—	1.6
Z-9, 12 : AC	99.0	—	0.8

[a] From Bjostad *et al.*, 1984*a* and *b*.

The sex attractant pheromone of the female CL moth is a mixture containing the six compounds listed in Table 3.1. A blend's ability to modulate the behavior of males depends to a large degree on how closely it approximates the natural mixture. The major constituent of the pheromone blend, (Z)-7-dodecenyl acetate (Z-7, 12:AC), was essential for flight initiation and maintenance of upwind flight. However, when presented alone it failed to elicit significant amounts of sustained upwind flight, with the result that less than 20% of the males tested reach the source. A blend containing only the other five components of the blend was even less effective than Z-7, 12:AC alone and failed to elicit any of the typical upwind behavioral responses, including oriented flight in the wind tunnel. However, when these materials were combined so as to mimic the composition of the effluvia released by females, more than 95% of the insects tested produced the full sequence of male behaviors with a speed and vigor similar to that elicited by calling females. Although these results demonstrated that the full behavioral response required the five minor components of the CL blend in combination with the primary component, it was also clear that any one of the five could be removed from the blend without significantly altering its potency (Linn *et al.*, 1984, 1986). Thus, although the major component is necessary to initiate the behavior, it is not possible to make a similar assignment of a specific behavioral response to any one of the minor components.

The sex attractant pheromone of the RBLR moth is a mixture containing the seven compounds listed in Table 3.1. Its ability to modulate the behavior of males depends to a large degree on the relative amount of each of these components in the blend. Especially important for optimal behavioral activity is the precise ratio between the principal component of the blend, (Z)-11-tetradecenyl acetate (Z-11,14:AC), and its geometric isomer, (E)-11-tetradecenyl acetate (E-11,14:AC). These two compounds must be present in a precise Z:E ratio of 100:9 in order to achieve maximal behavioral activation of the male (Baker, Carde & Roelofs, 1976; Bjostad *et al.*, 1984*b*). The relative proportionality among the remaining five identified minor components may be less critical, but is is clear that their presence in the blend is crucial for rapid and complete orientation behaviors (Bjostad *et al.*, 1984*b*; Linn *et al.*, 1986). It is also important to note at this point that the most abundant constituent (Z-11,14:AC) in the RBLR pheromone blend is largely ineffective in eliciting upwind flight in male RBLR moths, quite unlike the situation in the CL where the most abundant single component (Z-7,12:AC) of the blend induced more than 80% of the males to take flight and resulted in nearly 20% of them proceeding 1.5 m upwind in the flight tunnel to reach the pheromone source. However, the optimal ratio of Z- and E-11, 14:AC does elicit behavioral responses in male RBLR moths that are qualitatively similar to those elicited by Z-7,12:AC alone in male CL moths (Linn *et al.*, 1986). This behavioral sensitivity to a particular ratio of compounds is not unique to the RBLR but has been found in several other species and has been suggested as a mechanism for maintaining species isolation (Minks, Roelofs, Ritter & Persoons, 1973; Roelofs & Carde, 1974; Silk & Kuenen, 1986; Bailey, McDonough & Hoffmann, 1986; Linn & Roelofs, 1983).

It should be clear from the above description that, although both species rely on multicomponent pheromones for long-distance attraction, the mechanisms responsible for achieving this goal and maintaining the required degree of species specificity need not be the same. The experiments reported here continue our attempts to unravel blend perception in the peripheral olfactory system. Because of the large number of different single and multicomponent stimuli that can be derived from the diverse components identified in each species' pheromone blend, and the requirement that each be evaluated at several different intensity levels, it is not practical to attempt to evaluate all the mixtures that might be of interest. Therefore, the results presented here describe, in the main, the responses to some of the mixtures that include principal components of the pheromone blend. It is hoped that this comparative approach will help to define a portion of the available range of solutions to this important problem in olfactory communication science.

Here, we evaluate the sensitivity of primary olfactory receptor neurons in both CL and RBLR males with standard extracellular single sensillum recording and impulse sorting techniques (O'Connell, 1975, 1985; O'Connell, Kocsis & Schoenfeld, 1973; O'Connell, Grant, Mayer & Mankin, 1983). The extracellular recordings obtained in both species indicate that trichoid sensilla are innervated by at least two spontaneously active receptor neurons. Each produces typical biphasic action potentials whose amplitude and temporal characteristics permit their ready identification (O'Connell *et al.*, 1973). The larger amplitude action potential and the receptor neuron that produces it were labeled as 'A' and the smaller as 'B'. These assignments were always made without regard to the response specificities of the receptor neurons under study. Although there are several distinct classes of olfactory sensilla on the antenna of both species, several of which are distinguished by unique morphological and physiological characteristics (Grant & O'Connell, 1986; O'Connell *et al.*, 1983), we will consider here only that class of trichoid sensilla that contains receptor neurons which are responsive to the principal component of the respective pheromone blends.

II. Results

A. *Trichoplusia ni*

The average magnitude of the electrical response obtained from 12 pairs of individual receptor neurons stimulated with selected single and multiple component stimuli is displayed in Figure 3.1. In all, five (I–V) of the seven behaviorally relevant compounds have been evaluated as individual stimuli. In each case, the concentrations evaluated were proportional to the amounts measured in female CL pheromone glands. The absolute level of Z-7, 12:AC (0.005 µg) chosen, was one that would produce a half-maximal response from the average A-receptor neuron. Z-7,12:OH, which is not produced by the female, was evaluated at an intensity equal to that used for Z-7,12:AC. As demonstrated previously (O'Connell, 1985), the average A-receptor neuron is reliably responsive to low doses (0.0005 µg) of Z-7,12:AC, whereas the average B-receptor neuron is responsive to comparable amounts of Z-7,12:OH. Neither of these receptor neurons is particularly responsive to the other single components evaluated. These included 12:AC (0.0005 µg), Z-5,12:AC (0.0005 µg) and Z-7,14:AC (0.00005 µg). Of the five binary mixtures evaluated, three elicited responses that were significantly different

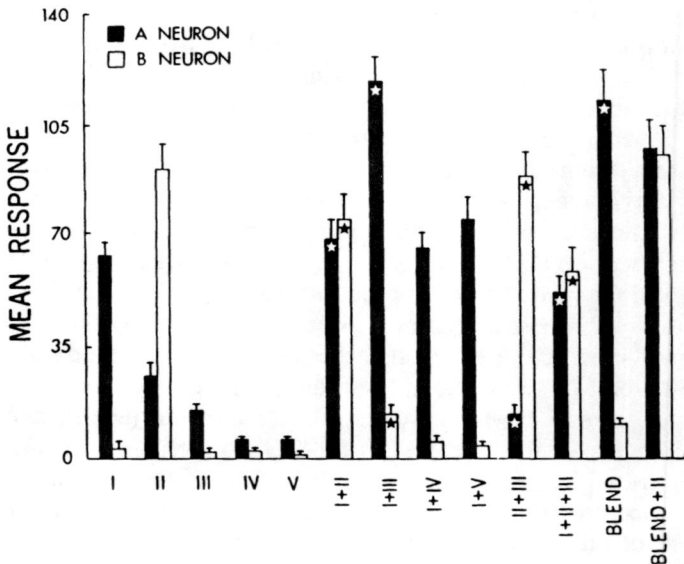

Fig. 3.1. The average (mean ± SEM) response (in impulses/10 seconds) to stimulation with the indicated single and multiple component stimuli. In all cases, a minimum of 12 pairs of male CL moth A- and B-receptor neurons contributed to each mean (range 12 to 26). The compounds and the doses evaluated are: I = Z-7, 12:AC, (0.005 µg); II = Z-7, 12:OH, (0.005 µg); III = 12:AC, (0.0005 µg); IV = Z-5, 12:AC, (0.0005 µg); V = Z-7, 14:AC, (0.00005 µg); Blend = all the compounds listed for the CL moth in Table 3.1. Statistically significant differences, $p < .02$, two tailed, Wilcoxon matched-pairs signed-ranks test (Siegel, 1956), between the observed and expected responses to multicomponent stimuli are starred. Cartridges containing mineral oil were ineffective stimuli for both A-receptor (1 ± 1 imp/10 s) and B-receptor (2 ± 1) neurons. (From O'Connell, Beauchamp & Grant, 1986.)

(starred bars in Fig. 3.1) from those expected if one assumed that pheromone mixtures should elicit responses that are simple sums of the responses elicited by their constituents. Addition of only 10% 12:AC to the principal component of the CL pheromone, Z-7, 12:AC, enhanced the amount of neural activity elicited in both A- and B-receptor neurons. This synergy is impressive, especially in the light of the relatively narrow tuning of these receptor neurons and the general failure of 12:AC to be an effective, single stimulus for either neuron. Dodecyl acetate (12:AC) seems unique in this regard because the other two minor pheromone components examined (Z-5, 12:AC and Z-7, 14:AC) did not significantly alter the responses elicited in either receptor neuron when they were individually combined with Z-7,12 : AC. The responses elicited in both receptor neurons by a blend containing all six of the identified female-produced compounds were, on average, indistinguishable from those elicited by the binary mixture containing only Z-7,12 : AC and 12 : AC. In addition, the average magnitude of the A-receptor neuron discharge elicited by this blend was significantly larger than would be expected from the additive model.

The responses elicited by mixtures that contain Z-7,12:OH, the behavioral inhibitor of male flight responses, were, on average, reliably smaller than those predicted by summing the responses obtained with the individual components of the mixture. For example, the three components blend (Fig. 3.1) elicited responses which were, on average, smaller than those expected from the additive model and, in fact, smaller than those obtained with the most effective single component. The one exception to the rule involved the mixture containing all seven behaviorally active compounds. In this case, both receptor neurons produced responses that were, on average, equal to those expected from the additive model (Fig. 3.1). Thus, the minor components of the blend do exert some influence over the responsiveness of these receptor neurons, at least to the extent that they prevent the significant reduction in discharge expected when Z-7,12:OH is added to a mixture.

From this brief summary, it should be apparent that individual olfactory receptors may produce responses to certain odor mixtures that are significantly different from the sum of the responses elicited by the mixtures' individual components. It should also be obvious from discussions elsewhere in this volume that there are a great number of mechanistic possibilities, largely drawn from the enzymological literature, that could account for both the positive and negative interactions observed. In any case, it is not possible to construct a simple explanation for these observations that involves only one kind of stimulus interaction with the CL receptor neuron membrane, and that simultaneously accounts for both the observed responses to single pheromone components and the various interaction effects observed with various components of the CL pheromone blend.

B. *Argyrotaenia velutinana*

In each of the male RBLR trichoid sensilla sampled, the electrical activity of two receptor neurons was observed. The average A-receptor neuron responded reliably to low doses (0.0001 µg) of Z-11, 14:AC, but was largely unresponsive to low doses of all of the other single component stimuli identified in the RBLR pheromone blend (Table 3.1). The average B-receptor neuron responded reliably to low doses (0.0001 µg) of

E-11, 14:AC and, again, was largely unresponsive to low doses of the other single component stimuli (Akers & O'Connell, in press). In both receptor neurons, small responses were observed in response to stimulation with much larger amounts (three to four orders of magnitude) of several of the other individual components of the female pheromone.

As noted earlier, this experiment was designed to evaluate the responses elicited by stimuli that contained blends of Z- and E-11,14:AC. In particular, we wished to determine if the different Z:E ratios elicited responses in trichoid receptor neurons that would correlate with the behavioral responses showing a peak in male behavior at the 100:9 ratio (Baker *et al.*, 1976). The average responses of A-receptor neurons (Fig. 3.2) and B-receptor neurons (Fig. 3.3), observed in 31 individual sensilla, are plotted so as to highlight potential interactions among the components as the amount of E-11, 14:AC, added to various fixed amounts of Z-11, 14:AC, was increased.

At each of the dosage levels of Z-11, 14:AC employed, the average of A-receptor neuron response remained constant, irrespective of the percentage of E-11, 14:AC added to the blend (Fig. 3.2). We consistently failed to observe a substantial alteration in neural output at the optimal 100:9 ratio at any of the three dosage levels. There was a small decrease

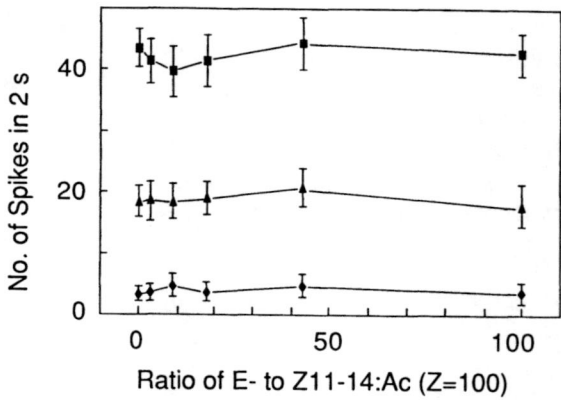

Fig. 3.2. The average (mean ± 95% confidence interval of the mean) responses elicited in 31 male RBLR moth A-receptor neurons by stimulation with various mixtures of Z- and E-11,14:AC. Each curve was obtained in response to stimuli containing the indicated amounts of E-11,14:AC added to one of three different fixed amounts of Z-11,14:AC (from Akers & O'Connell, 1988). The absolute amount of Z-11,14:AC in the various sets of stimuli ranged, in decade steps, from 0.01 µg in the top curve to 0.0001 µg in the bottom curve.

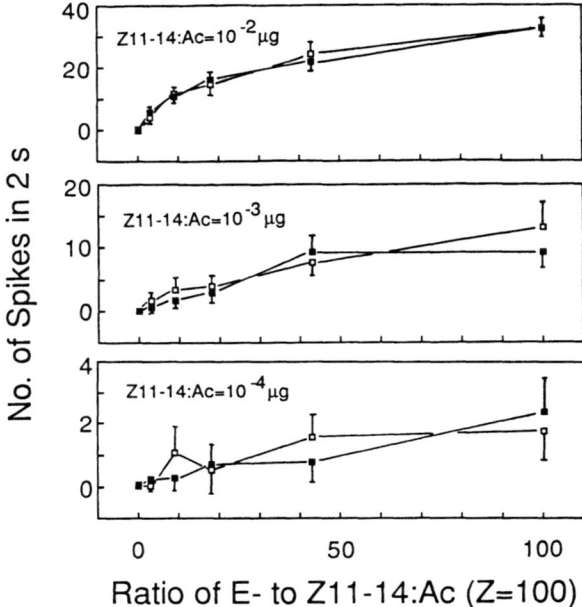

Fig. 3.3. The average (mean ± 95% CI) responses of the 31 companion B-receptor neurons elicited simultaneously by the same stimuli referred to in Figure 3.2 (closed symbols). For comparison, we also display (open symbols) the average responses elicited in these same receptor neurons by stimulation with graded doses of E-11,14:AC alone. (From Akers & O'Connell, in press.)

in average neural output at this ratio for the highest dosage level of Z-11, 14:AC (0.01 μg) evaluated. However, the mean responses across all of the blends, within any one dosage level, were not significantly different from one another. Since E-11,14:AC by itself fails to excite the average A-receptor neuron and since, in combination with Z-11, 14:AC (a compound that is excitatory), it also fails to alter the response expected on the basis of the content of active component, it seems likely that positive and negative interactions between Z- and E-11, 14:AC that might alter the response properties of A-receptor neurons do not occur.

The average responses of the companion B-receptor neurons to stimulation with these same blends are illustrated in Fig. 3.3 (closed symbols) To provide a ready comparison we have also plotted the average responses elicited in these same receptor neurons by stimulation with graded amounts of E-11, 14:AC alone (open symbols). Again, there is no indication of a significant alteration in the responses elicited by the 100 : 9 ratio. As expected, the responses observed at any one dosage level of Z-11,14:AC increased in proportion to the amount of E-11, 14:AC added to the blend. The average response function generated by stimulation with graded amounts of E-11,14:AC alone (open circles) and that generated by stimulation with the various binary mixtures of Z- and E-11, 14:AC (filled circles) are indistinguishable. Since Z-11,14:AC by itself fails to excite the average B-receptor neuron and since, in combination with E-11, 14:AC (a compound that is excitatory), it also fails to alter the response expected on the basis of the content of active component, it seems likely that positive and negative interactions between Z- and E-11, 14:AC that might alter the response properties of B-receptor neurons do not occur.

From these observations in the RBLR moth it seems fairly clear that there is no correlation between the responses elicited by stimulation with various blends of Z- and E-11, 14:AC in pheromone receptor neurons and the behavioral consequences of such stimulation. In other words, the interactive effects of certain ratios of Z- and E-11, 14:AC on male behavior are not the result of interactive effects involving the two olfactory receptor neurons that appear to be responsible for their detection. Although not shown here, various multiple component stimuli, including the full 7-component blend, also fail to elicit responses that are different from those elicited by the appropriate single component stimuli (Akers & O'Connell, in press).

III. Discussion

It is clear from these two studies that the common problem of multicomponent pheromone perception may be solved in different ways by different species. Although both CL and RBLR males require multicomponent chemical signals for robust responses to calling females, the actual mechanisms by which the various components contribute to this behavior are not the same. In particular, the major component of the CL pheromone is by itself responsible for eliciting early stages of a male's response to females. In its absence, the five minor components of the blend are completely ineffective. In contrast, the major component of the RBLR pheromone is by itself largely inactive, eliciting behavioral responses in less than 10% of the males tested. However a binary mixture containing the 100:9 ratio of Z- and E-11, 14:AC elicits a substantially enhanced behavioral response in male RBLR which is qualitatively similar to that elicited by Z-7, 12:AC in male CL moths. Addition of the remaining minor components to the respective blends elicits vigorous behavioral responses, which result in more than 95% of the males reaching the pheromone source. To the extent that the details surrounding the behavioral responses of these two insects to their respective multicomponent pheromone blends are different from each other, it is not too surprising to find that the neural responses that these components elicit in primary olfactory receptor neurons are also different in the two species.

In the CL at least one of the minor components (12:AC) has a significant ability to modulate the amount of neural activity produced in pheromone receptor neurons when it is presented in a mixture containing the principal component of the pheromone blend. This finding is in contrast to that observed in the RBLR where none of the minor components, including the behaviorally relevant geometric isomer (E-11, 14:AC), modulated the amount of neural activity observed in pheromone receptor neurons. A second major conclusion from the present studies on the RBLR moth is that A-receptor neurons in trichoid sensilla detect low levels of Z-11, 14:AC and their companion B-receptor neurons detect comparably low levels of E-11, 14:AC. Both of these capabilities operate independently of each other and are not modulated by the addition of the other identified components of the female pheromone blend.

These observations lead to at least one additional issue. If neither of the receptor neurons in trichoid sensilla overtly detects the other single components of the pheromone blend that have demonstrated behavioral activity, and if these minor components collectively fail to modulate in any observable way the responses of the A- and B-receptor neurons to stimulation with Z- and E-11, 14:AC, then these minor components of the pheromone blend must be detected by other sensory structures on the antenna. A survey is currently under way to explore the response properties of trichoid sensilla on other parts of the antenna in an attempt to discover the means by which these minor pheromonal components are detected. With only a small number of sensilla sampled in each of the accessible regions of the antennal surface, receptor neurons with additional response properties upon stimulation with minor components of the RBLR blend have not been observed.

In general, if minor components of pheromone blends are to be important in modulating pheromone-guided behavior, one is bound to conclude that the organisms possess sensory receptors that are uniquely responsive, in some fashion, to these components. Although this seems to be the case for the principal components of the pheromone blends in the two species considered here, it remains true that there continue to be minor, behaviorally relevant, components which do not seem to have receptor neurons devoted to their detection in antennal trichoid sensilla. If one is convinced that these minor components must have their own specialized channel of communication, then this apparent anomaly may be resolved by assuming that important minor components of the pheromone blend are detected and processed by receptor neurons in only a small fraction of the total number of the trichoid sensilla on the antenna (Lofstedt, Van Der Pers, Lofqvist, Lanne, Appelgren, Bergstrom & Thelin, 1982; Van Der Pers & Lofstedt, 1986). In this 'rare sensillum' argument, the failure to find a trichoidal receptor neuron responsive to a behaviorally relevant minor component of blend does not suggest that other strategies have been employed for the detection and processing of multicomponent stimuli, but rather indicates that too few sensilla have been examined in order to find the 1% or 2% that may be devoted to the compound in question. Implicit in this argument is the notion that the number of receptor neurons responsive to a particular component is directly proportional to its abundance in the blend. Although there is no way, short of actually finding them, to prove that such rare sensilla exist, and moreover that they participate in the detection and processing of minor pheromone components, one could agree that this arrangement might be appropriate for minor components that affect close-range behaviors. However, it seems an unlikely explanation for the minor components in the CL and RBLR pheromone evaluated here, which are thought to modulate behavior at a distance. For these materials, one might well imagine that a fairly large complement of receptor neurons should be necessary, in order to compensate for their initial low abundance in the pheromone bouquet and the distance over which they are required to act. In other words, it seems more likely to us that behaviorally important pheromone components would be processed by enough receptor neurons to ensure that there is a large neural safety factor associated with their detection and processing. This seems to be the case in the RBLR, where there are substantial numbers of receptor neurons devoted to the detection of E-11, 14:AC, a component that is important in long-distance behavioral responses of males and is present in the pheromone gland at less than one tenth the abundance of the major component. Similar distributions have been observed in the turnip moth, where 70% of the sensilla sampled contained receptor neurons tuned to Z-5, 10:AC, a behaviorally relevant component of the female pheromone blend whose abundance equals 6% of the total materials produced (Lofstedt *et al.*, 1982). Because receptor neurons tuned to other minor components of the pheromone blends of various species have not been found, this suggests to us that other neural mechanisms or other sensilla classes are being employed.

Acknowledgments

These studies have been supported by the Alden Trust and by US National Institute of Neurological and Communicative Disorders and Strove (NINCDS) grant NS-14453 awarded to Robert J. O'Connell and NINCDS National Research Service Award, NS-07913 awarded to R. Patrick Akers.

References

Akers, R. P., & O'Connell, R. J. (in press). The contribution of olfactory receptor neurons to the perception of pheromone component ratios in male redbanded leafroller moths. *Journel of Comparative Physiology, A* **163**.

Bailey, J. B., McDonough, L. M., & Hoffmann, M. P. (1986). Western avocado leafroller, *Amorbia cuneana* (Walsingham), (Lepidoptera: Torticidae): discovery of populations utilizing different ratios of sex pheromone components. *Journal of Chemical Ecology* **12**, 1239–1245.

Baker, T., Carde, R. T., & Roelofs, W. (1976). Behavioral responses of male *Argyrotaenia velutinana* to components of its sex pheromone. *Journal of Chemical Ecology* **2**, 333–352.

Bjostad, L. B., Linn, C. E., Du, J. W., & Roelofs, W. L. (1984a). Identification of new sex pheromone components in *Trichoplusia ni* predicted from biosynthetic precursors. *Journal of Chemical Ecology* **10**, 1309–1323.

Bjostad, L., Linn, C., & Roelofs, W. (1984b). Identification of new sex pheromone components in *Trichoplusia ni* and *Argyrotaenia velutinana* predicted from biosynthetic precursors. In T. E. Acree & D. M. Soderlund, eds, *Semiochemistry flavors and pheromones*, pp. 199–203. Walter de Gruyter, Berlin.

Grant, A. J., & O'Connell, R. J. (1986). Neurophysiological and morphological investigations of pheromone-sensitive sensilla on the antenna of male *Trichoplusia ni*. *Journal of Insect Physiology* **32**, 503–515.

Linn Jr, C. E., & Roelofs, W. L. (1983). Effect of varying proportions of the alcohol component on sex pheromone blend discrimination in male Oriental fruit moths. *Physiological Entomology* **8**, 291–306.

Linn Jr, C. E., Bjostad, L. B., Du, J. W., & Roelofs, W. L. (1984). Redundancy in a chemical signal: behavioral responses of male *Trichoplusia ni* to a 6-component sex pheromone blend. *Journal of Chemical Ecology* **10**, 1635–1658.

Linn Jr, C. E., Campell, M. G., & Roelofs, W. L. (1986). Male moth sensitivity to multicomponent pheromones: critical role of female-released blend in determining the functional role of components and active space of the pheromone. *Journal of Chemical Ecology* **12**, 659–668.

Lofstedt, C., Van Der Pers, J. N. C., Lofqvist, J., Lanne, B. S., Appelgren, M., Bergstrom, G., & Thelin, B. (1982). Sex pheromone components of the turnip moth, *Agrotis segetum*: chemical identification, electrophysiological evaluation and behavioral activity. *Journal of Chemical Ecology* **8**, 1305–1321.

Minks, A. K., Roelofs, W. L., Ritter, F. J., & Persoons, C. J. (1973). Reproductive isolation of two tortricid moth species by different ratios of a two-component sex attractant. *Science* **180**, 1073–1074.

O'Connell, R. J. (1975). Olfactory receptor responses to sex pheromone components in the redbanded leafroller moth. *Journal of General Physiology* **65**, 179–205.

O'Connell, R. J. (1985). Responses to pheromone blends in insect olfactory receptor neurons. *Journal of Comparative Physiology, A,* **156**, 747–761.

O'Connell, R. J., Kocsis, W. A., & Schoenfeld, R. L. (1973). Minicomputer identification and timing of nerve impulses mixed in a single recording channel. *Proceedings of the Institute of Electrical and Electronics Engineers* **61**, 1615–1621.

O'Connell, R. J., Grant, A. J., Mayer, M. S., & Mankin, R. M. (1983). Morphological correlates of differences in pheromone sensitivity in insect sensilla. *Science* **220**, 1408–1410.

O'Connell, R. J., Beauchamp, J. T., & Grant, A. J. (1986). Insect olfactory receptor responses to components of pheromone blends. *Journal of Chemical Ecology* **12**, 451–467.

Roelofs, W. L., & Carde, R. T. (1974). Oriental fruit moth and lesser appleworm attractant mixtures refined. *Environmental Entomology* **3**, 586–588.

Siegel, S. (1956). *Nonparametric statistics for the behavioral sciences.* McGraw-Hill, New York.

Silk, P. J., & Kuenen, L. P. S. (1986). Spruce budworm (*Choristoneura fumiferana*) pheromone chemistry and behavioral responses to pheromone components and analogs. *Journal of Chemical Ecology* **12**, 367–383.

Van Der Pers, J. N. C., & Lofstedt, C. (1986). Signal-response relationship in sex pheromone communication. In T. L. Payne, M. C. Birch & C. E. J. Kennedy, eds, *Mechanisms in insect olfaction*, pp. 235–241. Oxford University Press, Oxford.

4

Olfactory discrimination of mixtures: behavioral, electrophysiological and theoretical studies using the spiny lobster *Panulirus argus*

Charles D. Derby, Marie-Nadia Girardot,
Peter C. Daniel and Jacqueline B. Fine-Levy

*Department of Biology, Georgia State University,
Atlanta, Georgia, USA*

I.   Introduction

This chapter describes our attempts to understand how olfactory systems are organized to discriminate the quality of odorants. Our previous work has emphasized discrimination of quality of single-odorant compounds (Derby & Ache, 1984a; Girardot, Fine & Derby, 1986; Derby & Atema, 1988; Fine-Levy, Derby & Daniel, 1987). We describe here the neural basis for quality coding of mixtures of odorant compounds. Our approach is to couple behavioral and neurophysiological studies, asking first, what odorant mixtures can the animal behaviorally discriminate; and second, what is the neural code for these behaviorally discriminable odorant mixtures. While this parallel behavioral-neurophysiological approach has been used successfully to study gustatory discrimination by mammals (Nowlis, Frank & Pfaffman, 1980: Smith, Van Buskirk, Travers & Bieber, 1983; Smith, this volume; Frank, this volume) and blowflies (Maes & Bijpost, 1979; Maes & Ruifrok, 1986), to our knowledge this is the first attempt to do so for olfactory discrimination by any species.

The spiny lobster *Panulirus argus* was used in our studies of olfactory discrimination. The peripheral olfactory organ of spiny lobsters is the pair of antennules, or first antennae (Ache & Derby, 1985). Contained in

65

*Copyright © 1989 by Academic Press Australia.
All rights of reproduction in any form reserved.*

these antennules are approximately 800 000 primary olfactory receptor cells, which are modified ciliary cells that project from the periphery to the brain.

The chemical stimuli used in these experiments are multicomponent artificial mixtures (Table 4.1), whose compositions are based on chemical analyses of tissue extracts of the potential food of spiny lobsters (Carr & Derby, 1986b). These four mixtures are similar in that they all contain most of the same components and have the same total concentration, yet they differ in the relative amounts of these components.

A second topic addressed in this chapter concerns the significance of mixture interactions, which are commonly found in behavioral and neurophysiological studies of spiny lobsters and other crustaceans. A theoretical analysis is presented of how mixture interactions may affect discrimination and perception of mixtures by enhancing or obscuring the

Table 4.1. Composition of artificial mixtures of crab, mullet, oyster and shrimp, at 500 μM

Compound[a]	Concentration (μM) in mixture			
	Crab	Mullet	Oyster	Shrimp
AMINO ACIDS[b]				
Alanine (Ala)	25.2	16.1	45.2	31.6
β-Alanine (β-Ala)	0	0.168	10.6	0
α-Aminobutyrate (α-ABA)	0	0.194	0	0
Arginine (Arg)	26.9	1.68	3.81	13.1
Asparagine (Asn)	2.29	2.06	1.41	1.49
Aspartate (Asp)	0.424	2.03	6.07	1.23
Cysteine (Cys)	1.41	0	0.778	0
Glutamate (Glu)	3.34	2.55	6.26	2.46
Glutamine (Gln)	34.7	3.11	4.78	9.38
Glycine (Gly)	133	18.7	29.5	139
Histidine (His)	1.62	41.0	0.593	0.482
Hydroxyproline (Hyp)	1.88[c]	0.381	1.52	0
Isoleucine (Ile)	1.26	0.876	0.220	1.57
Leucine (Leu)	2.88	1.56	0.480	2.83
Lysine (Lys)	2.67	7.14	2.04	0.723
Methionine (Met)	4.61	0.436	0.148	1.54
3-Methylhistidine (3mHis)	0	0.734	0	0
Ornithine (Orn)	0	1.63	1.15	0
Phenylalanine (Phe)	1.09	0.514	0.037	0.804
o-Phosphoserine (pSer)	0	0.540	0	0
Proline (Pro)	65.5	2.87	15.2	17.0
Serine (Ser)	3.12	3.00	4.15	2.00
Taurine (Tau)	22.1	70.0	204	49.1
Threonine (Thr)	4.75	2.94	0.703	0.946
Tryptophan (Trp)	1.54	0	0	0

difference in the quality of a mixture versus that of some components of that mixture.

II. Behavioral discrimination

Two conditioning techniques were employed to study behavioral discrimination: an associative learning technique based on classical conditioning of an aversive response (Fine-Levy *et al.*, 1987; Fine-Levy, Girardot, Derby & Daniel, 1988), and a non-associative learning technique based on a habituation paradigm (Daniel & Derby, in press). From these conditioning studies, the perceived similarity of any two chemicals can be determined from the degree of generalization of the conditioned response to the non-conditioned stimuli: the greater the generalization, the more similar the chemicals.

Table 4.1. (continued)

Compound[a]	Concentration (μM) in mixture			
	Crab	Mullet	Oyster	Shrimp
Tyrosine (Tyr)	1.19	0.220	0.333	1.37
Valine (Val)	3.74	1.66	0.519	3.42
NUCLEOTIDES, -SIDES, AND RELATED COMPOUNDS				
Adenosine 5'-monophosphate (AMP)	0.294	0.108	3.67	7.25
Adenosine 5'-diphosphate (ADP)	2.14	0.071	0.296	2.12
Adenosine 5'-triphosphate (ATP)	10.3	0	0	0.589
Guanosine 5'-monophosphate (GMP)	0	0.365	0.481	0
Inosine 5'-monophosphate (IMP)	1.86	25.2	0.704	2.97
Xanthosine 5'-monophosphate (XMP)	0	1.06	0	0
Hypoxanthine (Hyx)	0.388	0.441	0	0.339
Inosine (Ino)	0.812	6.06	0.444	0.089
QUATERNARY AMMONIUM COMPOUNDS				
Betaine (Bet)	43.5	38.0	125	75.0
Homarine (Hom)	5.52	0	12.3	10.3
Trimethylamine oxide (TMAO)	39.8	36.2	0	30.4
ORGANIC ACIDS				
L-lactate (1-Lact)	51.2	210	3.37	40.9
D-lactate (d-Lact)	0	0	0.556	0
Succinate (Succ)	0	0	13.2	0
TOTAL CONCENTRATION (μM)	500	500	500	500

From Carr & Derby, 1986*b*.
[a] Abbreviation is in parentheses.
[b] All amino acids are the L-isomer.
[c] From W. E. S. Carr (personal communication).

A.   Aversive conditioning

There are three phases to this experiment: preconditioning, conditioning, and postconditioning. In preconditioning, the chemical stimuli, which consisted of two concentrations (500 and 50 µM) of each of the four mixtures (Table 4.1), were presented every 15 minutes. The responses of untrained animals were recorded for three minutes after stimulus presentation. Fourteen behaviors were monitored, three of which (see below) were incorporated into the analysis because they showed significant changes following conditioning. That these behaviors are largely mediated by olfaction is demonstrated from behavioral experiments in which anosmic lobsters do not show the behaviors upon chemical stimulation. During the conditioning phase, the unconditioned, aversive stimulus was a dark object moved rapidly toward the animal. This aversive stimulus reliably elicited an avoidance response (tail flip). Four groups of three animals were used in this study. Each group was trained to one of the four mixtures. Training involved pairing the aversive stimulus with presentation of either concentration of only one of the mixtures. Training to both concentrations of a single mixture increases the likelihood that the cue to which the animals attend is stimulus quality rather than quantity. In this differential training paradigm, both concentrations of the other three mixtures were also presented each day but were not paired with the aversive stimulus. Training lasted for five to seven days. In the postconditioning phase, responses to the four mixtures were recorded as during preconditioning.

The effect of training to each mixture was quantitatively evaluated by comparing the pre-to the postconditioning responses. Three behaviors showed significant training effects; these were search, grab at stimulator, and backwards walking or running. The effects of training can be described as passive avoidance (a decrease in search or grab at stimulator) and active avoidance (an increase in backwards walking or running). The conditioning effect can be measured based on each of these three behaviors. The presentation of the results can be simplified by combining the effect on all three behaviors, described as an aversion index.

As an example, the results for animals conditioned to shrimp mixture are shown in Figure 4.1. The greatest aversions were to the two concentrations of the conditioned chemical, shrimp mixture. The next greatest aversion was to crab mixture. Oyster mixture and mullet mixture produced some aversion, but significantly less than did crab mixture. Therefore, crab mixture is perceived by the lobsters as more similar to shrimp mixture than is either oyster or mullet mixture.

The relative similarities among all of the mixtures can be summar-

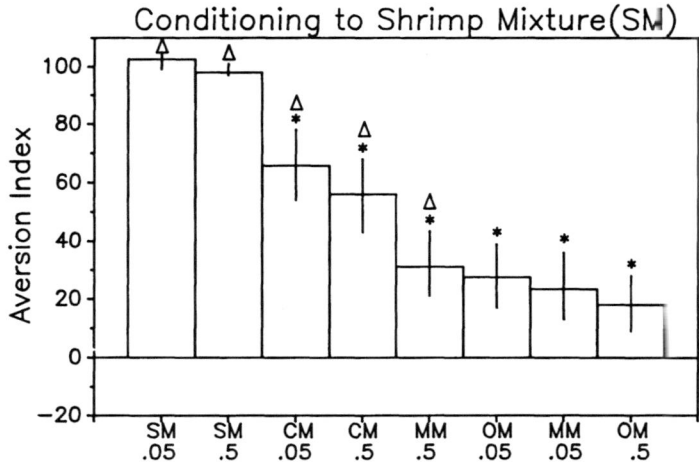

Fig. 4.1. Aversions to mixtures for lobsters differentially and aversively conditioned to shrimp mixture. Aversion index is derived from conditioned changes in three behaviors: search, grab at stimulator, and backwards walking or running. Aversion values (mean +/− standard error for three animals) are shown for shrimp, crab, mullet, and oyster mixtures, (SM, CM, MM and OM, respectively), each at two concentrations, 0.5 and 0.05 mM. Asterisks denote aversion values significantly different from the aversion values for shrimp mixture (MANOVA, $p < .05$. Delta symbols indicate significant aversions based on comparison of 95% confidence limits with a zero aversion index. (From Fine-Levy, Girardot, Derby & Daniel, 1988.)

ized by incorporating the results of all four conditioning groups in a multivariate analysis. Figure 4.2 shows such a result using multidimensional scaling, based on Euclidean distances between stimuli. A two-dimensional solution is used, since 98% of the total variance in the data is accounted for by these two dimensions. The degree of similarity between chemicals is inversely proportional to the distance between them in this two-dimensional space. Three main observations can be made from this analysis. First, changes in stimulus intensity have little effect on stimulus quality. Second, oyster and mullet mixtures are obviously distinct from each other and from crab and shrimp mixtures. Third, shrimp and crab mixtures are the most similar types of mixtures, but they are still separated along dimension 2.

B. Habituation

Habituation techniques were also used to determine the degree of similarity between crab mixture and each of the other three mixtures (Daniel & Derby, in press). Lobsters were habituated to crab mixture

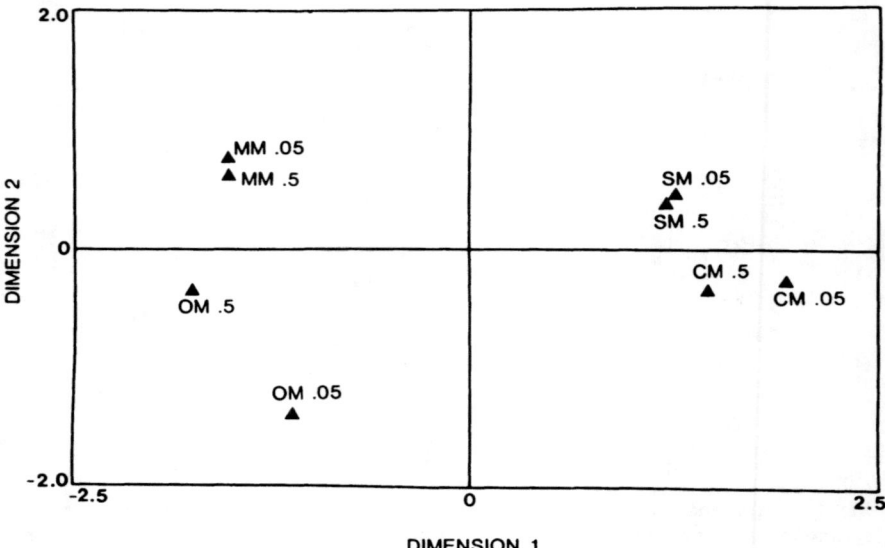

Fig. 4.2. Relative similarities among mixtures as determined from aversive conditioning studies. The locations of crab, shrimp, mullet and oyster mixtures, each at two concentrations (0.5 and 0.05 mM), are shown in a two-dimensional space derived from multidimensional scaling. This analysis is based on aversion values from four groups of three animals, with members of each group being aversively conditioned to both concentrations of only one of the four mixtures. See text for explanation. (Crab mixture, CM; shrimp mixture, SM; mullet mixture, MM; oyster mixture, OM) (From Fine-Levy, Daniel and Derby, submitted.)

by presenting 5 mL volumes of alternating concentrations (500 and 50 μm) of that mixture every five minutes . Habituation, defined as failure to respond to 10 consecutive presentations of the habituated stimulus, normally occured by three hours. This relatively long time for habituation is probably related to the fact that these mixtures are extremely relevant and biologically important stimuli for these animals.

Generalization of the habituated response to the non-habituated stimuli is shown for animals habituated to crab mixture in Figure 4.3. The degree of generalization is derived from a habituation index. The habituation index was greatest for the habituated stimulus, crab mixture (90%), followed by shrimp mixture (65%), oyster mixture (49%), and mullet mixture (47%). The habituation index value for crab mixture was significantly different from those for oyster and mullet mixtures.

These results using a habituation technique indicate that crab mixture and shrimp mixture are more similar to each other than to oyster or mullet mixture, and are therefore consistent with the results of aversive conditioning.

III. Neural discrimination

In order to produce the behavioral discriminations described above, the nervous system must receive and code information contained within chemical stimuli. To achieve an understanding of the underlying mechanisms we have been studying how olfactory receptor cells in the antennules, and olfactory interneurons in the brain, code information contained in single compounds and in chemical mixtures. This chapter deals only with quality coding by the antennular receptor cells.

Spiny lobsters can behaviorally discriminate among certain single compounds contained within their food, including taurine, AMP, and glutamate (Fine-Levy *et al.*, 1987). When tested with these and other single compounds, the olfactory receptor cells of spiny lobsters are often found to be extremely narrowly tuned, responding strongly or only to one of the test compounds. Receptor cell types responsive only to taurine, glutamate, betaine, glycine, AMP, or ATP have been described for spiny lobsters (Fuzessery, Carr & Ache, 1978; Derby & Ache, 1984a; Derby, Carr & Ache, 1984a; Gleeson & Ache, 1985; Carr, Gleeson, Ache & Milstead, 1986). These findings fueled our speculation that these narrowly tuned receptor cells could function as a labeled line coding

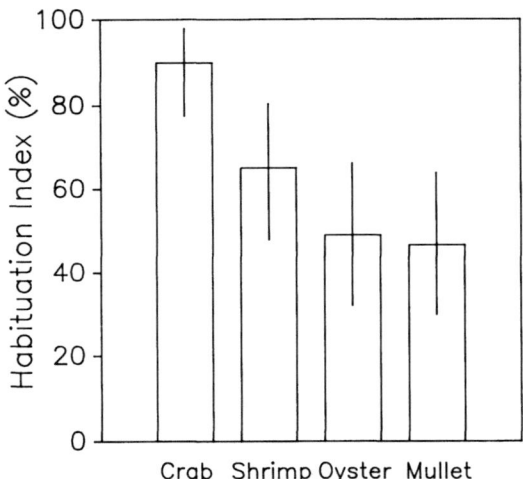

Fig. 4.3. Relative similarities between crab mixture and the other mixtures as determined from habituation studies. Habituation index values are shown for lobsters habituated to crab mixture. Habituation index is defined as [(pre-habituation response — post-habituation response)/pre-habituation response] $\times$ 100. Values are mean $+/-$ 95% confidence limits for six animals habituated to 500 and 50 μM crab mixture. The values for oyster mixture and mullet mixture are significantly different from crab mixture (MANOVA, $p < .05$). (From Daniel & Derby, in press.)

system in the discrimination of single compounds (Derby & Ache, 1984*a*; Derby & Atema, 1988). However, it has also been demonstrated that these receptor cells can differentiate among these same single chemicals by using an across-neuron pattern (ANP) code, a population code in which stimulus quality is indicated by the pattern of activity across a population of receptor cells (Derby & Ache, 1984*a*; Derby, Hamilton & Ache,1984*b*; Carr & Derby, 1986*a*; Girardot *et al.*, 1986). Both types of codes appear to be candidates for coding quality of chemicals for which narrowly tuned receptor cells exist.

A different picture of coding by receptor cells emerges when the responses of olfactory cells to mixtures of chemicals are considered (Girardot & Derby, 1988). In this study, the four artificial mixtures used in the behavioral study (Table 4.1) were tested, each at three concentrations (500, 50, and 5 µM). Responses of olfactory receptor cells were recorded from an excised antennule preparation (Derby & Ache, 1984*a*; Gleeson & Ache, 1985). Action potentials from single cells were recorded extracellularly from axons, using a fine-tipped suction electrode. The response to chemical stimulation was quantified as the number of impulses in the first five seconds of response.

The responses to mixtures for 30 olfactory receptor cells are included in this study. Response specificity can be quantified using the breadth of tuning metric of Smith and Travers (1979), where specificity of cells ranges from 0 (maximal narrow tuning) to 1 (maximal broad tuning). Based on their responses to mixtures, these cells are characterized as broadly tuned, with mean breadth values of 0.90, 0.92, and 0.95 for concentrations of 5, 50, and 500 µM, respectively. Thus, receptor cells that are narrowly tuned for single chemicals are broadly tuned for mixtures. Obviously, a strictly labelled line code would not allow this population of receptor cells to differentiate among these mixtures.

To test whether a population code could be used in the discrimination of behaviorally discriminable mixtures, the differences in ANPs elicited by the mixtures were analyzed using multivariate analyses (Girardot & Derby, 1988). The results of one such analysis, multidimensional scaling based on Euclidean distances between stimuli, are shown in Figure 4.4. In this three-dimensional solution (in which 94% of the total variance in the data is accounted for), each type of mixture is represented by a triangular space (stimulus space), delineated by the locations of the three ANPs for the three concentrations of that mixture. Four important points are evident from this figure. First, changes in concentration by as much as 100-fold have relatively minor effects on the ANPs and therefore on stimulus quality. This is indicated by the fact that ANPs for the three concentrations of a mixture type are relatively close to each other. Second, oyster mixture is most distinct from the other

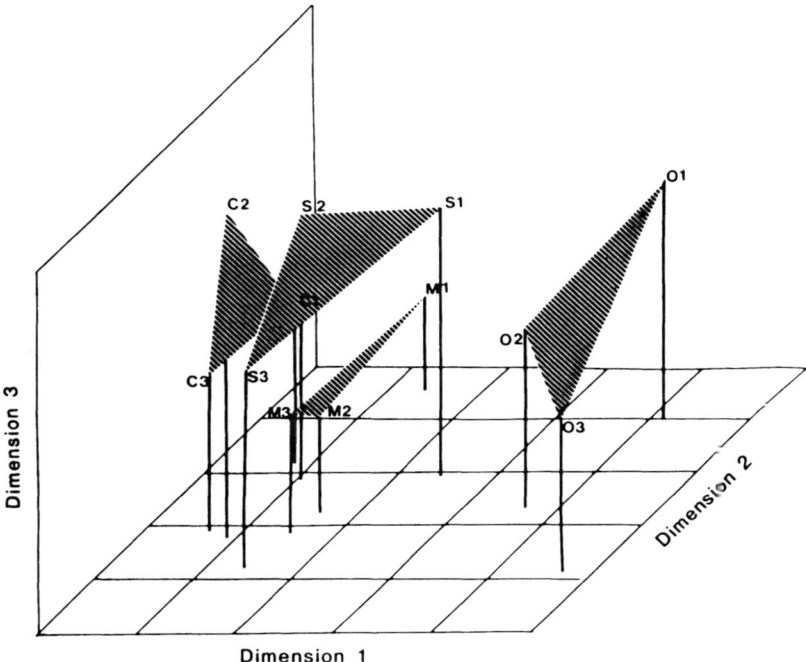

Fig. 4.4. Relative similarities among mixtures as determined from neural studies. The locations of crab, shrimp, mullet and oyster mixtures (C, S, M and O, respectively), each at three concentrations (1, 2, 3 = 5, 50, and 500 µM, respectively), are shown in a three-dimensional space derived from multidimensional scaling. This analysis is based on the responses of 30 olfactory receptor cells to the four mixtures. Each point represents the across-neuron pattern for one stimulus, and each shaded triangle represents the stimulus space for a two-log unit concentration range of one mixture type. (From Girardot & Derby, 1988.)

mixtures. Third, mullet mixture is also clearly distinct from crab and shrimp mixtures, although to a lesser degree than is oyster mixture. Fourth, crab and shrimp mixtures are more similar to each other than to any other mixtures, but even their stimulus spaces do not overlap and are therefore somewhat distinct.

These results based on neural responses are in complete agreement with the results of the behavioral experiments. This strongly suggests that in this olfactory system, at least at the level of olfactory receptor cells, the quality of complex mixtures is encoded by ANPs.

IV.   Similarity in compositions of mixtures

The differences in the compositions of the mixtures in Table 4.1 obviously do not reside in their total concentrations, the number of components, or the types of components, but rather in the *relative* amounts of the components. The degree of similarity in the overall compositions of the mixtures can be derived from multivariate analysis of the concentrations of the components. The results of one such technique, cluster analysis, are represented in a dendrogram (Fig. 4.5). The smaller the cluster distance for any two stimuli or clusters of stimuli, the more similar are their overall compositions. Hence, crab and shrimp mixtures are relatively similar in composition, while mullet and oyster mixtures are relatively different from each other as well as from crab and shrimp mixtures. This is a satisfying result in that it agrees with the behavioral and neural results. Thus, the ability of the olfactory receptor cells to detect the relative differences in the amounts of the components of the mixtures allows these animals to discriminate behaviorally among them.

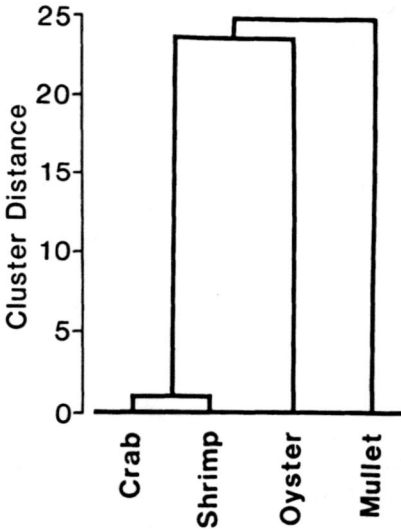

Fig. 4.5.   Relative similarities among the compositions of the mixtures. This dendrogram represents the results of a cluster analysis, based on the relative amounts of the components of the mixtures as given in Table 4.1. See text for explanation. (From Fine-Levy, Girardot, Derby & Daniel, 1988.)

However, not all the compounds in a mixture contribute equally to its unique identity. Another multivariate technique, principal components analysis, was used to identify which compounds best differentiate any two mixtures. For a given pairwise comparison of mixtures, two principal components are produced, one of which describes the degree to which each compound contributes to the difference between the two mixtures. Which and how many chemicals most contribute to the difference between the two mixtures were determined by sequentially removing those chemicals with the largest principal component values. This process is continued until enough chemicals have been removed to reach a correlation coefficient for the mixtures that is greater than 0.90. Figure 4.6 summarizes the results of these analyses for all six comparisons. This figure shows only those compounds that must be removed in order to reach the 0.90 correlation level between the mixtures being compared, thus representing only those compounds that make the mixtures different. Each comparison is depicted by either a triangle (for shrimp-mullet, shrimp-oyster and mullet-oyster comparisons) or quadrangle (for crab-mullet, crab-shrimp and crab-oyster comparisons) formed of solid lines. A dashed line separates each comparison into two spaces, each occupied by the compounds that are characteristic of each mixture. The compounds are also ranked in order of their principal component value, thus giving an assessment of each compound's relative contribution to the differences between the mixtures. For example, the comparison of shrimp and mullet is represented by the triangle on the left side of the figure. When all the compounds in shrimp and mullet mixtures are included in the principal components analysis, the two mixtures are appreciably different in composition, with a relatively low correlation coefficient of 0.585. Twelve compounds make shrimp and mullet different. Histidine (His), at the mullet end of the traingle, is the most characteristic of mullet relative to shrimp, while homarine (Hom), at the shrimp end of the triangle, is the most characteristic of shrimp relative to mullet.

In addition, our knowledge of the spiny lobster's olfactory system can be used to further refine our list of components that we predict are important in the discrimination of these mixtures. The electrophysiological study of Derby & Ache (1984*b*) identified compounds in crab mixture that are either stimulants or interactants (suppressants or synergists) to the olfactory system. These contributory compounds are indicated in Figure 4.6. Thus homarine, ADP, ATP and arginine are biologically active 'shrimp-like' compounds relative to mullet, while AMP is the only biologically active 'mullet-like' compound relative to shrimp among those tested for activity. From an initial list of 41 compounds, we can isolate just five compounds that are the most likely candidates involved in discrimination between the shrimp and mullet mixtures.

Based on this analysis, our predictions are that lobsters may be able to use as few as five to eight compounds to discriminate between any two food mixtures. These predictions can take the form of testable hypotheses. In spite of these reductionistic predictions, we believe that the addition of more compounds to a mixture should add to its uniqueness. This probably explains why the ANPs for the four tissue extracts are more distinct from each other than are the ANPs for the four related but simpler artificial mixtures (Girardot & Derby, 1988).

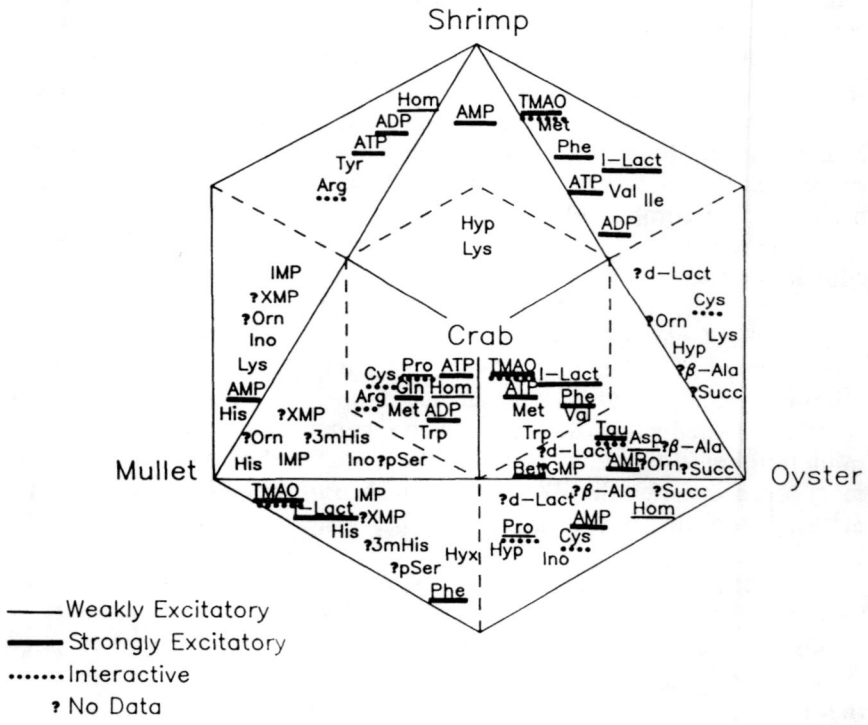

Fig. 4.6. Components of the mixtures responsible for the dissimilarities in compositions of the mixtures. Principal components analysis was used to define which components are most responsible for making the compositions of the mixtures different from each other. See text for explanation. (Daniel & Derby, unpublished data.)

V. Mixture interactions and their influence on the perception and discrimination of odorant mixtures

Mixture interactions have been operationally defined as chemosensory phenomena where the response to a mixture cannot be predicted based on knowledge of the responses to its individual components. Critical to the identification of mixture interactions is how to predict the expected response to the mixture. As our knowledge of the chemosensory system under question increases, our predictions can become more refined and therefore our identification and understanding of mixture interactions can improve.

Mixture interactions are commonly described in studies of olfaction and gustation with crustaceans. Suppression and synergism have been identified from behavioral studies of shrimp (Carr & Derby, 1986a,b) and California spiny lobsters (Zimmer-Faust, Tyre, Michel & Case, 1984), and from electrophysiological studies of olfactory receptor cells and olfactory interneurons in Florida spiny lobsters (Derby & Ache, 1984b; Derby & Atema, 1988; Ache, this volume) and American lobsters (Atema, this volume).

Mixture interactions change the message carried by sensory neurons. Since these neurons carry information about the chemical environment around the animal, alterations in these messages, as by mixture interactions, might have important ramifications on neural coding and chemosensory discrimination. The following section is a theoretical consideration of these influences.

Imagine a hypothetical olfactory system composed of a total of 40 receptor cells, 20 of each of two types, with types defined according to their response spectra for three stimuli: compound 1, compound 2, and a mixture of 1 and 2. The first cell type responds to component 1 and not to component 2, while the second cell type responds to component 2 but not 1. A mixture of components 1 and 2, under the condition of no mixture suppression, stimulates both types of cells. The magnitudes of the responses for each cell are standardized to that cell's maximum response. The relationship between the ANPs generated by component 1, component 2, as well as the two-component mixture *under different degrees of mixture suppression*, as revealed through multidimensional scaling, is shown in Figure 4.7. The responses of the two cell types are shown in parentheses [(cell type 1, cell type 2)] under each stimulus. The similarity of the stimuli can be evaluated from the distance between them in the space, such that stimuli that are situated near each other in space have relatively similar ANPs. Thus, the influence of mixture suppression on

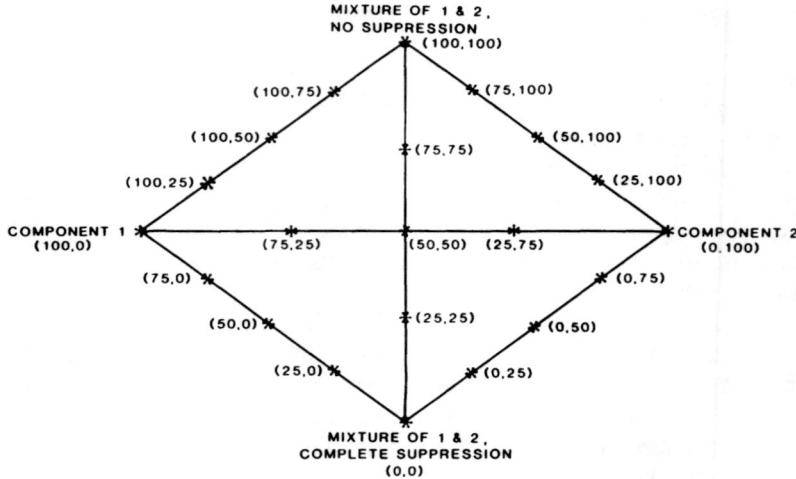

Fig. 4.7. Effect of mixture interactions on perception and discrimination of chemical stimuli. Locations of two single compounds (component 1 and component 2) and of the mixture of the two components under different absolute and relative amounts of mixture suppression are shown in the two-dimensional space derived from multidimensional scaling. The analysis is based on the responses of 40 cells, 20 responsive to component 1 and not component 2, and 20 responsive to component 2 and not component 1. For each stimulus, the responses of cell type 1 and cell type 2 are given in parentheses, respectively. Each point thus represents the across-neuron pattern for that stimulus. See text for explanation. (Derby & Girardot, unpublished data.)

the quality of that mixture relative to the quality of its components is revealed in this figure. The extent of this effect depends on the absolute and relative amounts of suppression, which are represented by the vertical and horizontal dimensions, respectively.

The vertical dimension describes the effect of changing the absolute amount of suppression such that there is reciprocal suppression between the two components (i.e. component 1 suppresses the response of cell type 2 to component 2, and component 2 suppresses the response of cell type 1 to component 1). For example, the responses to the mixture under no suppression and under complete suppression are represented by (100,100) and (0,0), respectively. Some increase in the absolute amount of mixture suppression results in the mixture being some similar to either component, and this similarity is greatest when the sum of the responses of the two types of cells is the same for the mixture and the individual components. In the case of this two-component mixture, this occurs when the reciprocal suppression is 50% (response values of [50,50]). Reciprocal suppression that is greater or less than 50% causes the mixture to be relatively more different from either component. This effect is one of *intensity*, in that the same effect can be produced by changing the concentration of the two components in an equally effective dose manner. Nonetheless, changes along the vertical dimension have a relatively minor effect on discrimination of the mixture from its components as compared with changes in this mixture along the horizontal dimension.

The horizontal dimension expresses the effect of changes in the relative amount of suppression between the two components of the mixture, such that the total amount of suppression in the two types of cells is constant. The main point drawn from this dimension is that relatively greater suppression by a component makes the mixture *more* similar to that *suppressing* component. The extreme example of this is 100% unilateral suppression. For example, when component 1 causes 100% suppression of the response of cell type 2 to component 2 and when the response of cell type 1 is not suppressed at all by component 2, the response to the mixture is (100,0), which is the same as the response to component 1 alone. Component 1 and the mixture under this suppression have exactly the same ANP and therefore the same quality. The corollary of this relationship is that unilateral suppression causes the mixture to become *less* similar to the *suppressed* component. Thus, mixture suppression can greatly enhance the contrast between the quality of a mixture and some of its components and obscure the difference between the mixture and other components. In this regard, the effect of changes in the relative amount of suppression is greater than that of changes in the absolute amount of suppression. This type of analysis was extended to the discrimination of three-component mixtures as well as to chemosensory systems composed of neurons with broad and overlapping response spectra, and the same rules as described for the two-component mixture were found to apply.

Data on the gustatory system of mammals (Travers & Smith, 1984; Smith, this volume) support the assertions of the model. ANPs for most two-component mixtures of tastants, when analyzed by multidimensional scaling, lie intermediate between the two components. An exception is a mixture of quinine and sodium chloride. This mixture lies closer to sodium chloride. This probably explains the psychophysical observations in humans (Bartoshuk, 1979) and behavioral observations in hamsters and rats (Nowlis *et al.*, 1980) that the bitter taste of quinine is suppressed by sodium chloride.

VI.   Summary and conclusions

Spiny lobsters are good subjects for the study of olfactory discrimination. They can be conditioned using either associative or non-associative techniques, and these techniques can be used to determine which odorants are perceived by them as smelling similar and which as dissimilar. Lobsters can discriminate among the members of a set of single chemicals, as well as among the members of a set of mixtures that contain the same compounds but in different relative amounts. Most of the olfactory receptor cells of spiny lobsters respond to some degree to all mixtures. Nonetheless, this population of neurons can enable the animal to distinguish between the mixtures on the basis of across-neuron patterns, which are different for the different mixtures. The quality of the across-neuron patterns is relatively stable, in spite of changes in total concentration of the mixtures. The assertion that the quality of odorant mixtures is coded by across-neuron patterns is supported by behavioral and physiological studies that are in agreement with each other, and with an analysis of the compositions of the mixtures as to the degree of similarity among the four mixtures: crab and shrimp mixtures are relatively similar to each other, while oyster and mullet mixtures are relatively different form each other and from crab and shrimp mixtures.

Mixture interactions are very common in olfaction by spiny lobsters and, because these interactions result in an alteration of the messages carried by neurons, they can have a pervasive effect on discrimination of odorants. The example described in this chapter demonstrates that mixture suppression can result in an enhancement of differences between a mixture and that mixture's suppressed components, and obfuscation of differences between that mixture and its suppressing components.

Acknowledgments

Support for this research was provided by US National Institute of Neurological and Communicative Disorders and Stroke (NINCDS) grant NS-22225 and a grant from the Whitehall Foundation, Palm Beach, Florida. We thank David Blaustein for photographic assistance.

References

Ache, B. W., & Derby, C. D. (1985). Functional organization of olfaction in crustaceans. *Trends in Neuroscience* **8**, 356–360.

Bartoshuk, L. M. (1979). Taste interactions in mixtures of sucrose with NaCl and sucrose with QHCl (abstract). *Society for Neuroscience* **5**, 125.

Carr, W. E. S., & Derby, C. D. (1986*a*). Chemically stimulated feeding behavior in marine animals: the importance of chemical mixtures and the involvement of mixture interactions. *Journal of Chemical Ecology* 12, 987–1009.

Carr, W. E. S., & Derby, C. D. (1986b). Behavioral chemoattractants for the shrimp, *Palaemonetes pugio*: identification of active components in food extracts and evidence of synergistic mixture interactions. *Chemical Senses* 11, 49–64.

Carr, W. E. S., Gleeson, R. A., Ache, B. W., & Milstead, M. L. (1986). Olfactory receptors of the spiny lobster: ATP-sensitive cells with similarities to P2-type purinoceptors of vertebrates. *Journal of Comparative Physiology* 158, 331–338.

Daniel, P. C., & Derby, C. D. (in press). Behavioral olfactory discrimination of mixtures in the spiny lobster (*Panulirus argus*) based on a habituation paradigm. *Chemical Senses* 13.

Derby, C. D., & Ache, B. W. (1984a). Quality coding of a complex odorant in an invertebrate. *Journal of Neurophysiology* 51, 906–924.

Derby, C. D., & Ache, B. W. (1984b). Electrophysiological identification of the stimulatory and interactive components of a complex odorant. *Chemical Senses* 9, 201–218.

Derby, C. D., & Atema, J. (1988). Chemoreceptor cells in aquatic invertebrates: peripheral filtering mechanisms in decapod crustaceans. In J. Atema, R. R. Fay, A. N. Popper & W. N. Tavolga, eds, *Sensory biology of aquatic animals*, pp. 365–386, Springer, New York.

Derby, C. D., Carr, W. E. S., & Ache, B. W. (1984a). Purinergic olfactory receptor cells of crustaceans: response characteristics and similarities to internal purinergic cells of vertebrates. *Journal of Comparative Physiology* 155, 341–349.

Derby, C. D., Hamilton, K. A., & Ache, B. W. (1984b). Processing of olfactory information at three neuronal levels in the spiny lobster. *Brain Research* 300, 311–319.

Fine-Levy, J. B., Derby, C. D., & Daniel, P. C. (1987). Chemosensory discrimination: behavioral abilities of the spiny lobster. *Annals of the New York Academy of Science* 510, 280–283.

Fine-Levy, J. B., Girardot, M.-N., Derby, C. D., & Daniel, P. C. (1988). Differential associative conditioning and olfactory discrimination in the spiny lobster *Panulirus argus*. *Behavioral and Neural Biology* 49, 315–331.

Fuzessery, Z. M., Carr, W. E. S., & Ache, B. W. (1978). Antennular chemosensitivity in the spiny lobster. *Panulirus argus*: studies of taurine sensitive receptors. *Biological Bulletin* 154, 226–240.

Girardot, M.-N., & Derby, C. D. (1988). Neural coding of quality of complex olfactory stimuli in lobsters. *Journal of Neurophysiology* 60, 303–324.

Girardot, M.-N., Fine, J. B., & Derby, C. D. (1986). Coding of odorant quality by the olfactory system of the lobster: behavioral and neural analysis of discrimination of quality of single chemicals and chemical mixtures (abstract). *Society for Neuroscience* 12, 1352.

Gleeson, R. A., & Ache, B. W. (1985). Amino acid suppression of taurine-sensitive chemosensory neurons. *Brain Research* 335, 99–107.

Maes, F. W., & Bijpost, S. C. A. (1979). Classical conditioning reveals discrimination of salt taste quality in the blowfly *Calliphora vicina*. *Journal of Comparative Physiology* 133, 53–62.

Maes, F. W., & Ruifrok, A. C. C. (1986). Neural coding of salt taste quality in the blowfly *Calliphora vicina*. II. Ensemble coding. *Journal of Comparative Physiology* 159, 89–96.

Nowlis, G. H., Frank, M. E., & Pfaffmann, C. (1980). Specificity of acquired aversions to taste qualities in hamsters and rats. *Journal of Comparative Physiological Psychology* 94, 932–942.

Smith, D. V., & Travers, J. B. (1979). A metric for the breadth of tuning of gustatory neurons. *Chemical Senses* 4, 215–229.

Smith, D. V., Van Buskirk, R. L., Travers, J. B., & Bieber, S. L. (1983). Coding of taste stimuli by hamster brainstem neurons. *Journal of Neurophysiology* **50**, 541–558.
Travers, S. P., & Smith, D. V. (1984). Responsiveness of neurons in the hamster parabrachial nuclei to taste mixtures. *Journal of General Physiology* **84**, 221–250.
Zimmer-Faust, R. K., Tyre, J. E., Michel, W. C., & Case, J. F. (1984). Chemical mediation of appetitive feeding in a marine decapod crustacean: the importance of suppression and synergism. *Biological Bulletin* **167**, 339–353.

5

Adaptation and mixture interactions in chemoreceptor cells: mechanisms for diversity and contrast enhancement

Jelle Atema, Paola Borroni, Bruce Johnson,
Rainer Voigt and Linda Handrich

Boston University Marine Program, Marine Biological Laboratory
Woods Hole, Massachusetts, USA

I. Introduction

Chemoreceptor cells in the receptor organs of various animals interface with organ-specific chemical stimulus environments. As do other receptor cells (Hartline, Wagner & MacNichol, 1952; Costalupes, Young & Gibson, 1984), they serve to enhance contrast between biologically relevant signals and various forms of environmental noise, and in higher animals to provide diversity in response patterns. Contrast enhancement can be accomplished by spectral as well as temporal/intensity differentiation of stimuli. Diversity of stimulus representation is provided by a large population of receptor cells, each with unique characteristics of response to spectral and temporal aspects of the stimulus. For chemoreceptors, spectral contrast refers to discrimination of different compounds and mixtures, whereas temporal contrast refers to discrimination of the rates at which different spectral signals change in concentration. Although we have begun to interpret our data on temporal contrast discrimination, this chapter will be restricted almost entirely to spectral contrast. While further along than temporal pattern studies, even the study of spectral contrast in chemical senses is only in its infancy. As this chapter will show, there may be important and unexpected temporal components involved in enhancement of spectral contrast.

Contrast and diversity are created by molecular mechanisms that turn receptor cells into spectral and temporal filters. Enhancement of

Copyright © 1989 by Academic Press Australia.
All rights of reproduction in any form reserved.

contrast and diversity may be aided by mixture effects and cross-adaptation in which competitive binding and differential coupling of receptors and transduction mechanisms, such as second messengers, result in unique responses for each cell. Temporal filtering results from adaptation and disadaptation rates of receptor cells (e.g. Block, Segall & Berg, 1982, 1983; Segall, Block & Berg, 1986). These rates vary greatly among cells, adding to each cell's uniqueness as it responds to the natural mixtures that determine the behavior of many animals. Based on the great diversity of neural activity patterns generated across its receptor cells, the central nervous system (CNS) appears capable of 'choosing' patterns. It is the task of the CNS to select patterns based on previous experience. Certain patterns are no doubt genetically fixed or preferred due to natural and sexual selection. This chapter reviews recent results from experiments in our laboratory on filter properties of lobster chemoreceptor cells leading to a spectral filter model. We will argue that spectral filtering in chemoreceptors may be based on (variable) distribution of differently tuned receptor sites across receptor cells.

In the lobster, *Homarus americanus*, we have defined receptor cell response properties of four different chemoreceptor organs (two for smell = lateral and medial flagella of the antennule; and two for taste = chelated walking legs and maxillipeds) and responses of the whole animal in different behavioral tasks. We also have behavioral evidence for some of the functions of the different receptor organs as well as physical/chemical information about the nature of the different stimulus distributions around these organs (for reviews, see Atema 1985, 1988). This information allows us to test receptor cells, receptor organs and whole animals with biologically relevant stimuli. The use of realistic stimuli can avoid years of dead-end research. Stimuli must be considered both in terms of qualitative (spectral) and quantitative (dynamic/temporal) realism.

As a large aquatic invertebrate, the lobster has a number of features that make it a favorable model for the study of chemoreceptive contrast. Its size facilitates behavioral observation and manipulation as well as physiological access to receptor cells and organs. The aquatic environment has physical/chemical advantages over the aerial medium both for stimulus delivery and quantification, and for scaling of flumes and odor plume models (Vogel, 1983; Atema, 1987). Invertebrate models also have advantages over vertebrates in that one can record easily from receptor cells. In addition, invertebrates often show intricate signal filtering in the periphery, whereas vertebrates use neural processing. For instance, lobsters have relatively narrowly tuned chemoreceptor cells for common amino acids, ammonium and other compounds. This feature makes it initially easier to understand the rules by which mixture stimuli are processed at the level of receptor cells.

II. Methods

The physiological results described in this chapter are based on data from extracellular recordings of primary chemoreceptor cells. Typically, an appendage is excised and placed in a recording chamber, where the cut end is bathed in Ringer solution and the distal end with chemoreceptor sensilla is superfused in a flow of artificial sea water (ASW). Hook or suction electrodes are used to record action potentials from small nerve bundles, resulting in clear recordings of one or two cells. Response magnitude is measured as number of spikes per unit time or per response. Stimuli are highly reproducible 1 s pulses (Fig. 5.1A) of certain concentrations of amino acids or other compounds injected into the ASW flow. Stimuli are separated by one to five minutes depending on stimulus concentration to allow for recovery from adaptation (Voigt & Atema, 1987). Backgrounds are presented by changing ASW flow to background flow (Fig. 5.1B). For further methodological details see Johnson, Voigt, Borroni & Atema (1984).

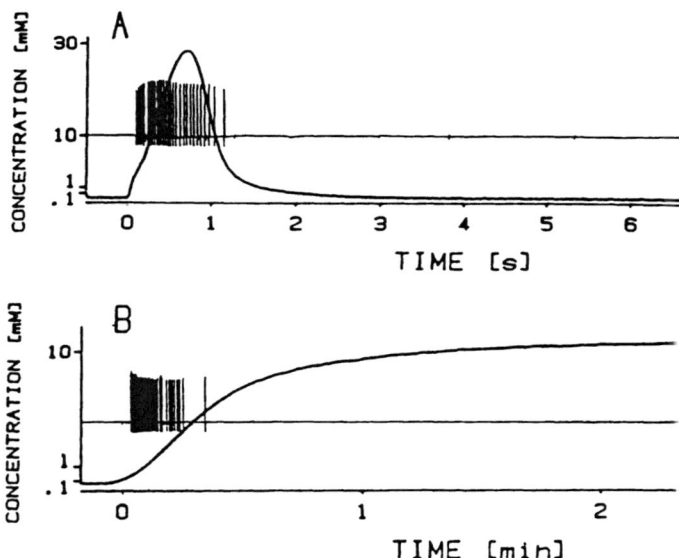

Fig. 5.1. A. Spiking response of a lobster leg ammonium cell superimposed on the concentration profile of a 1 s standard pulse of ammonium (10^{-4} M NH_4 in ASW). B. Spiking response of an ammonium cell to a slowly rising NH_4 concentration: no activity before or after self-adaptation. Time courses of stimulus and background concentrations are determined by conductivity change in separate experiments; therefore the temporal alignment of concentration profile and response is estimated (accuracy within seconds). (From Borroni & Atema, in press.)

III.   Results

Spectral tuning

The majority of receptor cells in the chemoreceptor organs of lobsters are narrowly tuned: the cells respond best to any one compound out of a series of related compounds such as amino acids (Derby & Atema, 1982; Johnson & Atema, 1983; Johnson *et al.*, 1984; Tierney, Voigt, & Atema, 1988; Corotto & Atema, 1987). This divides the cells into 'spectral' populations defined by best compound. Some cells have second(third, etc.)-best compounds to which they respond less, often far less, at equimolar stimulus concentrations. Within most spectral populations, the second(etc.)-best compounds are different from cell to cell. The picture emerging from hundreds of receptor cells thus far studied shows that each organ is made up of a large number of spectral subpopulations. Spectrally homogeneous subpopulations can be subdivided further by excitability and various other aspects of the stimulus-response functions of the individual cells. Other criteria, such as suppression and enhancement by mixtures and adaptation and disadaptation rates lead to further separation of receptor cells. The result is that no one single receptor cell is identical to another, thus creating enormous diversity across a population of thousands (taste) to hundreds of thousands (smell) receptor cells.

In this chapter we will use examples from four spectral populations responding to ammonium (NH_4), glutamate (Glu), hydroxyproline (Hyp) and taurine (Tau). It is important to state that the designation 'ammonium cells' is shorthand for ammonium-sensitive cells and not necessarily ammonium-best cells. The same is true for other cell types. Since it is impossible to test a cell with all known chemical compounds and impractical to test with a large number of compounds before testing anything else, the spectral classification of a cell is often limited to mere sensitivity to a search stimulus. For example, when searching with ammonium stimuli certain cells respond. These cells are called ammonium-sensitive cells. Only when a cell has been tested with a number of single compounds can one discover if it is indeed an ammonium-best cell. Even then, 'best' is still restricted to the compounds tested. Experience has shown that ammonium-sensitive cells are almost always ammonium-best. The same is true for taurine cells. But this is not necessarily the case for other populations, e.g. hydroxyproline cells. In this paper we will distinguish specifically between ammonium-sensitive cells and ammonium-best cells, and similarly for the other spectral populations.

We will summarize by cell population recent results of studies on self- and cross-adaptation and on mixture suppression and enhancement leading to a tentative receptor cell model of spectral (and temporal) filtering.

1. Ammonium cells

Ammonium-sensitive cells demonstrate an important characteristic: complete self-adaptation to prolonged exposure of ammonium ions. In a slowly rising ammonium concentration the cells fire initially and then become and remain silent long before the concentration reaches its maximum (Fig. 5.1B). Pulses of higher concentrations of ammonium (signals) superimposed on a constant ammonium background (noise) can cause a response again, implying that the receptor cell is desensitized rather than depleted of some component needed for response. The instantaneous dynamic response of ammonium cells can be described by a family of stimulus response functions, parallel to each other and originating from thresholds fixed by the background concentration of ammonium (Fig. 5.2); i.e. the signal to noise ratio determines the mean response of these cells (Borroni & Atema, 1987; in press).

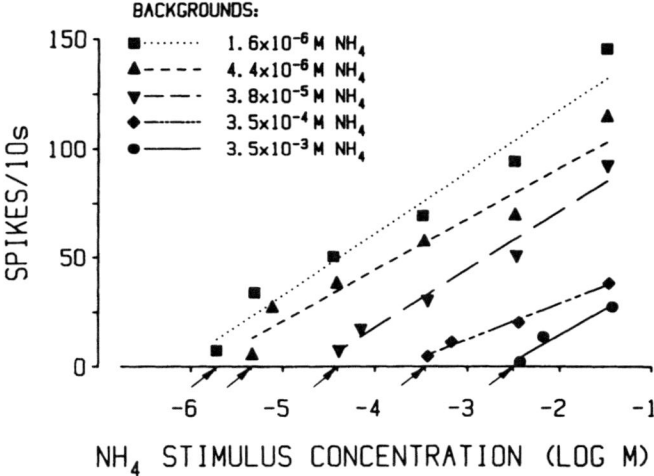

Fig. 5.2. Complete self-adaptation and parallel shift of NH_4 response function of lobster leg ammonium cells (N = 14) when stimulated with ascending concentration series of 1 s pulses in five different NH_4 backgrounds (indicated by arrows). (After Borroni & Atema, in press.)

Table 5.1. Mean relative response to NH_4, Glu, Bet and Hyp and tuning breadth (H) of three lobster leg ammonium-best cell subpopulations

	NH_4	Glu	Bet	Hyp	H
NH_4	100	0	0	0	0.00
NH_4-Bet	100	13	3	1	0.28
NH_4-Glu	100	61	19	15	0.66

The diversity index (H) ranges from 0 ('most narrowly tuned', response to only one stimulus) to 1 ('not tuned', equal response to all stimuli).

The population of ammonium-best cells can be subdivided by second-best compounds into pure ammonium cells, ammonium-betaine cells, and ammonium-glutamate cells. The tuning breadth of a subpopulation is an important determinant of its cross-adaptability (Borroni & Atema, 1987). Pure ammonium cells (narrowly tuned, Table 5.1) did not alter their response to a concentration series of ammonium pulses in low or high backgrounds of betaine or glutamate (Bet): i.e. they showed no cross-adaptation even at background concentrations 300 times the ammonium stimulus concentration (Fig. 5.3A). Ammonium-betaine cells (broadly tuned, Table 5.1) showed cross-adaptation (39% response suppression at the highest stimulus concentrations) to high but not to low betaine background (Fig. 5.3B). Evidently, the lower betaine background concentration was below threshold for cross-adaptation. Ammonium-glutamate cells (most broadly tuned, Table 5.1) cross-adapted to both high and low glutamate backgrounds (Fig. 5.3C). Surprisingly, the degree of cross-adaptation (35% response suppression at the highest stimulus concentrations) was equal in the two glutamate backgrounds, suggesting that the adapting mechanism of the ammonium cells was already saturated by the lower Glu background. In all cases, cross-adaptation was essentially different from self-adaptation (Fig. 5.2), in that the slopes of stimulus-response functions in cross-adaptation diminished significantly. This reflects the fact that cross-adaptation was most effective at high stimulus concentrations, while hardly affecting the weaker stimuli.

The conclusion from the cross-adaptation experiments on ammonium-best cells is the following. Only compounds that excited a cell also cross-adapted it. A cell's sensitivity to other compounds (reflected in its tuning breadth) determined the degree of cross-adaptation. Cross-adapting mechanisms had a threshold and saturated easily.

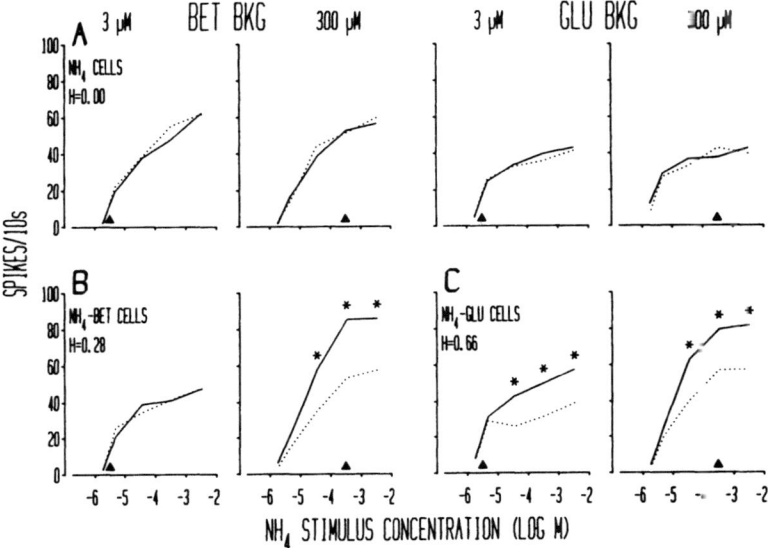

Fig. 5.3. Response functions of three different ammonium-best cell subpopulations from lobster legs in ASW (solid line) and in cross-adapting backgrounds (broken line). **A.** Pure ammonium cells are extremely tuned (H = 0.00) and not affected even narrowly by high (300μM) betaine or glutamate backgrounds (BKG). **B.** Ammonium-betaine cells are more broadly tuned (H = 0.28) and suppressed in high but not low (3 μM) betaine backgrounds. **C.** Ammonium-glutamate cells are even more broadly tuned (H = 0.66) and nearly equally suppressed in both high and low glutamate backgrounds. See Table 5.1 for response breadths. Arrowheads indicate approximate background concentrations. Asterisks indicate a significant difference between responses in different backgrounds ($p < .05$; paired t-test). (After Borroni & Atema, unpublished data.)

2. Glutamate cells

Glutamate sensitive cells were tested with a series of identical glutamate pulses to determine their recovery from self-adaptation to previous pulses (Voigt & Atema, 1987). Self-adaptation is a fast process (less than a few seconds), leading in most of the cells observed in lobsters to complete cessation of response. Disadaptation (recovery from self-adaptation), however, is a slower process often requiring minutes for full recovery. The population of glutamate cells can be subdivided into several groups of cells, each with different adaptation, disadaptation and excitation properties. Details of this work fall outside the scope of this

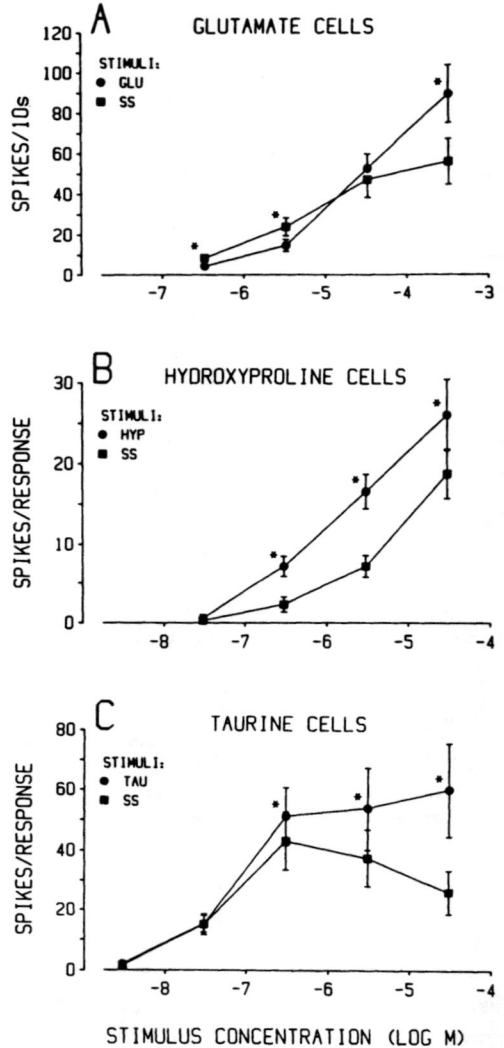

Fig. 5.4. Mixture effects in glutamate (A, N = 15), hydroxyproline (B, N = 21) and taurine (C, N = 17) cells. Glutamate cells were from lobster legs, Hyp and Tau cells from antennules. Population stimulus-response functions are for the best compound and the mixture, both in ASW background. Hyp cells show a parallel shift in the dose-response function. Glu and Tau cells show a decreased slope of the function as a result of mixture stimulation. Mixture (SS) composition: equimolar amounts of taurine, hydroxy-L-proline, L-glutamate, L-glutamine, L-arginine, L-alanine, L-lysine, L-aspartate, L-leucine, L-proline, ammonium chloride, betaine, glycine, sucrose and ethanol. Mixture curve plotted at the concentration of each single compound within the mixture. Asterisks indicate a significant difference between responses to single compound and mixture ($p < .05$, two-tailed Wilcoxon signed-ranks test). (After Johnson *et al.*, 1985.)

chapter, but contribute significantly to response diversity referred to in the Introduction and Discussion. Moreover, since mixture effects are probably at least in part related to cross-adaptation mechanisms, the time courses of adaptation and disadaptation cannot be ignored in mixture studies.

To explore mixture effects, glutamate cells were tested with pairs of glutamate and mixture stimulus pulses at increasing concentrations (Johnson, Borroni & Atema, 1985). The mixture contained 15 compounds, each in equimolar amounts, including glutamate, hydroxyproline and taurine. As in cross-adapting ammonium cells, the stimulus response functions began to deviate significantly at the higher concentrations, where the mixture caused lower responses than glutamate alone (Fig. 5.4A). This 'mixture suppression' was seen regularly in most cells at high concentrations. At lower concentrations, however, most cells showed enhanced responses to the mixture. Responses of individual cells showed great variability. These results suggest that more narrowly tuned cells have an instantaneous cross-adaptation mechanism that becomes increasingly effective at higher stimulus concentrations, whereas at lower concentrations different mixture compounds excite the cell with little or no change in cell sensitivity. The sensitivity of individual glutamate cells to mixture components was not tested in these experiments, which relied on population tuning data from Johnson *et al.* (1984), where glutamate cells were shown to be narrowly tuned. Mixture effects, both suppression and enhancement, were clearly established for the glutamate cell population.

3. *Hydroxyproline cells*

Hydroxyproline (Hyp)-sensitive cells were investigated in some detail for various reasons, including their surprisingly low maximum firing rate (Johnson, Merrill, Ogle & Atema, 1987). From that study it became clear that there are several subpopulations of Hyp-sensitive cells. One of these is a rather homogeneous subpopulation of Hyp-best cells characterized by low firing rate, no spontaneous activity and narrow tuning. Other subpopulations are more broadly tuned.

The same mixture that suppressed glutamate cells also suppressed Hyp-sensitive cells (Fig. 5.4B). When tested with the same paired procedure as used in the Glu cell tests above, the Hyp cell population showed a consistently and significantly lower mean response function for the mixture stimuli than for the Hyp stimuli (Johnson *et al.*, 1985). No mixture enhancement was found in the more broadly tuned Hyp-sensitive population.

In a second series of tests on mixture effects, Hyp-best cells were presented with 3×10^{-6} M Hyp, the 15-compound equimolar mixture

(SS) including Hyp, and the same mixture without Hyp (SS-Hyp), all in artificial seawater (ASW) background (Johnson, Voigt & Atema, in press). Then the background was switched from ASW to 10^{-5} M SS-Hyp (concentration of each mixture component 10^{-5} M). This test paradigm allowed us to compare responses to Hyp and SS-Hyp when presented simultaneously (i.e. as SS) in ASW (mixture suppression protocol) and when presented as a Hyp pulse in a SS-Hyp background (cross-adaptation protocol). In the latter case, the cells could be considered cross-adapted (for at least three minutes) to SS-Hyp. The results showed first that SS-Hyp stimulated this Hyp-sensitive population quite

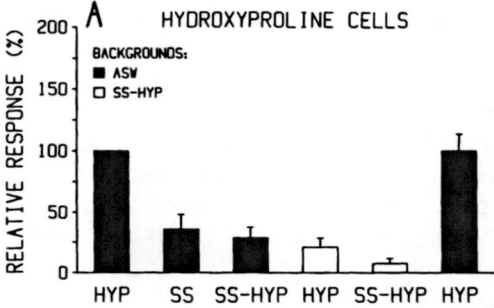

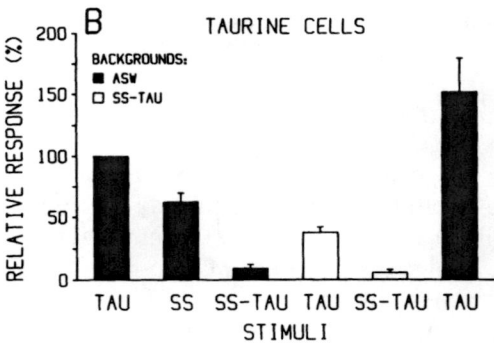

Fig. 5.5. Comparison of mixture suppression (bars 1 and 2) and cross-adaptation (bars 1 and 4) in Hyp (A, N = 8) and Tau (B, N = 9) cell populations from lobster antennules. (Bars 1 to 6 run from left to right, respectively.) Cell responses are normalized to its response to best compound. Significant mixture suppression, i.e. difference between first and second bar ($p < .05$; two-tailed Wilcoxon signed-ranks test), was observed in both Hyp and Tau cells. Significantly greater cross-adaptation than mixture suppression, i.e. difference between second and fourth bar ($p < .05$; two-tailed Wilcoxon signed-ranks test) occurred for Tau, but not for Hyp. SS-Hyp stimulus was 30% above SS-Hyp background (10^{-5} M), as was SS-Tau. See text for details and protocol. (After Johnson, Voigt & Atema, in press.)

well, although less than Hyp alone (Fig. 5.5A, bar 3 vs. 1). Based on the ammonium cell results discussed earlier, we would expect therefore that SS-Hyp also can cause significant cross-adaptation. As expected, when SS-Hyp and Hyp were then presented together (i.e. SS stimulus in ASW), the response was significantly suppressed (Fig. 5.5A, bar 1 vs. 2), a 'classical' case of mixture suppression. This suppression was not significantly greater when the SS-Hyp compounds were given three minutes to cross-adapt the Hyp cells (Fig. 5.5A, bar 4 vs. 2), i.e. the SS-Hyp compounds exerted their cross-adapting effects immediately. This is different from Tau cells, as will be seen below. Both the adapted and the suppressed responses were fully reversible (Fig. 5.5A, bar 6).

Finally, Hyp-best cells were also tested with binary equimolar mixtures of Hyp and each of the remaining 14 single components of the mixture to identify suppressing and enhancing compounds (Johnson *et al.*, in press). The surprising result was that all cells differed in the degree to which individual compounds affected the response to Hyp. No consistent patterns emerged, except that suppression was more common than enhancement. Although compounds such as glutamine and glycine *suppressed* Hyp responses in most cells, they *enhanced* Hyp responses in other cells. Only arginine suppressed Hyp responses in all cells. The taurine-Hyp mixture caused reduced responses in a few cells and increased responses in a few other cells when compared with the responses of the same cells to Hyp alone; most cells showed no change. This extends the overall pattern of receptor cell diversity.

4. Taurine cells

Taurine (Tau) sensitive cells are among the best studied in lobster chemoreception (Ache, 1982). In *H. americanus* they occur in all four chemoreceptor organs studied so far; they are especially prominent in antennules. The very low natural background levels of taurine may make it a suitable signal for the detection of distant odor sources (Fuzessery, Carr & Ache, 1978; Johnson & Atema, 1986). We will restrict our treatment of Tau cells here to the same three experiments described above for Hyp cells. Like Glu cells, Tau cells tend to be narrowly tuned.

When tested with paired stimuli of Tau and SS, Tau cells showed mixture suppression of the population response function, but only at higher concentrations (Fig. 5.4C). At low stimulus concentrations, individual cells showed great variability in their responses; many cells actually showed enhanced responses to the mixture. The results and interpretation for the Tau population are the same as for the Glu cells described above; perhaps such features are seen best in narrowly tuned populations.

When tested in ASW with SS, Tau cells showed less mixture suppression than Hyp cells (Fig. 5.5B vs. Fig. 5.5A, bars 1 vs. 2). Further, the SS-Tau mixture was only slightly stimulatory by itself and less than SS-Hyp for Hyp cells (Fig. 5.5B vs. Fig. 5.5A; bar 3). These results are not inconsistent with the ammonium cell results, which showed compounds that stimulate a cell can also suppress it. After a background switch from ASW to 10^{-5} M SS-Tau, the now pre-adapted Tau cells gave lower responses to Tau than when presented with SS (i.e. Tau and the SS-Tau mixture simultaneously) in ASW (Fig. 5.5B, bar 4 vs. 2). Tau and Hyp cells thus show an important difference in the time course of adaptation: where Hyp cells show the same response reduction to Hyp whether SS-Hyp is presented simultaneously or as a cross-adapting background, Tau cells are significantly more suppressed when presented with Tau in a SS-Tau background than by simultaneous exposure to Tau and SS. This implies that the SS-Tau compounds act slower on the Tau cells than the SS-Hyp compounds on Hyp cells. This difference may indicate that two different processes are involved, and/or that more stimulatory compounds (SS-Hyp) immediately saturate the cell's cross-adaptation process (i.e. the more broadly tuned Hyp cells) whereas less stimulatory compounds can affect the cross-adapting process gradually (i.e. the more narrowly tuned Tau cells).

In binary presentations, Tau cells and Hyp cells were equally diverse. An interesting comparison is the response of Hyp and Tau cells to a binary equimolar Hyp/Tau mixture. Most Tau cells were enhanced by that binary mixture, where Hyp cells were not much affected (Johnson *et al.*, in press).

IV.   Discussion

A.   *Receptor cell model*

Contrast and diversity are provided by spectral and temporal filter properties of the receptor cells. Our results from studies on self- and cross-adaptation and on mixture suppression and enhancement lead to the following filter model of (lobster) chemoreceptor cells. The purpose of such an 'extracellular' model is to describe filter properties of whole cells, and to provide constraints for biochemical and biophysical studies aimed directly at elucidating cellular mechanisms. Any statement regarding intracellular mechanisms remains speculative until tested.

The model proposes that each cell contains a unique blend of receptor sites. The more dominant a particular site, the more narrowly the cell is tuned to the stimuli that bind to that site. Furthermore, each site is tuned to a best compound. Different sites have different tuning breadth, i.e. different affinities and/or efficacies for different compounds. However, each site is probably genetically constrained (low variability). Thus, spectral tuning of a cell is determined at two levels: tuning breadth of its receptor sites and its particular, cell-specific blend of sites. This allows for various degrees of competitive inhibition (or agonism) at each of the receptor sites of a cell. It also allows for sites competing for transduction mechanisms, such as second messenger cascades. This, in turn, would lead to the possibility of excitation summation ('enhancement') at low stimulus concentrations and response suppression at high stimulus concentrations. The fewer and the more specific the sites, the more narrowly the cell is tuned.

We use the term 'mixture suppression' for cases for instantaneous cross-adaptation. Such effects are consistent with competition for receptor sites, although other mechanisms can be imagined. In extreme cases of narrow tuning, we may be dealing with cells completely dominated by one receptor site. This site may have different affinities and efficacies for different compounds. Greatest efficacy for ammonium, taurine or glutamate would make it an ammonium, taurine or glutamate cell. Other compounds with lesser *efficacy* would be seen as second(etc.)-best compounds in whole-cell recordings. Different *affinities* for such second-best compounds would result in more or less efficient mixture suppression.

We use the term cross-adaptation when a time element is involved in desensitizing the cell. In cells dominated by more than one site, such as perhaps the ammonium-glutamate or the ammonium-betaine cells, we might see cross-adaptation in addition to mixture suppression. The different sites may converge on common transduction mechanisms, thus allowing the cell's sensitivity for its best compound to be reduced by prior exposure to a second-best compound *beyond* the response reduction caused by mixture suppression, (see taurine cell results, Fig. 5.5B). However, when the best and the second-best compounds are presented simultaneously, we can see excitation summation, i.e. response enhancement compared to the response to the best compound alone. Thus, in cells where we see mixture enhancement at lower mixture concentrations (i.e. several individual taurine and glutamate cells, see also Fig. 5.4A), different sites may sum their effects on transduction mechanisms, while not yet showing much mixture suppression: at the lower stimulus concentrations each site is primarily affected by its best compound with little interference from the less efficient second-best compounds. (The same additive effects occur with certain binary mixtures in Tau and Hyp cells). At higher mixture concentrations the compounds become more efficient in acting as second-best compounds at 'each other's' best sites, leading to overall response reduction: mixture suppression overwhelms excitation summation.

Self-adaptation is competitive, but it may take place at different sites. An ammonium-best cell may have, besides specific ammonium sites, other sites where ammonium is a second-best compound. Self-adaptation would desensitize all sites proportionally, and reduce each site's ability to compete for subsequent transduction mechanisms.

This model is consistent with mixture suppression and enhancement, and with self- and cross-adaptation results obtained in our laboratory as reviewed in this chapter. It also provides for clearly falsifiable statements that form the basis for experimental work of the present and future. Thus far, we have substantial experimental results for only a small part of this model. Events that are subthreshold for spiking responses may well be involved, but can obviously not be measured directly in extracellular recordings.

Observed diversity in temporal filtering (adaptation and disadaptation) properties of receptor cells (Voigt & Atema, 1987) may be an integral part of mixture effects. It may be due to boundary layer integration and to differences in components of the second messenger cascade (chemoreception: Pace, Hanski, Salomon & Lancet, 1985; Sklar, Anholt & Snyder, 1986; Shirly, Robinson, Dickinson, Aujla & Dodd, 1986; Huque & Bruch, 1986; vision: Lamb, 1986; Fein, 1986). Our results point in the direction that spectral diversity is created

1. by variable distribution of perhaps hundreds of different receptor proteins across hundreds of thousands of receptor cells;
2. by different time courses of excitation and adaptation for different compounds that act on any one cell.

The latter automatically could provide the basis for temporal filtering diversity. Both spectral and temporal filter properties may result from the same cellular processes.

B.   Diversity and contrast

The various examples given above clearly illustrate that the more criteria used to interrogate a cell, the more diversity turns up among cells. This enormous diversity seems advantageous for a general detection system, such as olfaction, where many food odors and individual body odor mixtures must be discriminated. Indeed, lobsters can distinguish between body odors (and urine) of male and female lobsters (Atema & Cowan, 1986) and between the body odors of two different species of mussels (Derby & Atema, 1981). Also, specific mixtures are more potent behavioral stimuli than single-mixture components or partial mixtures (Borroni, Handrich & Atema, 1986). To code for such subtly different mixtures, it is likely that each odor, be it a single compound or a mixture, will generate a unique output pattern of activity across all the receptor cells. Only in specialized detection systems, such as we find in insect pheromone receptors, the demands of communication may have selected for extreme sensitivity to a few signal compounds at the cost of diversity. This remains to be tested, however.

Discrimination of different odor mixtures must normally be done against environmental noise backgrounds. This requires detection of contrast between a complex signal and a complex background. Detection of chemical contrast is evident in behavioral experiments with lobsters and hermit crabs. In backgrounds of mussel extract, hermit crabs did not respond to mussel extract stimuli until, as expected, the stimulus exceeded the background; similar results were obtained with fish extract stimuli in fish extract backgrounds. However, hermit crabs were capable of detecting a mussel extract stimulus against high backgrounds of a fish extract and vice versa (Sammon & Atema, 1987). Similarly, lobsters showed parallel shifts in behavioral response functions (Fig. 5.6A), when stimulated with a 22-compound synthetic mixture based on mussel extract presented against different background concentrations of that same mixture (Handrich & Atema, 1987). Self-adaptation to a complex mixture may lower cell responses across the board without significantly distorting the activity pattern across the receptor cells. Such behavioral results resemble the effects of self-adaptation in ammonium cells (Fig. 5.2). As in single ammonium cells, signal-to-background ratios determine the animal's behavioral response to mixtures. However, in elevated ammonium backgrounds, not uncommon in nature, the response functions for the mixture stimuli have a different slope (Fig. 5.6B), indicating non-linear effects of ammonium. Increased background concentration of ammonium may distort firing patterns across the receptor cells, making the mixture less recognizable (Handrich & Atema, 1987).

It is unlikely that behavioral capabilities of contrast detection can be explained directly by peripheral sensory filter properties. However, in order to understand the workings of the CNS, it is important to know what kind of information is extracted by the receptors. In addition, detailed knowledge of receptor cell filter properties serves to guide research into the underlying molecular and biochemical processes of chemoreception. In turn, both molecular and behavioral/CNS processes will help to determine the spectral and temporal filter characteristics of

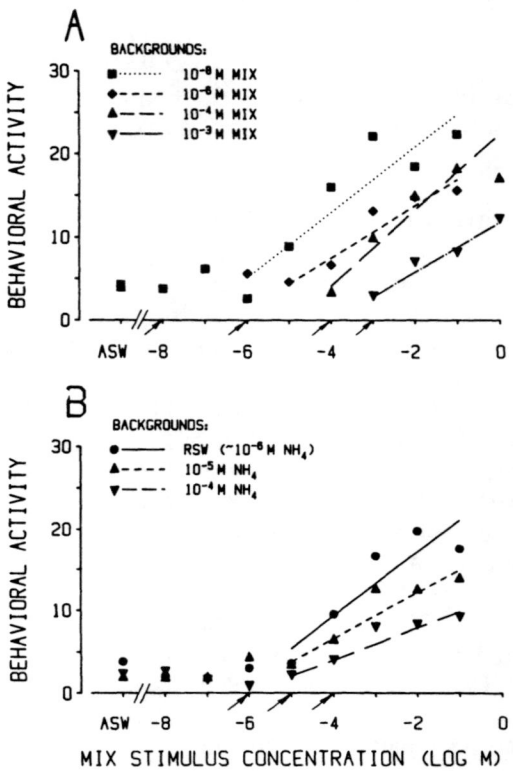

Fig. 5.6. A. Parallel shift of behavioral response functions to mixture stimuli (MIX = mixture of 22 compounds, mostly amino acids, in proportions found in mussel extract) caused by increasing MIX backgrounds. B. In increasing ammonium background concentrations, response functions to MIX stimuli decrease in slope. Arrows indicate approximate background concentration levels. RSW = raw sea water. ASW = artificial sea water. (After Handrich & Atema, 1987.)

peripheral receptor cells, as they contribute to contrast enhancement of behaviorally important mixtures against mixture backgrounds that fluctuate in space and time (Atema, 1985, 1988).

Acknowledgments

We thank Drs B. W. Ache, J. Brand, B. Bryant, and R. Payne for discussion and critical review of the manuscript. This work was supported by grants from Whitehall Foundation, Palm Beach, Florida, and the US National Science Foundation (BNS 85–12585).

References

Ache, B. W. (1982). Chemoreception and thermoreception. In H. L. Atwood & D. C. Sandeman, eds, *The biology of crustacea*, vol. 3, pp. 369–398. Academic Press, New York.

Atema, J. (1985). Chemoreception in the sea: adaptations of chemoreceptors and behavior to aquatic stimulus conditions. *Society of Experimental Biology Symposium* 39, 387–423.

Atema, J. (1987). Aquatic and terrestrial chemoreceptor organs: morphological and physiological designs for interfacing with chemical stimuli. 1 P. Dejours, L. Bolis, C. R. Taylor & E. R. Weibel, eds, *Terrestrial versus aquatic life: contrasts in design and function*, pp. 303–316. Lavinia Press, Padua.

Atema, J. (1988). Distribution of chemical stimuli. In J. Atema, A. N. Popper, R. R. Fay & W. N. Tavolga, eds, *Sensory biology of aquatic animals*, pp. 29–56. Springer-Verlag, New York.

Atema, J., & Cowan D. F. (1986). Sex identifying urine and molt signals in the lobster *Homarus americanus, Journal of Chemical Ecology* 12, 2065–2080.

Block, S. M., Segall, J. E., & Berg, H. C. (1982). Impulse responses in bacterial chemotaxis. *Cell* 31, 215–226.

Block, S. M., Segall, J. E., & Berg, H. C. (1983). Adaptation kinetics in bacterial chemotaxis. *Journal of Bacteriology* 154, 312–323.

Borroni, P. F., & Atema, J. (in press). Adaptation in chemoreceptor cells I. Self-adapting backgrounds determine threshold and cause parallel shift of response function. *Journal of Comparative Physiology, A.*

Borroni, P. F., & Atema, J. (1987). Self- and cross-adaptation of single chemoreceptor cells in the taste organs of the lobster, *Homarus americanus. Annals of the New York Academy of Sciences* 510, 184–186.

Borroni, P. F., Handrich, L. S., & Atema, J. (1986). The role of narrowly tuned taste cell populations in lobster (*Homarus americanus*) feeding behavior. *Behavioral Neuroscience* 100, 206–212.

Corotto, F., & Atema, J. (1987). Initial survey of the chemosensory response properties of lobster mouthparts: spectral populations and tuning breadth. *Biological Bulletin* 173, 436.

Costalupes, G. A., Young, E. D., & Gibson, D. J. (1984). Effects of continuous noise backgrounds on rate response of auditory nerve fibers in cat. *Journal of Neurophysiology* 51, 1326–1344.

Derby, C. D., & Atema, J. (1981). Selective improvement in responses to prey odors by the lobster, *Homarus americanus*, following feeding experience. *Journal of Chemical Ecology* 7, 1073–1080.

Derby, C. D., & Atema, J. (1982). Narrow-spectrum chemoreceptor cells in the walking legs of the lobster *Homarus americanus*: taste specialists. *Journal of Comparative Physiology* 146, 181–189.

Fein, A. (1986). Excitation and adaptation of Limulus photoreceptor by light and Inositol 1,4,5-triphosphate. *Trends in Neurosciences* 9, 110–114.

Fuzessery, Z. M., Carr, W. E. S., & Ache, B. W. (1978). Antennular chemosensitivity in the spiny lobster, *Panulirus argus*: studies of taurine sensitive receptors. *Biological Bulletin* 154, 226–240.

Handrich, L. S., & Atema, J. (1987). Effects of chemical noise on the detection of chemical stimuli. *Annals of the New York Academy of Sciences* 510, 342–344.

Hartline, H. K., Wagner, H. G., & MacNichol, E. F. (1952). The peripheral origin of neuron activity in the visual system. *Cold Spring Harbor Symposium on Quantitative Biology* **17**, 125–141.

Huque, T., & Bruch, R. C. (1986). Odorant- and guanine nucleotide-stimulated phosphoinositide turnover in olfactory cilia. *Biochemica Biophysica Research Communications* **137**, 36–42.

Johnson, B. R., & Atema, J. (1983). Narrow-spectrum chemoreceptor cells in the antennules of the American lobster, *Homarus americanus*. *Neuroscience Letters* **41**, 145–150.

Johnson, B. R., & Atema, J. (1986). Chemical stimulants for a component of feeding behavior in the common gulf-weed shrimp *Leander tenuicornis* (Say). *Biological Bulletin* **170**, 1–10.

Johnson, B. R., Voigt, R., Borroni, P. F., & Atema, J. (1984). Response properties of lobster chemoreceptors: tuning of primary taste neurons in walking legs. *Journal of Comparative Physiology* **155**, 593–604.

Johnson, B. R., Borroni, P. F., & Atema, J. (1985). Mixture effects in primary olfactory and gustatory receptor cells from the lobster. *Chemical Senses* **10**, 367–373.

Johnson, B. R., Merrill, C. L., Ogle, R. C., & Atema, J. (1987). Response properties of lobster chemoreceptors: tuning of olfactory neurons sensitive to hydroxy-proline. *Journal of Comparative Physiology* **162**, 201–211.

Johnson, B. R., Voigt, R., & Atema, J. (in press). Response properties of lobster chemoreceptor cells = response modulation by stimulus mixtures. *Physiological Zoology*.

Lamb, T. D. (1986) Photoreceptor adaptation: vertebrates. In H. Stieve, ed., *The molecular mechanism of photoreception*, pp. 267–286. Springer-Verlag, New York.

Pace, U., Hanski, E., Salomon, Y., & Lancet, D. (1985). Odorant-sensitive adenylate cyclase may mediate olfactory reception. *Nature* **316**, 255–258.

Sammon, L., & Atema, J. (1987). Detection of chemical contrast in hermit crabs. *Biological Bulletin* **173**, 438.

Segall, J. E., Block, S. M., & Berg, H. C. (1986). Temporal comparisons in bacterial chemotaxis. *Proceedings of the National Academy of Sciences* (USA) **83**, 8987–8991.

Shirley, S. G., Robinson, C. J., Dickinson, K., Aujla, R., & Dodd, G. H. (1986). Olfactory adenylate cyclase of the rat. Stimulation by odorants and inhibition by Ca^{2+}. *Biochemical Journal* **240**, 605–607.

Sklar, P. B., Anholt, R. R. H., & Snyder, S. H. (1986). The odorant-sensitive adenylate cyclase of olfactory receptor cells. *Journal of Biological Chemistry* **261**, 15 538–15 543.

Tierney, A. J., Voigt, R., & Atema, J. (1988). Response properties of chemoreceptors from the medial antennular filament of the lobster, *Homarus americanus*. *Biological Bulletin* **174**, 364–372.

Vogel, S. (1983). How much air passes through a silkmoth's antenna? *Journal of Insect Physiology* **29**, 597–602.

Voigt, R., & Atema, J. (1987). Signal-to-noise ratios and cumulative self-adaptation of chemoreceptor cells. *Annals of the New York Academy of Sciences* **510**, 692–694.

6

Central and peripheral bases for mixture suppression in olfaction: a crustacean model

Barry W. Ache

*C. V. Whitney Laboratory and Departments of Zoology and Neuroscience,
University of Florida, St Augustine, Florida, USA*

I. Introduction

Mixture interactions are complex, and attempts to formulate a set of
rules or formulae to predict the smell or taste of mixtures have yet to be
successful. Our inability to predict the smell or taste of mixtures is in part
because we have much to learn about the physiological basis of quality
coding. The experimental approach of interpreting physiology from
psychophysical experimentation is difficult. Attempts to assign suppression
in the perceived intensity of taste mixtures to the periphery or the central
nervous system (CNS), for example, have produced conflicting results
(e.g., Kroeze & Bartoshuk, 1985); more recent studies support both
peripheral and central contributions (Kroeze, this volume). Electrophysiological experimentation in animal models offers a more direct
approach to the underlying physiology, but with the obvious difficulty of
relating changes in neural activity to perception. Physiological recordings
from receptor cells in both vertebrates (Frank; Smith; both this volume)
and invertebrates (Atema, Borroni, Johnson, Voight & Handrick; Derby,
Girardot, Daniel & Fine-Levy; O'Connell & Akers; all this volume)
support the psychophysical findings that peripheral events contribute to
mixture interactions, but the molecular and/or cellular mechanisms
underlying peripheral interactions are unknown. Physiological investigation into central events that contribute to mixture interactions are just
beginning, again in both vertebrates (Bell, Laing & Panhuber, 1987) and
invertebrates (Derby *et al.,* this volume), and the mechanisms underlying
central interactions are also unknown.

PERCEPTION OF COMPLEX SMELLS
AND TASTES ISBN 0 12 042990 X

101

*Copyright © 1989 by Academic Press Australia.
All rights of reproduction in any form reserved.*

The present paper considers physiological events that potentially contribute to mixture interaction in an invertebrate model, the antennular (olfactory) pathway of crustaceans, in particular that of the spiny lobster.

II. The lobster model and background data

The important contributions that research on the horseshoe crab, *Limulus*, have made to our understanding of vision epitomizes the potential of using invertebrate models in sensory biology. In this light, work in this laboratory and others (Atema *et al.*; Carr, Trapido-Rosenthal & Gleeson; Derby *et al.*; all this volume) has focused on the antennular pathway of crustaceans to better understand the chemical senses. Antennular chemoreception is considered to be the 'olfactory' sense of decapod crustaceans such as lobsters (Atema, 1977) and, indeed, the crustacean antennular pathway, like the analogous pathway of insects, appears to have the same gross organizational plan as the vertebrate olfactory pathway. Recent research with insects (e.g., Christensen & Hildebrand, 1987) suggests the analogy with the vertebrate olfactory pathway extends well beyond the gross organizational plan to include similarities in the cellular organization of the olfactory brain. Ache and Derby (1985) review the anatomy and physiology of the crustacean 'olfactory' pathway and should be consulted for details outside the scope of the present paper. Of particular relevance to the present paper, however, are projection fibers with axons in the circumesophageal connectives that project from the brain to the ventral nervous system. Monitoring neural activity in these relatively high-order interneurons in response to chemical stimuli applied to the antennules gives a convenient physiological measure of the net result or 'output' of olfactory integration. These neurons, as well as those at lower levels of the pathway, can be recorded from directly in a perfused anterior end preparation (Ache & Sandeman 1980). Primary afferent activity can also be recorded in an excised, perfused antennule preparation (e.g., Gleeson & Ache, 1985).

Aquatic organisms such as lobsters facilitate studying mixture effects in that natural feeding cues tend to be mixtures of relatively simple organic molecules (Carr *et al.*, this volume). These mixtures can be duplicated easily and accurately in the laboratory. Using evoked activity in the circumesophageal or 'output' interneurons as an assay, Derby and Ache (1984) identified 31 components of such a feeding stimulus for the spiny lobster, an aqueous extract of crab muscle tissue, according to the type of contribution each component made to the overall stimulatory

Table 6.1. Identified chemical components of an aqueous extract of crab muscle tissue

| Stimulatory | Non-stimulatory | | Non-contributory |
	Suppressive	Synergistic	
AMP	Arginine	IMP	Asparagine
ADP	Cysteine	Threonine	Aspartic acid
ATP	Proline		Histidine
Alanine			Homarine
Betaine			Hypoxanthine
Glutamic acid[a]			Inosine
Glutamine			Isoleucine
Glycine			Leucine
Lactic acid			Lysine
Phenylalanine			Methionine
Taurine[b]			Serine
Trimethylamine			Tyrosine
oxide[a]			Valine

Based on chemical analysis by Carr, Netherton & Milstead (1984). The chemicals are grouped according to their type of contribution to the stimulatory capacity of the mixture to the olfactory pathway of the spiny lobster. (After Derby & Ache, 1984.)
[a] Also identified as synergistic.
[b] Also identified as suppressive.

capacity of a mixture of the 31 components. Table 6.1 summarizes the results of this study, in which the response evoked by each single component and the response evoked by submixtures of the remaining 30 components were compared to the response evoked by the complete mixture. The comparisons were based on a stimulus substitution model patterned after Hyman and Frank (1980) with a linear concentration/ response function of about 0.25 on semi-logarithmic coordinates. This function typified the response of the interneurons surveyed to the complete mixture. As with any stimulus substitution model, however, accuracy of prediction is highly dependent on selection of the appropriate concentration/response function. The predictions of this model were therefore tested by recombining the components by class (e.g., the three non-stimulatory suppressants) and confirming their collective action. Figure 6.1 illustrates a typical result obtained from one cell. While individual interneurons varied in the degree of suppression they expressed, suppression could be confirmed, and in fact was strong, at least at the single concentration of the complete mixture against which all comparisons were made. Synergism, however, was either too weak to be registered or was overpredicted by the model used. This study provided a

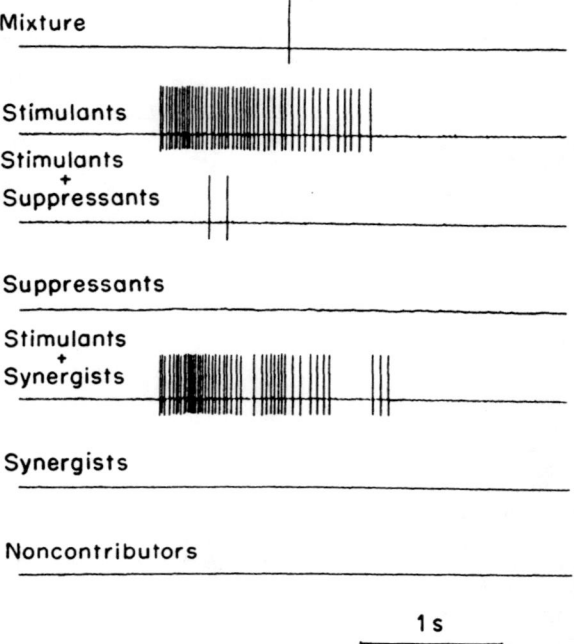

Fig. 6.1. Mixture suppression in a single 'output' interneuron in the brain of the spiny lobster. Responses were evoked by applying the chemicals listed in Table 6.1, combined by functional class, to the olfactory organ. The response evoked by combining all 31 components (trace 1) is far less than that evoked by the 12 stimulatory components (trace 2). The latter response, in turn, is almost entirely eliminated in the presence of the combined non-stimulatory suppressants (trace 3). In contrast, the combined non-stimulatory synergists fail to potentiate the response to the combined stimulants (trace 5). All traces are in temporal register. (After Derby & Ache, 1984.)

basis to investigate physiological events that might account for mixture suppression in the olfactory pathway of the lobster.

Derby, Ache and Kennel (1985) demonstrated that mixture suppression in the olfactory pathway has both peripheral and central components. The peripheral component was established by repeating the so-called confirmatory experiment mentioned above while recording from the olfactory receptor cells directly, as is easily done in crustaceans, and obtaining essentially similar results to those depicted in Figure 6.1. Evidence for the central component came from the results of a 'split-nose' experiment in which the initial confirmatory experiment was repeated, but with the stimulants and suppressants applied to different

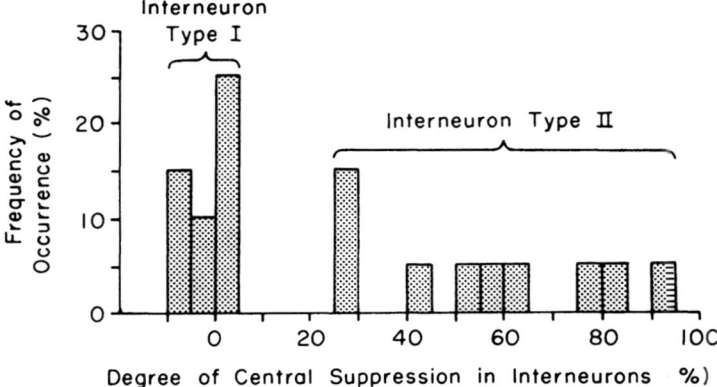

Fig. 6.2. Persistence of mixture suppression in 'output' interneurons ($n = 20$ in the brain of the spiny lobster when the combined stimulants and the combined nor-stimulatory suppressants (Table 6.1) are applied simultaneously, but to separate parts of the receptor field. Abscissa indicates per cent reduction (in 5% increments) in the response to the stimulants presented to one part of the field resulting from simultaneous presentation of suppressants to another part of the field. Half the interneurons tested, designated type II, showed > 25% suppression in this paradigm. (From Derby *et al.*, 1985.)

parts of the chemosensory field (typically the left and right lateral antennular filaments). Interneurons fell into two broad classes (Figure 6.2). One group (Type I) failed to show suppression when the stimulants and suppressants were presented co-temporally, but not co-spatially. A second group (Type II), however, showed ⩾25% suppression under these same conditions. The latter result argues strongly for a central component to mixture suppression, since suppression persisted even though the two components were not applied to the same receptor field. The validity of this interpretation is further supported by the observations that suppression was independent of side of presentation and that it persisted when stimulants and suppressants were applied to two separate antennular filaments on the same side of the organism (the left and right rami of one biramous antennule).

III. Peripheral suppression

A number of physiological and biochemical events could account for suppression at the periphery. These potentially include perireceptor events (Carr *et al.*, this volume) in addition to events inherent in the receptor cells, the object of the present paper. Ultrastructural analyses

(Grünert & Ache, 1988; Spencer & Linberg, 1986) indicate that the olfactory receptor cells of spiny lobsters are typical bipolar sensory cells with no morphological evidence of collateral processes or synaptic or ephaptic connections between sensory cells. In the absence of any information to the contrary, a reasonable working hypothesis is that these cells project independently to the CNS and, therefore, that peripheral interaction more likely reflects intra- rather than intercellular events. The latter possibility, however, cannot be entirely ruled out, and recent work with vertebrates (Gesteland & Adamek, 1987) describes an intercellular phenomenon that could contribute to peripheral mixture suppression.

In the lobster, at least two different events potentially suppress the output of receptor cells to mixtures. The concentration/response functions of receptor cells recorded in excised, perfused antennule preparations suggest that suppressants can compete with stimulants for common receptor sites (Gleeson & Ache, 1985; Ache, Gleeson & Thompson, in press).

As shown in Figure 6.3A, the concentration/response function of the response to a stimulatory component exhibits a parallel right shift in the presence of either one of two different suppressants; sufficiently high concentrations of the stimulatory component overcome the suppression.

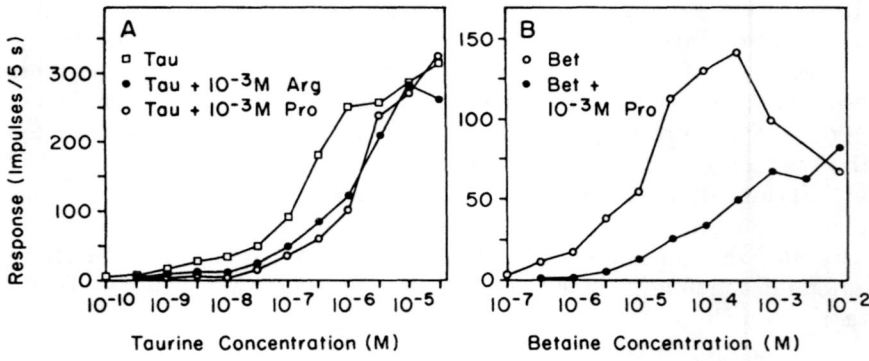

Fig. 6.3. Concentration/response functions of two olfactory receptor cells of the spiny lobster to single stimulants in the presence and absence of fixed doses of single suppressants. A. Suppression of a taurine(Tau)-sensitive cell by arginine (Arg) and proline (Pro). In this cell, increasing the stimulant concentration overcomes the suppression. **B.** Suppression of a betaine(Bet)-sensitive cell by proline. In this cell higher concentrations of the stimulant fail to overcome the suppression. A marked tachyphylaxis occurs at higher concentrations of the stimulant. (A. After Ache, Gleeson, & Thompson, 1987; B, After Ache *et al.*, in press.)

These findings, although physiological, are consistent with the expected kinetics of two agonists of different efficacies competing for a common receptor site according to the occupancy theory of receptor activation (e.g., Hollenberg, 1985). What makes this finding novel, even for stimulus substitution-based predictions, is that the suppressants are not themselves stimulatory to the cells tested. This holds true over a range of suppressant concentrations (Gleeson & Ache, 1985). To extend the pharmacological analogy, they are acting as full antagonists with no intrinsic activity (note in Fig. 6.3A that the concentration/response functions in the presence of the antagonist go to zero). Interestingly, these same suppressive compounds can be perfectly good stimulants for other receptor cells in the antennule. We now have similar results for five different suppressants interacting with three different stimulants (Gleeson & Ache, 1985; Ache *et al.*, in press), supporting the generality of this phenomenon in the lobster, at least among odorants of the same molecular class.

On the other hand, this same type of experiment suggests that not all binary pairings produce results that are consistent with the interpretation mentioned above (Ache *et al.*, in press). As indicated in Figure 6.3B, suppression cannot always be overcome by higher concentrations of stimulant, and the slope of the suppressed function can be reduced. This result was obtained for two different suppressants acting on two different stimulants. Care is always in order when interpreting biochemical mechanisms from physiological data, and these findings do not necessarily rule out competitive mechanisms. It is possible, for instance, for particularly 'sticky' antagonists to reduce transiently the number of binding sites available under conditions of full receptor occupancy (the 'hemi-equilibrium' state of Paton & Waud, 1964). On the other hand, the inability of higher concentrations of stimulant to overcome the suppression is consistent with the idea that other, possibly non-competitive mechanisms contribute to peripheral mixture suppression. These non-competitive interactions should be demonstrable physiologically, since it is possible to record intracellularly from lobster olfactory receptor cells using patch electrodes (Anderson & Ache, 1985).

As yet we have no evidence from intracellular recording that suppressants act by hyperpolarizing the receptor cells. As shown in Figure 6.4 for glycine suppression of taurine, while glycine markedly reduces the taurine-evoked depolarization of the receptor cell, glycine by itself does not hyperpolarize the membrane. This observation does not exclude the possibility that suppressants alter membrane conductance to ions whose equilibrium potential is at or near the membrane potential, either directly or indirectly by affecting one or more steps in a stimulant-evoked enzyme cascade.

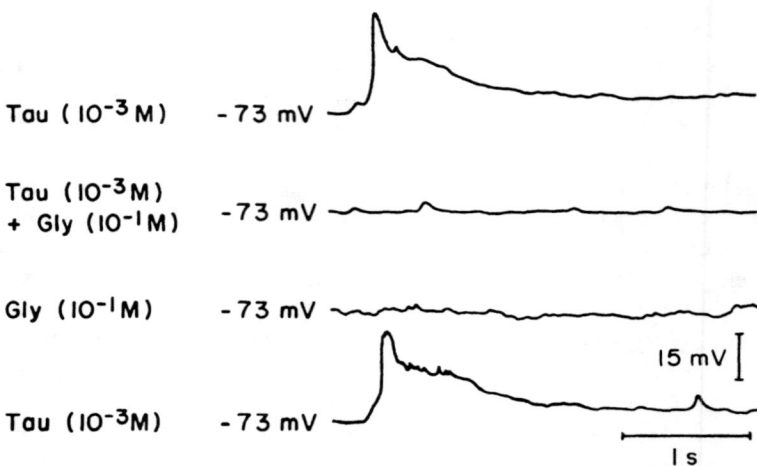

Fig. 6.4. Intracellular records from an olfactory receptor cell of the spiny lobster. Slow depolarization evoked by taurine (Tau, traces 1 and 4) is eliminated in the presence of glycine (Gly, trace 2). Glycine by itself is without effect (trace 3). Spiking was blocked by tetrodotoxin in these records. Membrane potential was set to −73 mv in each record. (Schmiedel-Jakob & Ache, unpublished data.)

Ion substitution experiments, in fact, suggest that multiple ionic currents contribute to the receptor potential. Table 6.2 shows data from 33 cells, for which one to three ions were substituted in the perireceptor environment prior to stimulating the cells with a mixture. No less than three different groupings are necessary to include the different patterns of ionic sensitivities observed. This suggests that either different ionic currents are inherent in different cells or, if all cells have the same currents, that they are being activated by different odorants. That different currents can be evoked in the same cell is suggested by conductance measurements made during the rising phase of mixture-evoked receptor potentials; both increases and decreases in conductance fuse into the final conductance state of the plateau phase of the receptor potential (Schmiedel-Jakob, Anderson & Ache, submitted). These multiple conductance states have been observed in numerous cells in response to mixture stimulation, but not in response to stimulation with single components. It remains to be shown, however, that different currents are evoked by the functionally different constituents of mixtures and, if so, that the currents activated by suppressants are consistent with a reduced probability of the cell to discharge. It is known from extracellular recording that crustacean chemoreceptors can be inhibited by some chemicals and excited by others (Derby & Harpaz, 1988).

Table 6.2. Grouped effects of ion substitution on the magnitude of the receptor potential in olfactory receptor cells of the spiny lobster

No. of cells	Ion substituted		
	Na^+	K^+	Ca^{++}
17	No effect	Reduced	Reduced
13	No effect	No effect	Reduced
14	Reduced	Reduced	No effect

From one to three ions were tested on 33 cells. The number of cells (column 1) is the number of the 33 cells whose ionic requirements are consistent with the profiles shown. Eleven cells are represented twice. (After Schmiedel-Jakob, Anderson & Ache, submitted.)

IV. Central mechanisms

As mentioned, the olfactory CNS of crustaceans is only recently receiving detailed attention. We still know little of the neural organization responsible for central mixture suppression. Derby *et al.* (1985) proposed the simplest model for central suppression consistent with our understanding of general patterns of neural organization in crustaceans (Fig. 6.5). Compounds identified as non-stimulatory suppressants in the interneuron assay excite at least some receptor cells (Derby & Ache, 1984; T. McClintock, unpublished data), providing a pathway to code for suppressants in the CNS. Olfactory afferents are thought to project ipsilaterally in crustaceans (Sandeman, 1982), eliminating the possibility that the central suppression identified by the 'split-nose' protocol is the result of afferents coding for suppressants directly inhibiting the central terminations of afferents coding for stimulants. The possibility that afferents coding for suppressants inhibit first-order interneurons cannot be eliminated, but it would be more consistent with the behavior of other crustacean sensory fibers if they made excitatory connections in the olfactory lobe. A reasonable working hypothesis, then, is that suppressant-evoked afference excites low-order inhibitory interneurons, which in turn inhibit excitatory connections between receptor cells and first-order interneurons and/or inhibit transmission between first- and higher-order interneurons, as shown in Figure 6.5.

This idea, if correct, should be testable by penetrating the olfactory brain with microelectrodes and characterizing the response of morphologically identified interneurons to various mixtures. Inhibition is well known to be a major component of information processing at the earliest synaptic levels of the olfactory pathway of vertebrates (Shepherd, 1977)

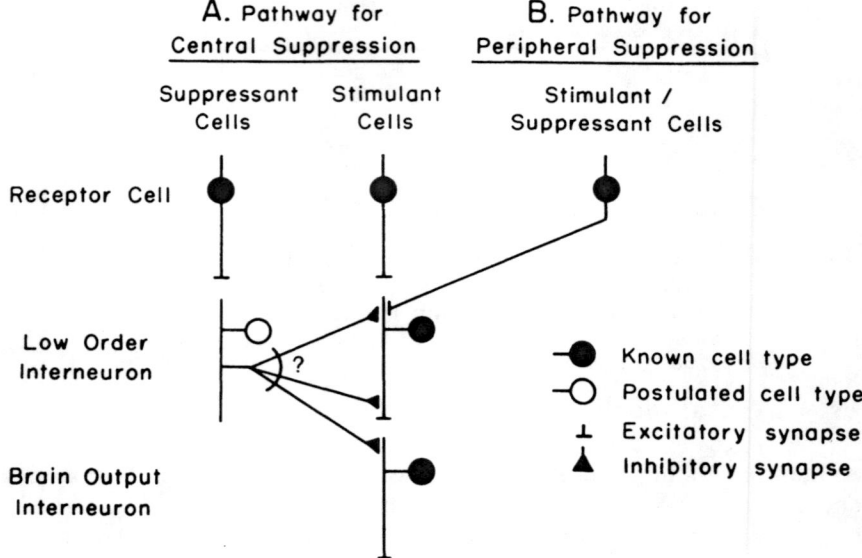

Fig. 6.5. Neural model for mixture suppression in the spiny lobster olfactory pathway. **A.** Central suppression: low-order interneurons in the olfactory lobe are excited by stimulant-evoked afference. These in turn either directly (as shown) or indirectly excite higher-order output interneurons. Parallel, suppressant-evoked afference excites other low-order interneurons which form inhibitory connections onto the output pathway. **B.** Peripheral suppression: afference reflecting peripheral suppression is reduced when the receptor cells are coactivated by suppressants. These cells form excitatory connections onto low-order interneurons in the output pathway. What for simplicity are pictured as single interneurons may well be multicellular networks. (After Derby *et al.*, 1985.)

and insects (Matsumoto & Hildebrand, 1981; Burrows, Boeckh & Esslen, 1982). There is evidence that this generalization extends to the olfactory lobe of crustaceans as well (Arbas, Humphreys & Ache, in press). It is necessary, however, to show if and how inhibition in the olfactory lobe contributes to mixture suppression.

Towards this end, Arbas *et al.* (in press) find a morphological class of non-spiking interneurons in the olfactory lobe of crayfish that characteristically receives an ongoing barrage of excitatory post-synaptic potentials (epsp's) (Fig. 6.6). The combined suppressants (as identified in the spiny lobster, Table 6.1) eliminate the epsp's (trace 2) and, when added to the combined stimulants (trace 3), eliminate the slow depolarization otherwise evoked in these cells by the stimulants (trace 1). This behavior is consistent with that predicted for low-order excitatory interneurons inhibited by suppressant-evoked afference. Cells that mediate this inhibition are as yet unknown, although Arbas *et al.* (in press) find a morphological class of spiking interneurons that is excited by the combined suppressants. Not all morphological classes of interneurons that arborize in the olfactory lobe show evidence of suppression with this same stimulus paradigm. The finding that compounds initially identified as suppressants in the lobster also behave as suppressants in crayfish suggests a possible conservation of receptor function, at least across closely related species.

V. Discussion and conclusion

Mixture suppression can be attributed to at least two levels of the olfactory pathway of the lobster. Two different inhibitory processes, one molecular and one cellular, potentially contribute to peripheral mixture

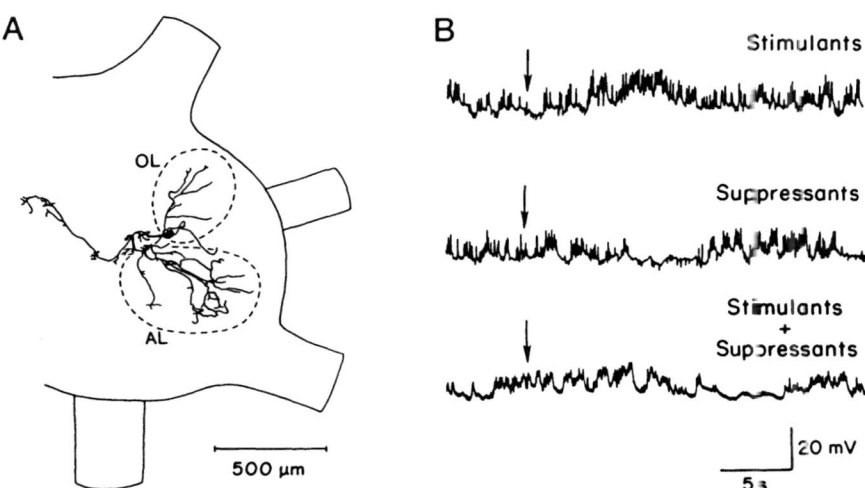

Fig. 6.6. Morphologically and physiologically identified olfactory interneuron in the CNS of the crayfish. **A.** Camera lucida drawing of a cobalt-stained cell that arborizes in the olfactory lobe (OL), the central terminus of the primary olfactory afferents, and the related accessory lobe (AL), suggesting it is a low-order olfactory interneuron. **B.** Intracellular traces recorded from the same cell prior to staining in response to stimulation of the olfactory receptors with the mixture components indicated (Table 6.1). Arrow indicates olfactometer on; the response latency, approx. 5 s, includes a delay for the stimulus to reach the olfactory organ. See text for details. (From Arbas *et al.*, in press.)

suppression. Central suppression presumably introduces yet a third inhibitory process, synaptic inhibition. Thus, mixture suppression in the lobster does not appear to be a unitary phenomenon that could readily be predicted by any simple rule or formula. It has yet to be established, of course, that any of these events actually contributes to the final percept of complex odorants by the organism and, if so, what the relative contribution of events at each level would be. Conditioning experiments (Fine–Levy, Girardot, Derby & Daniel, 1988; Derby *et al.*, this volume) suggest that lobsters can discriminate between naturally occurring odor mixtures derived from different foods, such as the extract of crab muscle tissue on which the present line of experimentation is based. Because suppressive interactions would result in unique across-fiber patterns for each particular odor mixture (Derby *et al.*, this volume), it is likely that these events contribute to the perception of complex odors in lobsters.

Inhibition broadens the scope for sensory coding over that available from excitation only. It is reasonable to expect inhibition to be an important component of coding at all levels of the olfactory pathway, even though the mechanism of the inhibition may be different, as is the case for the three processes discussed herein. It is also reasonable to expect the full range of inhibition to be utilized in order to maximize the coding potential available from inhibition. Because output pathways in the nervous system operate with discrete thresholds (e.g., spiking primary afferents and projection interneurons), the effect of sub-threshold levels of inhibition (or potentiation, for that matter) would not appear in response to single component activation of the pathway. Co-activated pathways, however, would express the bias placed on output elements by sub-threshold levels of inhibition. Unlike inhibition of supra-threshold activity, the reduced output caused by sub-threshold inhibition would not be factored into the expected response of a population of cells to a mixture that is calculated from the unitary action of its components. Predictions that are based on the summed responses evoked by individual components would be greater than those observed, indicating suppression.

Thus, what we observe as mixture suppression could be the natural consequence of sub-threshold levels of inhibition across the various coding elements in the olfactory pathway. This is not to exclude that other, perhaps non-physiological, processes contribute to mixture suppression. Sub-threshold inhibition, however, provides a potentially important physiological explanation for mixture suppression in lobster olfaction, if not in other organisms. As we more fully understand how odors are processed at all levels of the olfactory pathway, our ability to predict response to odor mixtures should increasingly match what we actually observe. When this can be done, perhaps we can say we understand the neural basis of quality coding.

The model proposed for mixture suppression in the lobster will certainly prove to be too simplistic as more types of cells are characterized, particularly in terms of its distinct, serially-organized 'levels' within the CNS. It does, however, provide a base from which to begin to unravel the inevitable complexities of studying information processing in neural networks. Such understanding may not happen soon, however, given the intractability of olfactory neuropil for revealing its organizational secrets, even in 'simple' systems such as the lobster.

Acknowledgments

The work summarized here owes much to my associates, past and present. Thanks are due Dr Charles Derby, who carried out the background work and modeling; Dr Richard Gleeson, for the first direct evidence for peripheral suppression; Ms Elizabeth Kennel and Ms Holly Thompson, for able technical assistance; Drs Edmund Arbas and Carol Humphreys, for intracellular recording in the CNS; Dr Ingrid Schmiedel-Jakob and Timothy McClintock, for intracellular recording from receptors; and Dr Ulrike Grunert for ultrastructural analysis of the receptor cells. I thank Drs Charles Derby and Richard Gleeson for critically reading the manuscript. Original research presented here was supported by the US National Science Foundation (BNS85–11256) and a grant from the Whitehall Foundation, Palm Beach, Florida.

References

Ache, B. W., & Derby, C. D. (1985). Functional organization of olfaction in crustaceans. *Trends in Neuroscience* **8**, 356–360.

Ache, B. W., & Sandeman, D. C. (1980). Olfactory-induced central neural activity in the Murray crayfish, *Euastacus armatus*. *Journal of Comparative Physiology* **140**, 295–301.

Ache, B. W., Gleeson, R. A., & Thompson, H. D. (1987). Mechanisms of interaction between odorants of olfactory receptor cells. *Annals of the New York Academy of Sciences* **510**, 145–147.

Ache, B. W., Gleeson, R. A., & Thompson, H. D. (in press). Mechanisms for mixture suppression in olfactory receptors of the spiny lobster. *Chemical Senses*.

Anderson, P. A. V., & Ache, B. W. (1985). Voltage- and current-clamp recordings of the receptor potential in olfactory receptor cells *in situ*. *Brain Research* **338**, 273–280.

Arbas, E. A., Humphreys, C. J., & Ache, B. W. (in press). Morphology and physiological properties of interneurons in the olfactory midbrain of the crayfish. *Journal of Comparative Physiology*.

Atema, J. (1977). Functional separation of smell and taste in fish and crustacea. In J. LeMagnen & P. MacLeod, eds, *Olfaction and taste VI*, pp. 165–174. IRL Press, London.

Bell, G. A., Laing, D. G., & Panhuber, H. (1987). Odour mixture suppression: evidence for a peripheral mechanism in human and rat. *Brain Research* **426**, 8–18.

Burrows, M., Boeckh, J., & Esslen, (1982). Physiological and morphological properties of interneurones in the deutocerebrum of male cockroaches which respond to female pheromone. *Journal of Comparative Physiology* **145**, 447–457.

Carr, W. E. S., Netherton III, J. C. & Milstead, M. L. (1984). Chemoattractants of the shrimp, *Palaemonetes pugio*: variability in responsiveness and the stimulatory capacity of mixtures containing amino acids, quaternary ammonium compounds, purines and other substances. *Comparative Biochemistry & Physiology* **77A**, 469–474.

Christensen, T. A., & Hildebrand, J. G. (1987). Functions, organization and physiology of the olfactory pathways in the lepidopteran brain. In A. P. Gupta, ed., *Arthropod brain: its evolution, development, structure and functions*, pp. 457–484. John Wiley, New York.

Derby, C. D., & Ache, B. W. (1984). Electrophysiological identification of the stimulatory and interactive components of a complex odorant. *Chemical Senses* **9**, 201–218.

Derby, C. D., & Harpaz, S. (1988). Physiology of chemoreceptor cells in the legs of the freshwater prawn, *Macrobrachium rosenbergii*. *Comparative Biochemistry & Physiology* **90A**, 85–91.

Derby, C. D., Ache, B. W., & Kennel, E. W. (1985). Mixture suppression in olfaction: electrophysiological evaluation of the contribution of peripheral and central neural components. *Chemical Senses* **10**, 301–316.

Fine-Levy, J. B., Girardot, M.-N., Derby, C. D., & Daniel, P. C. (1988). Differential associative-conditioning and olfactory discrimination in the spiny lobster, *Panulirus argus*. Behavior and Neural Biology **49**, 315–331.

Gesteland, R. C., & Adamek, G. D. (1987). Adaptation and mixture component suppression in olfaction. *Chemical Senses* **12**, 657.

Gleeson, R. A., & Ache, B. W. (1985). Amino acid suppression of taurine-sensitive chemosensory neurons. *Brain Research* **335**, 99–107.

Grünert, U., & Ache, B. W. (1988). Ultrastructure of the aesthetasc (olfactory) sensilla of the spiny lobster, *Panulirus argus*. *Cell & Tissue Research* **251**, 95–103.

Hollenberg, M. D. (1985). Receptor models and the action of neurotransmitters and hormones: some new perspectives. In H. I. Yamamura, S. J. Enna & M. J. Kuhar, eds, *Neurotransmitter receptor binding*, pp. 1–39. Raven Press, New York.

Hyman, A. M., & Frank, M. E. (1980). Effects of binary taste stimuli on the neural activity of the hamster chorda tympani. *Journal of General Physiology* **76**, 125–142.

Kroeze, J. H. A., & Bartoshuk, L. M. (1985). Bitterness suppression as revealed by split-tongue taste stimulation in humans. *Physiology & Behavior* **35**, 779–783.

Matsumoto, S. G., & Hildebrand, J. G. (1981). Olfactory mechanisms in the moth *Manduca sexta*: response characteristics and morphology of central neurons in the antennal lobes. *Proceedings of the Royal Society of London, Series B* **213**, 249–277.

Paton, W. D. M., & Waud, D. R. (1964). A quantitative investigation of the relationship between rate of access of a drug to a receptor and the ratio of onset or offset of action. *Archives of Experimental & Pathological Pharmacology* **248**, 124–143.

Sandeman, D. C. (1982). Organization of the central nervous system. In H. L. Atwood & D. C. Sandeman, eds, *The biology of crustacea*, vol. 3, pp. 1–61. Academic Press, New York.

Shepherd, G. M. (1977). The olfactory bulb: a simple system in the mammalian brain. In E. R. Kandel, ed., *Handbook of physiology*, vol. 1, part 2, pp. 945–968. American Physiological Society, Bethesda, Maryland.

Spencer, M., & Linberg, K. A. (1986). Ultrastructure of aesthetasc innervation and external morphology of the lateral antennule setae of the spiny lobster *Panulirus interruptus* (Randall). *Cell & Tissue Research* **245**, 69–80.

7

Receptor site specificity in taste and implications for mixture reception

Joseph G. Brand[1,2,3], Bruce P. Bryant[1],
D. Lynn Kalinoski[1] and Robert H. Cagan[4]

[1] *Monell Chemical Senses Center, Philadelphia, Pennsylvania, USA*
[2] *Veterans Administration Medical Center, Philadelphia, Pennsylvania, USA*
[3] *Department of Biochemistry, University of Pennsylvania, School of Dental Medicine,
Philadelphia, Pennsylvania, USA*
[4] *Colgate-Palmolive Company, Research and Development Division,
Piscataway, New Jersey, USA*

I. Introduction

Animals usually encounter mixtures of chemical stimuli. Understanding the ways in which these stimuli interact with peripheral chemoreceptors is a challenge whose resolution will help in ultimately describing how mixtures of stimuli affect perception and behavior. The biochemical events that occur at the periphery in taste and olfaction have been receiving increasing and more detailed investigation (see Cagan & Kare, 1981; Teeter & Brand, 1987). Considerable success in exploring receptor specificity at the peripheral level has been achieved for the amino acid taste system of the catfish (Krueger & Cagan, 1976; Cagan, 1981, 1986; Brand, Bryant, Cagan & Kalinoski, 1987; Kalinoski, Bryant, Shaulsky & Brand, submitted). United recently, our studies concentrated on the single-stimulus event — that is, on understanding the fundamental processes underlying stimulus recognition at the receptor level, including binding affinity, relative site density and initiation of transductive events. More recent studies have examined site specificity using binary mixtures.

*Copyright © 1989 by Academic Press Australia.
All rights of reproduction in any form reserved.*

Taste receptor binding sites exhibit a range of specificity, from highly specific to relatively non-specific. Salt taste receptors in mammals are an example of a specific site. The chloride salts of the cations sodium and lithium are the best known examples of purely salty tastes (Murphy, Cardello & Brand, 1981). The specificity of this receptor is derived from ion specific, amiloride-sensitive channels that mediate salty taste (Heck, Mierson & DeSimone, 1984; Brand, Teeter & Silver, 1985; DeSimone & Ferrell, 1985). As a consequence of this specificity, any agent or compound that can alter the activity or selectivity of the channels can affect salty taste perception (Schiffman, Lockhead & Maes, 1983; Schiffman, Simon, Gill & Beeker, 1986). Thus, mixtures of two stimuli, such as NaCl and a pharmacological agent active against the epithelial channel and which may also possess a taste, will modify the salty taste. This can be interpreted, providing one has knowledge of the receptor event, as a purely peripheral event. Without an understanding of the specificity and pharmacology of the receptor process, an erroneous interpretation of the mixture response could easily be made.

At the other end of the specificity spectrum is the bitter modality. Hundreds of compounds taste bitter. Depending upon the nature of the specificities of the receptor/transductive events, the tastes of mixtures of these stimuli may be expected to be either additive (for example, where two or more stimuli interact with different receptors) or competitive (where two or more stimuli interact with the same receptor). Cross-adaptation studies suggested overlapping classes of bitter compounds, but distinct classes corresponding to distinct receptor entities were not observed (McBurney, 1974). Predicting the nature of these interactions is possible only with a more complete understanding of these peripheral events. In a study a decade ago, Brand, Zeeberg and Cagan (1976) observed no differences in binding of the bitter stimulus quinine to bovine taste and non-taste tissues. This study found no evidence for bitter receptor macromolecules unique to taste bud tissues that recognize the bitter stimulus quinine. The bitter modality may achieve its specificity at a step subsequent to the initial interaction with the peripheral plasma membrane. Many bitter stimuli are phosphodiesterase inhibitors (Cagan, 1976; Kurihara, 1972). Perhaps this inhibition acts as part of the recognition event and, through enzyme effects, could influence the perception of mixtures. Only with a greater understanding of peripheral binding and transductive events can rational conclusions be made regarding taste mixture interactions.

We have been studying receptor binding and, more recently, the possible steps in transduction (Huque, Brand, Rabinowitz & Bayley,

1987; Kalinoski, LaMorte & Brand, 1987) in the taste receptor system of the catfish, *Ictalurus punctatus*. This animal possesses taste receptors for several amino acids and is sensitive to them at concentrations approaching nanomolar (Caprio, 1975). Based on electrophysiological evidence, Caprio had earlier suggested the existence of at least two populations of receptors: one sensitive to neutral amino acids including L-alanine and the other specific to L-arginine (Caprio & Robinson, 1978; Caprio, 1982). These electrophysiological results were supported by behavioral data (Stewart, Bryant & Atema, 1979). The receptor site for L-alanine has been extensively investigated (Krueger & Cagan, 1976; Cagan, 1979; Zelson & Cagan, 1979; Brand *et al.*, 1987). The site binds L-alanine with a micromolar K_D and recognizes other neutral amino acids. The site that binds L-arginine (Cagan, 1981, 1986) has recently been characterized (Kalinoski *et al.*, submitted). As anticipated by previous electrophysiological and biochemical data, this site is more specific than the one(s) for L-alanine. Based on binding competition data, Cagan (1986) suggested the existence of at least three site types for amino acids in the taste system of the catfish. In this construct, L-alanine can bind to two separate sites, one recognizing L-alanine, L-threonine, L-serine, and possibly D-alanine and β-alanine. The other recognizes L-serine, L-alanine, glycine, D-alanine and β-alanine. These data also confirmed the separate identity of the L-arginine site.

We have used both biochemical and neurophysiological techniques to characterize the amino acid taste receptor sites in catfish. Using both of these techniques, we have delineated the enantiomeric specificity of the alanine receptor sites and further defined the structural specificity of the L-arginine receptor site. In order to define the selectivity of these sites, several types of biochemical binding inhibition studies using binary mixtures, and neurophysiological cross-adpatation studies were performed. We describe the findings here and discuss the implications of these results for mixture interactions at the receptor level.

II. Materials and methods

Channel catfish, *Ictalurus punctatus* were housed as described (Cagan, 1979; Brand *et al.*, 1987). Animals used for neurophysiological studies were held for no longer than seven days in holding tanks with recirculated, charcoal-filtered water at 25°C, while those used for binding studies were held for up to two weeks in recirculated water at 8°C. Where binding studies are described, Fraction P2 was used. This fraction was prepared by scraping the taste-bud rich epithelium from the barbels and the top of the head, from lips to the operculi. Fraction P2 was then prepared by differential centrifugation (Krueger & Cagan, 1976). This fraction is enriched in taste ligand binding activity. Radiolabeled amino acids were purchased from commercial suppliers: L-[³H]arginine from ICN Pharmaceuticals, L-[³H]alanine from New England Nuclear and D-[¹⁴C]alanine from Amersham and ICN. Other chemicals were reagent grade purchased from Sigma (St. Louis) or Fisher Scientific.

Binding activity to Fraction P2 by L-[³H]alanine and D-[¹⁴C]alanine was determined as previously described (Krueger & Cagan, 1976; Brand *et al.*, 1987). Binding studies of L-[³H]arginine to Fraction P2 were carried out in 0.01 M Tris-HCl, pH 7.8, that contained 1 mM $CaCl_2$, as described (Krueger & Cagan, 1976; Cagan, 1979) with some modifications. To a sample of 1.0 mL of Fraction P2 were added 0.05 mL radioactive ligand plus either 0.05 mL water (for total counts) or 0.05 mL of 0.5 M L-arginine (for non-specific binding). The samples were incubated on ice for one hour. Incubation was terminated by the rapid addition, with vortexing, of 1 mL ice-cold, 1 mM L-arginine. A 1 mL sample was rapidly filtered through a Millipore membrane (0.45 μ) and quickly washed with 10 mL of the ice-cold Tris-Ca buffer. The dilution of isotope achieved by this method reduced non-specific binding of L-[³H]arginine of the filters by 15-fold. The dilution step could be performed in less than 10 s. All assays were performed in triplicate, which varied by less than 5%. Protein was determined by the method of Bradford (1976) or Lowry, Rosebrough, Farr and Randall (1951) using bovine serum albumin as a standard.

For neurophysiological studies, fish were immobilized with Flaxedil (0.3 mg per kg, IM) and kept anesthetized during the course of the experiment with MS-222 (1:10,000) applied in the respiratory water flow. The maxillary barbel was positioned in a plastic tube through which a background aqueous medium, designated as 'artificial pond water' (APW) (0.3 mM NaCl, 0.2 mM KCl, 0.2 mM $CaCl_2$, 0.2 mM $NaHCO_3$), flowed at 10 to 15 mL per minute. Stimulus solutions were prepared fresh daily by diluting frozen stock solutions of 10^{-2} M amino acid with APW. Test stimuli were applied to the barbel by injecting 1 mL volumes into the background flow of APW. The barbel was rinsed with APW for three to four minutes between test stimuli. Responses were measured by integrating multiunit neural activity (time constant, 1.0 s) from teased bundles of the facial nerve (Caprio, 1975). Only bundles that responded well to both 10^{-6} M L-alanine and 10^{-6} M L-arginine were used for experiments.

Table 7.1. Cross-adaptation of neural responses to L-alanine and D-alanine

Nerve bundle	Adapting stimulus Test stimulus	D-alanine L-alanine	L-alanine D-alanine
1		0.0	0.0
2		51.7	0.0
3		19.3	14.9
4		30.5	42.9
5		0.0	0.0
6		53.0	22.0
7		14.2	36.4

Values are responses to the test stimulus following adaptation to the adapting stimulus; i.e. L-alanine after adaptation to D-alanine (middle column), or D-alanine after adaptation to L-alanine (right-hand column), and are expressed as a percentage of the unadapted response to the test stimulus. This unadapted response is defined as 100%. Each value is the mean of triplicate determinations. D-alanine concentration was 5×10^{-4} M for all bundles tested. L-alanine concentration for each bundle was adjusted to give a response of equal magnitude to that given by D-alanine. The concentration of L-alanine with different nerve bundles varied from 2×10^{-8} to 5×10^{-7} M. The data are from seven fish, with a single bundle being used from each. (From Brand, Bryant, Cagan & Kalinoski, 1987, with permission.)

III. Results and discussion

Site specificities

Previous results suggested that there may be overlap in enantiomeric specificity for L- and D-alanine (Cagan, 1986). Cross-adaptation experiments (Table 7.1) supported this observation since adaptation was observed, although the degree varied depending upon the bundle selected. The incomplete cross-adaptation seen in some of the bundles points to the existence of at least two classes of receptor populations sensitive to either or both L- and D-alanine.

Both L-alanine and D-alanine bind to Fraction P2, with L-alanine exhibiting a K_D of 1.5 µM, while D-alanine exhibited a K_D of 2 5 µM for the high affinity state. Interestingly, D-alanine also showed considerably more binding than L-alanine at higher concentrations (> 500 µM), suggesting heterogeneity in the D-alanine binding sites. When L-alanine and D-alanine were used in the same binding experiment in an attempt to define the nature of this heterogeneity, it was found that while L-alanine readily displaced D-alanine, this displacement was not uniform across all concentrations (Fig. 7.1). The data suggest that some D-alanine sites were

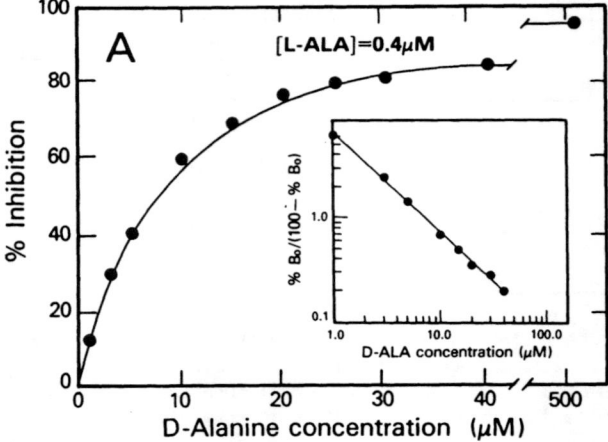

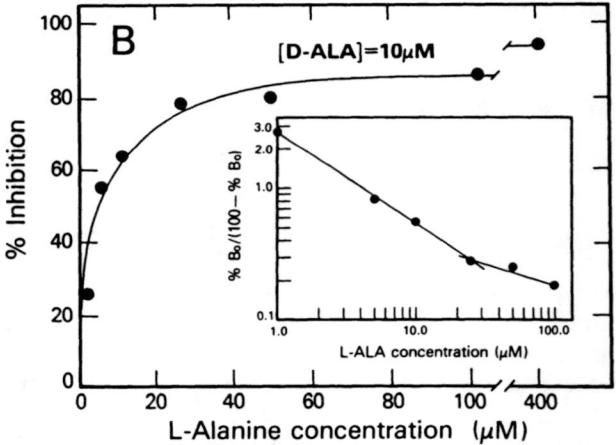

Fig. 7.1.   Per cent inhibition of binding of each enantiomer of alanine versus concentration of the other enantiomer. **A**. Binding of L-[³H] alanine (0.4µM) to Fraction P2 is inhibited by increasing concentrations of D-alanine up to 500 µM. The inset shows a modified Hill plot where Bo represents the total counts bound in the absence of the enantiomeric competitor, and % Bo represents the percentage of the total counts that are bound in the presence of the unlabeled enantiomeric competitor. **B**. Binding of D-[¹⁴C] alanine (10 µM) to Fraction P2 is inhibited by increasing concentrations of unlabeled L-alanine up to 400 µM. The inset shows the modified Hill plot. (From Brand, Bryant, Cagan & Kalinoski (1987), with permission.)

less accessible to L-alanine than others. In fact, at high concentrations of D-alanine, for example at 200 μM, L-alanine could displace D-alanine only to a level of 70% (Brand *et al.*, 1987). Thus, the inhibitory ability of each enantiomer toward the other is concentration-dependent in a manner that suggests site heterogeneity.

The manner in which mixtures of L- and D-alanine interact with receptors depends upon their relative concentrations. At low concentrations, they would be more likely to compete one with the other. In fact, double-label experiments (using D-[^{14}C] alanine and L-[^{3}H] alanine) at concentrations below 60 μM revealed that the total number of available sites for both is not greater than that for either one alone at equivalent concentrations (Brand *et al.*, 1987). However, at higher concentrations (> 200 μM) both biochemical and neurophysiological experiments indicated site independence. For most nerve bundles sampled, a 1:1 mixture of equally stimulatory concentrations of L-alanine and D-alanine evoked a larger magnitude of response than did either component alone. The mean of responses to the mixtures was 12.4% greater than the mean of responses to either component alone (Brand *et al.*, 1987).

In contrast to the binding sites for L-alanine, which appear to be relatively non-specific (Caprio, 1982; Cagan, 1986), our recent biochemical and neurophysiological work has confirmed the more specific nature of the binding site for L-arginine. Binding of the taste stimulus L-arginine to Fraction P2 was reversible, saturable and protein-dependent. Binding affinity has been estimated at 2×10^{-8} M for a high affinity state and 1×10^{-6} M for a lower affinity state. At pH 7.8, an apparent single rate constant for association at 4°C was 4.7×10^{5} M^{-1} min^{-1}. Dissociation was more complex, yielding two rate constants of 1.8 min^{-1} and 8.3×10^{-3} min^{-1}. The rather slow dissociation constants and high affinities that were determined by equilibrium kinetics for L-arginine, and for L-alanine (Krueger & Cagan, 1976), may be indicative of a system whose primary function is specific low-threshold detection. The threshold of such a system is, of course, limited by ambient concentrations of stimuli. At these concentrations, stimulus-receptor binding is expected to be at or near equilibrium. The slow dissociation rates of this type of receptor could render it less well adapted for making fine temporal (and hence spatial) resolutions of chemical stimuli.

The receptor site for L-arginine exhibits specificity for L-arginine and its close structural analogs. For example, L-arginine methyl ester and L-α-amino-β-guanidino propionic acid (L-AGPA) and to a lesser extent D-arginine, cross-adapt L-arginine in a neurophysiological assay (Bryant, Harpaz, & Brand, 1987; Kalinoski *et al.*, submitted). Certain other

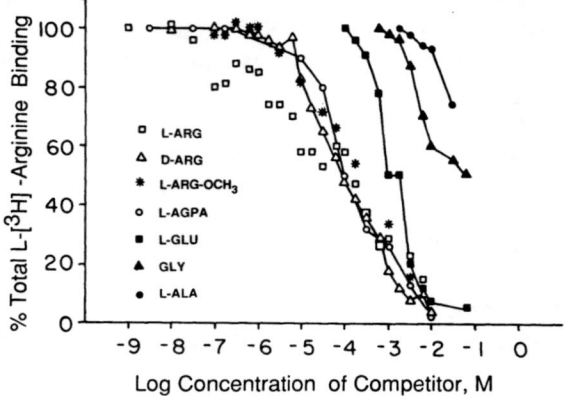

Fig. 7.2   Per cent total binding of L-[^{3}H] arginine at 10^{-6} M versus log concentration of various competitive amino acids. L-[^{3}H] arginine was incubated with Fraction P2 in the presence of each competitor at the stated concentrations for the one hour incubation period on ice. Binding was measured using an ultrafiltration assay as detailed under *Materials and methods*. The competitors examined were, L-arginine (□), D-arginine (△), L-arginine methyl ester (*), L-α-amino-β-guanidino propionic acid (L-AGPA) (o), L-glutamate (■), glycine (▲), and L-alanine (•). The data are averages from two separate experiments. (From Kalinoski, Bryant, Shaulsky, & Brand, submitted.)

derivatives, particularly those with a blocked amino group, are not only poor cross-adaptors but also poor stimuli. Of other amino acids that are stimulatory, all are poor cross-adaptors of L-arginine, for example glycine, L-alanine and L-glutamate (Kalinoski *et al.*, submitted). These results from cross-adaptation studies of L-arginine by analogs and by other amino acids are paralleled by binding competition studies (Fig. 7.2). The analogs L-AGPA, L-arginine methyl ester and D-arginine, that cross-adapted L-arginine well, were also better able to inhibit L-arginine binding to Fraction P2. While L-arginine in homologous inhibition studies exhibited two affinity states, none of the competing amino acids or analogs (with the exception of the poor stimulus L-lysine) expressed this heterogeneity. The effect of L-glutamate (Fig. 7.2) is especially interesting, since not until it was present at a concentration of 3×10^{-4} M was significant inhibition of 10^{-6} M L-[^{3}H] arginine observed, yet within two orders of magnitude total inhibition was achieved. On the other hand, inhibition of L-arginine binding by L-alanine was essentially non-existent.

Implications for mixtures

This chapter has focused on competitive binding effects of amino acids at two characterized taste receptor binding sites. One site, that recognizing the amino acid stimulus L-alanine, was shown by previous studies to be relatively non-specific. Occupation was permitted for L-alanine, L-threonine, L-serine, glycine and D-alanine, but competitive binding studies at two concentrations suggested at least two subclasses of this type of site (Cagan, 1986). At higher concentrations of D-alanine, a separate D-alanine active pathway was also noted (Brand *et al.*, 1937). Thus, results of stimulation by mixtures of these amino acid stimuli at lower, physiologically relevant concentrations may be explained by simple competitive effects; until the receptor sites are saturated, the stimuli act in an associative manner at the same receptors, and therefore through the same neural pathways. An exception to this could occur at higher concentrations, where additional selective sites and possibly additional neural pathways may become active, such as those for D-alanine.

In the case of a more selective stimulus, such as L-arginine, mixtures of L-arginine and L-alanine would be primarily additive. The L-arginine receptor site is relatively specific, exhibiting narrow conformational requirements. For example, carboxy-terminal modifications are tolerated, but not amino terminal; side chain length is not excessively critical, yet enantiomeric specificity is seen (Bryant *et al.*, 1987; Kalinoski *et al.*, submitted).

One intriguing result from the competitive binding studies with L-arginine concerns L-glutamate competition. At low concentrations of L-glutamate, no competition for L-arginine binding is observed. But at higher concentrations competition is obvious, and within two orders of magnitude it is complete. Thus, the effect of a mixture of L-arginine and L-glutamate would be expected to be concentration-dependent as opposed to, for example, a mixture of L-arginine and L-alanine. In the latter case, both excite separate receptors throughout the concentration range. While the binding results help to define the immediate peripheral event, subsequent steps at the receptor cell level (second messenger events) and at the neural level may act to converge the signals, giving rise to summations not anticipated by binding events alone.

One aspect of the receptor process not yet dealt with in depth in our work, or elsewhere, is the nature of the transduction steps. We hypothesize that the signal can be transduced within the cell through a variety of mechanisms including stimulus/receptor associated channels and direct linking with second messenger events. Evidence has been obtained for the latter in taste since active adenylate cyclase and phosphoinositide systems have been described (Huque *et al.*, 1987; Kalinoski *et al.*, 1987). If these systems were differentially sensitive to stimulus concentration and differentially sensitive to unique receptors, the possibilities for mixtures eliciting complex responses would be enhanced greatly.

Acknowledgments

This work was supported in part by US National Institutes of Health (NIH) Research Grants NS-15740, NS-23622 and NS-22620 from US National Institute for Neurological Communicative Disorders and Stroke, by US Basic Research Services Grant SO7-RR05825-07 from NIH, by US National Research Service Award Fellowship 1 F32 NS-07809 to Dr B. P. Bryant, and by the US Veterans Administration.

References

Bradford, M. M. (1976). A rapid and sensitive method for the quantitation of microgram quantities of protein utilizing the principle of protein-dye binding. *Analytical Biochemistry* 72, 248–254.

Brand, J. G., Zeeberg, B. R., & Cagan, R. H. (1976). Biochemical studies of taste sensation V. Binding of quinine to bovine taste papillae and taste bud cells. *International Journal of Neuroscience* 7, 37–43.

Brand, J. G., Teeter, J. H., & Silver, W. L. (1985). Inhibition by amiloride of chorda tympani responses evoked by monovalent salts. *Brain Research* 334, 207–214.

Brand, J. G., Bryant, B. P., Cagan, R. H., & Kalinoski, D. L. (1987). Biochemical studies of taste sensation XIII. Enantiomeric specificity of alanine taste receptor sites in catfish, *Ictalurus punctatus*. *Brain Research* 416, 119–128.

Bryant, B. P., Harpaz, S., & Brand, J. G. (1987). Structure-activity relationships in the arginine receptive taste pathways of the channel catfish, *Ictalurus punctatus* (abstract). *Chemical Senses* 12, 643–644.

Cagan, R. H. (1976). Biochemical studies of taste sensation II. Labeling of cyclic AMP of bovine taste papillae in response to sweet and bitter stimuli. *Journal of Neuroscience Research* 2, 363–371.

Cagan, R. H. (1979). Biochemical studies of taste sensation VII. Enhancement of taste stimulus binding to a catfish taste receptor preparation by prior exposure to the stimulus. *Journal of Neurobiology* 10, 207–220.

Cagan, R. H. (1981). Recognition of taste stimuli at the initial binding interaction. In R. H. Cagan & M. R. Kare, eds, *Biochemistry of taste and olfaction*, pp. 175–203. Academic Press, New York.

Cagan, R. H. (1986). Biochemical studies of taste sensation XII. Specificity of binding of taste ligands to a sedimentable fraction from catfish taste tissue. *Comparative Biochemistry & Physiology* 85A, 355–358.

Cagan, R. H., & Kare, M. R. (1981). *Biochemistry of taste and olfaction*. Academic Press, New York.

Caprio, J. (1975). High sensitivity of catfish taste receptors to amino acids. *Comparative Biochemistry & Physiology* 52A, 247–251.

Caprio, J. (1982). High sensitivity and specificity of olfactory and gustatory receptors of catfish to amino acids. In T. J. Hara, ed., *Chemoreception in fishes*, pp. 109–134. Elsevier, Amsterdam.

Caprio, J., & Robinson, J. J. (1978). Adaptation and cross-adaptation of the peripheral olfactory and gustatory systems of the catfish to amino acids (abstract). *Society for Neuroscience* **4**, 86.

DeSimone, J. A., & Ferrell, F. (1985). Analysis of amiloride inhibition of chorda tympani taste response of the rat to NaCl. *American Journal of Physiology* **249**, R52–R61.

Heck, G. L., Mierson, S., & DeSimone, J. A. (1984). Salt taste transduction occurs through an amiloride-sensitive sodium transport pathway. *Science* **223**, 403–405.

Huque, T., Brand, J. G., Rabinowitz, J. L., & Bayley, D. L. (1987). Phospholipid turnover in catfish barbel (taste) epithelium with special reference to phosphatidylinositol-4, 5-bisphosphate (abstract). *Chemical Senses* **12**, 666–667.

Kalinoski, D. L., LaMorte, V., & Brand, J. G. (1987). Characterization of a taste-stimulus sensitive adenylate cyclase from the gustatory epithelium of the channel catfish (abstract). *Society for Neuroscience* **13**, 1405.

Krueger, J. M., & Cagan, R. H. (1976). Biochemical studies of taste sensation IV. Binding of L-[^{3}H]alanine to a sedimentable fraction from catfish barbel epithelium. *Journal of Biological Chemistry* **259**, 88–97.

Kurihara, K. (1972). Inhibition of cyclic 3',5'-nucleotide phosphodiesterase in bovine taste papillae by bitter stimuli. *Federation of European Biochemical Societies, Letters* **27**, 279–281.

Lowry, O. H., Rosebrough, N. J., Farr, A. L., & Randall, R. J. (1951). Protein measurement with the Folin phenol reagent. *Journal of Biological Chemistry* **193**, 265–275.

McBurney, D. H. (1974). Are there primary tastes for man? *Chemical Senses & Flavour* **1**, 17–28.

Murphy, C., Cardello, A. V., & Brand, J. G. (1981). Tastes of fifteen halide salts following water and NaCl: anion and cation effects. *Physiology & Behavior* **26**, 1083–1095.

Schiffman, S. S., Lockhead, E., & Maes, F. W. (1983). Amiloride reduces the taste intensity of Na and Li salts and sweeteners. *Proceedings of the National Academy of Sciences, USA* **80**, 6136–6140.

Schiffman, S. S., Simon, S. A., Gill, J. M., & Beeker, T. G. (1986). Bretylium tosylate enhances salt taste. *Physiology & Behavior* **36**, 1129–1137.

Stewart, A., Bryant, B. P., & Atema, J. (1979). Behavioral evidence for two populations of amino acid receptors in catfish taste (abstract). *Biological Bulletin* **157**, 396.

Teeter, J. H., & Brand, J. G. (1987). Peripheral mechanisms of gustation: physiology and biochemistry. In T. E. Finger & W. L. Silver, eds, *Neurobiology of taste and smell*, pp. 299–329. John Wiley and Sons, New York.

Zelson, P. R., & Cagan, R. H. (1979). Biochemical studies of taste sensation VIII. Partial characterization of alanine-binding taste receptor sites in catfish, *Ictalurus punctatus*, using mercurials, sulfhydryl reagents, trypsin and phospholipase C. *Comparative Biochemistry & Physiology* **64B**, 141–147.

8

Processing of mixtures of stimuli with different tastes by primary mammalian taste neurons

Marion E. Frank

Department of BioStructure and Function, University of Connecticut, Health Center, Farmington, Connecticut, USA

I. Introduction

Responses of single mammalian sensory neurons to taste mixtures have been studied in two different ways, which involve use of different kinds of mixture stimuli. In what is essentially a neuroethological approach (Camhi, 1984), the stimuli have been complex mixtures of chemicals present in potential food items of the species. For adult golden hamsters (*Mesocricetus auratus*), mixture stimuli might be extracts of sunflower seed, apple or raisin. Examples of this approach are seen in the work of Boudreau and colleagues on the primary afferents of a number of non-primate mammals (e.g. Boudreau, Anderson & Oravec, 1975; Boudreau, Oravec & Hoang, 1982), and in the work of Rolls and Scott and their colleagues on neurons in the central nervous system of primates (Rolls, Yaxley, Sienkiewicz & Scott, 1985; Scott, Yaxley, Sienkiewicz & Rolls, 1986). It assumes that the taste system is 'designed' to deal with the complex mixtures encountered in nature and may best be applied to the end-product of neural processing at higher levels of the nervous system. The approach is analytic, starting with complex mixtures and working down to the elementary chemical constituents.

In a more traditional sensory neurophysiological approach, the stimuli have been mixtures of pure chemicals that are known to affect the taste system of the species in question. This use of contrived mixtures represents a synthetic approach to taste mixture study. Examples are seen in the work of Hyman and Frank (1980*b*) and Travers and Smith (1984)

Copyright © 1989 by Academic Press Australia.
All rights of reproduction in any form reserved.

on primary afferent and brain stem neurons of hamsters. It is this synthetic approach that is taken here. A central question addressed in this report is whether the effects of mixtures on primary taste afferents are predictable from the known effects of the component chemicals. The answer is that the effects of mixtures of taste stimuli with different tastes are not. A systematic exploration of the effects of contrived stimulus mixtures shows that mixture interactions occur at an early stage in the gustatory pathways.

A synthetic approach to the study of mixtures is commonly used by psychophysicists dealing with human perception, who have defined the theoretical framework of the approach (e.g. Cameron, 1947; Moskowitz, 1973; Bartoshuk, 1975; Bartoshuk & Gent, 1985). Chemicals with similar or dissimilar tastes may be studied. Studies of mixtures of stimuli with similar tastes typically address the number of receptor classes required to explain the similar perception with stimulation by diverse chemical compounds. Examination of responses of single chorda tympani neurons of hamsters to mixtures of a sugar and 'sweet' (see below) amino acid (Hyman & Frank, 1980*b*; Frank, 1985*b*) is an example of a study of mixtures of similar tastes. In contrast, the study of stimuli with different tastes has the objective of discovering mixture interactions or emergent properties of taste mixtures. Mixture interactions may result in alterations in the quality or intensity of the component stimuli. Examination of responses of single parabrachial brain stem neurons of hamsters to mixtures of a 'sweet' sugar and 'salty' sodium salt is an example of a study of mixtures with different tastes (Travers & Smith, 1984). Studies of the abilities of hamsters to identify components of mixtures reported here suggest that mixture interactions seen in responses of primary taste afferents alter the intensity but not the quality of taste perceptions.

II.   The study of synthetic mixtures

Establishing which stimuli have distinct or similar tastes is essential for the interpretation of synthetic mixture studies. Human beings can be questioned about the taste quality of particular chemicals, and those stimuli that are described by the same adjective(s) are considered similar (Bartoshuk & Cleveland, 1977; Kelling & Halpern, 1983; Halpern, 1987). But the adjectives used for taste qaulity are at least somewhat dependent upon the cultural milieu in which a person lives and the richness of the language that culture uses to describe food (O'Mahony & Ishii, 1987). It is also possible that cultural learning establishes distinctions among 'taste', which rely on chemosensory systems other than gustation. Olfactory and gustatory perceptions interact, with olfactory components of multisensory stimuli adding to perceived 'tastes' (Murphy, Cain & Bartoshuk, 1977; Murphy & Cain, 1980). Furthermore, certain 'taste' losses in humans can be shown to be purely olfactory (Gent, Goodspeed, Zagraniski & Catalanotto, 1987).

Attributing similarity or distinctiveness to taste stimuli for non-humans whose nervous systems can be directly tapped is another issue. Humans can directly describe sucrose and D-phenylalanine as sweet, whereas hamsters cannot. A simple technique, developed by Nowlis (Nowlis & Frank, 1977; Nowlis, Frank & Pfaffmann, 1980), allows hamsters to 'communicate' to humans about their taste perceptions. After establishing a 'conditioned food averson' (Braveman & Bronstein, 1985), the degree that the animal 'confuses' one chemical with others can be assessed by measuring the 'generalization' of the aversion to other stimuli. With this 'language', hamsters 'described' sucrose and D-phenylalanine (see above) as similar-tasting (Nowlis *et al.*, 1980).

In the following, mixture interactions in the series of transduction/processing events occurring in taste buds are addressed. Nerve impulses from chorda tympani neurons, primary afferents that innervate anterior lingual taste bud receptors, were recorded. The stimuli were binary mixtures of sucrose, sodium chloride (NaCl) and hydrochloric acid (HCl). Figure 8.1 presents patterns of generalization of taste aversions across 12 taste stimuli and indicates that sucrose, NaCl and HCl have different tastes to hamsters.

Earlier studies had determined the lowest molarities at which reliable aversions could be established for each compound (Nowlis *et al.*, 1980). These molarities were used for development and subsequent testing of the aversions. An aversion to sucrose (upper pattern) generalized to other 'sweeteners', fructose (fruc) and saccharin (sacch); but the hamsters did not detect any similarity between sucrose (sucr) and the salts and acids tested. An aversion to NaCl (central pattern) generalized only to sodium nitrate (NaNO$_3$), indicating its reliance on a detectable concentration of sodium ions. An aversion to HCl (lower pattern) generalized to acids, non-sodium salts and quinine hydrochloride (quin); apparently, the animals detected similarities among these seven stimuli but did not find the taste of HCl and sodium salts or sweeteners similar. Earlier work had shown that aversions to fructose or saccharin generalize to sucrose, an aversion to NaNO$_3$ generalizes to NaCl, and aversions to acetic acid (H-acet), citric acid (H-citr), ammonium chloride (NH$_4$Cl), quinine hydrochloride, potassium chloride (KCl) and magnesium sulfate (MgSO$_4$) generalize to HCl (Nowlis *et al.*, 1980).

In the following studies, sucrose, NaCl and HCl are considered stimuli that have distinctly different tastes for hamsters, and they are combined in binary mixtures. Neural data on the effects of NH$_4$Cl are

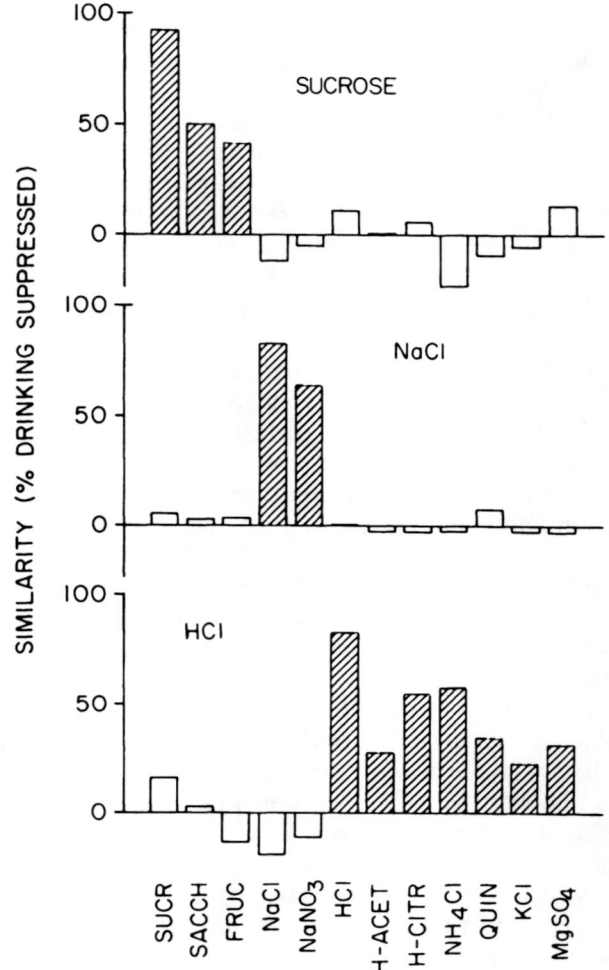

Fig. 8.1 Generalization patterns for aversions learned to sucrose, sodium chloride and hydrochloric acid in hamsters. One group of 12 hamsters learned an aversion to sucrose, a second group of 12 learned an aversion to NaCl, and a third group of 12 learned an aversion to HCl (See Frank 1985*a* for procedural details). Each pattern shows the 'mean per cent suppression of drinking' to the conditional stimulus and the 11 other test stimuli listed along the abscissa. The percentage that drinking was reduced relative to controls is plotted. Striped bars designate mean suppressions that are more than two standard errors greater than zero ($p < .05$). 'Per cent suppression' can be taken as a measure of similarity between a conditional and test stimulus (Nowlis & Frank, 1981*a*). The stimuli were 0.1 M sucrose, 0.001 M Na saccharin, 0.3 M fructose, 0.1 M NaCl, 0.1 M NaNO$_3$, 0.01 M HCl, 0.01 M acetic acid, 0.003 M citric acid, 0.03 M NH$_4$Cl, 0.001 M quinine HCl, 0.3 M KCl, and 0.1 M MgSO$_4$.

also considered in evaluating the substitution of one stimulus (NH_4Cl) for another with a similar taste (HCl) in the analysis (Fig. 8.1, lower pattern). Citric acid, a naturally occurring acid, is substituted for similar tasting HCl (Fig. 8.1, lower pattern) in studies of the effects of mixtures of stimuli with different tastes on intake.

III.   Neural effects of mixing stimuli with different tastes

Three types of response profiles (S, N, and H) are seen among the single neurons of the hamster chorda tympani (Frank, 1973; Hyman & Frank, 1980b; Frank, Bieber & Smith, 1988). Neurons with the S profile respond vigorously to sucrose and other sweeteners but weakly, if at all, to salts or acids. Most neurons with N response profiles respond vigorously to sodium salts, but weakly to sweeteners and non-sodium salts or acids. Neurons with S or N response profiles may be 'specialists', each identifying one type of nutrient. In contrast, neurons with H profiles are 'generalists', responding to a large number of salts (sodium and non-sodium), acids and other compounds. Six S, four N, and six H response profiles across single concentrations of sucrose, NaCl, HCl and NH_4Cl are given in Table 8.1. The response was measured by counting the number of nerve impulses occurring during its first five seconds. The 0.1

Table 8.1.   Hamster chorda tympani responses to mixtures and components

	S						N				H					
	S1	S2	S3	S4	S5	S6	N1	N2	N3	N4	H1	H2	H3	H4	H5	H6
Sucrose	68	113	56	117	68	186	23	16	6	0	5	0	10	0	1	1
NaCl	6	0	0	0	0	14	80	64	72	94	51	13	7	11	15	6
HCl	0	0	0	0	0	10	37	20	32	44	199	57	27	68	50	124
NH_4Cl	9	0	0	0	11	15	29	7	11	16	117	56	26	57	44	90
Sucr-NaCl	61	105	45	94	34	140	78	50	35	69	58	30	11	23	12	28
Sucr-HCl	22	35	14	20	40	82	—	—	41	34	224	68	30	101	52	135
Sucr-NH_4Cl	61	52	14	45	40	173	37	15	14	27	122	59	18	51	45	75
NaCl-HCl	0	0	0	0	0	15	44	50	45	88	247	81	54	105	72	147
NaCl-NH_4Cl	0	7	2	0	0	24	59	38	49	43	131	48	47	88	40	84
HCl-NH_4Cl	0	1	0	0	0	17	24	29	16	48	267	109	62	144	81	171

Values in the table are numbers of nerve impulses elicited during 5 s of stimulation. Stimuli are 0.1 M sucrose, 0.01 M NaCl, 0.003 M HCl, and 0.05 M NH_4Cl and their binary mixtures. The data are derived from Hyman, unpublished doctoral dissertation. The Rockefeller University, New York 1978. S, N, and H identify functional types of nerve fibers (see text). S1-S6, N1-N4, and H1-H6 are individual examples of the three types of fibers. Dashes indicate data were not obtained. Values for responses to the most effective stimulus and mixtures containing that stimulus are shown in bold type for each fiber type. Those mixtures elicit smaller responses than sucrose does in every S fiber ($p < .00001$), smaller responses than NaCl does in every N fiber ($p < .001$), but larger responses than HCl does in every H fiber ($p < .001$). Probabilities are based upon two-tailed binomial tests.

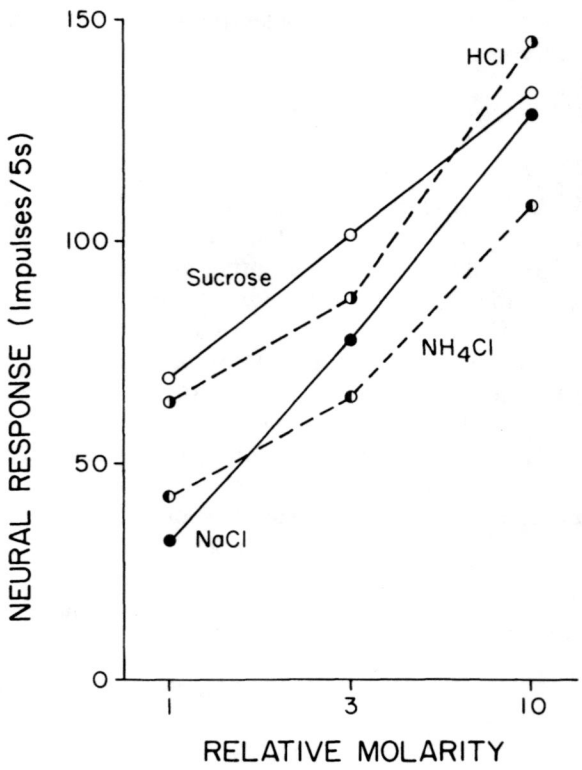

Fig. 8.2 Neural responses of hamster chorda tympani fibers with S, N, or H response profiles to three concentrations of highly effective stimuli. Mean responses (neural impulses/5 s) for six fibers with S profiles (open circles) to 0.03, 0.1, and 0.3 M sucrose; responses for four fibers with N profiles (filled circles) to 0.003, 0.01, and 0.03 M NaCl; and responses for six fibers with H profiles (half-filled circles) to 0.001, 0.003, and 0.01 M HCl, as well as 0.016, 0.05, and 0.16 M NH$_4$Cl are plotted. Relative concentration '3' is the concentration used in the study of binary mixtures, for which data are given in Table 8.1. Transformed 'stimulus-response functions' were used to obtain the 'response predicted if stimuli substitute' presented in Table 8.2, as explained by Hyman and Frank (1980*b*). Data from Hyman, unpublished doctoral dissertation. The Rockefeller University, New York, 1978.

M sucrose, 0.01 M NaCl, 0.003 M HCl, and 0.05 M NH$_4$Cl elicited equal-sized, filtered, multi-unit responses from the entire chorda tympani nerve (Hyman & Frank, 1980a).

Figure 8.2 presents mean responses to this 'center' concentration, which was used in binary mixtures, and two concentrations, differing by a factor of 10, above and below it. Concentration-response functions recorded for sucrose from neurons with S profiles, for NaCl from neurons with N profiles, and for HCl and NH$_4$Cl from neurons with H profiles are shown. The 'center' sucrose concentration fell above, but the 'center' NaCl concentration fell well below, the middle of the effective ranges for the stimuli (Hyman & Frank, 1980a, b). In both cases, the relation between log concentration and response is quite linear. Unfortunately, few chorda tympani neurons with N response profiles are clearly activated by 0.01 M NaCl and they do not encompass the range of reactiveness of N profiles observed when 0.03 M NaCl is used (Frank, 1973; Frank et al., 1988). Four highly reactive chorda tympani neurons with N profiles were recorded. Highly reactive taste afferents respond proportionally to other less effective stimuli (Hyman & Frank, 1980b; Frank, Contreras & Hettinger, 1983; Frank, 1985b; Frank, 1985c; Frank et al., 1988). Note this effect in the most reactive neurons with S and H profiles: 'S6' and 'HI'.

Every hamster chorda tympani neuron recorded with either an S or an N response profile is activated less when a most effective stimulus is mixed with a stimulus with a different distinct taste than it is by a most effective stimulus alone. This can be seen by comparing the values in the rows for neurons with S or N profiles (in bold print) in Table 8.1. The result is a reduction in the signal sent to the central nervous system regarding specific nutrients. The mean effects for sucrose, NaCl, HCl and their three binary mixtures are seen graphically in the upper and central patterns in Figure 8.3. The greatest reduction in signal sent by neurons with S profiles is seen when HCl is added to sucrose. That signal is about one-third of the signal elicited by sucrose alone. When NaCl or NH$_4$Cl are added, the signal is reduced about half as much. The signals sent by neurons with N profiles are reduced to about two-thirds of the signal elicited by NaCl alone when sucrose, HCl or NH$_4$Cl are added.

In contrast to the lessening of the signals sent by specialist taste neurons, every generalist neuron with an H profile is activated more when the most effective stimulus (HCl) is mixed with a stimulus with a different distinct taste than it is by HCl alone. This can be seen by comparing the values in the rows for neurons with H profiles (in bold print) in Table 8.1. Responses for the HCl-NH$_4$Cl mixture are not in bold print because these stimuli have similar tastes (Fig. 8.1). The result is a

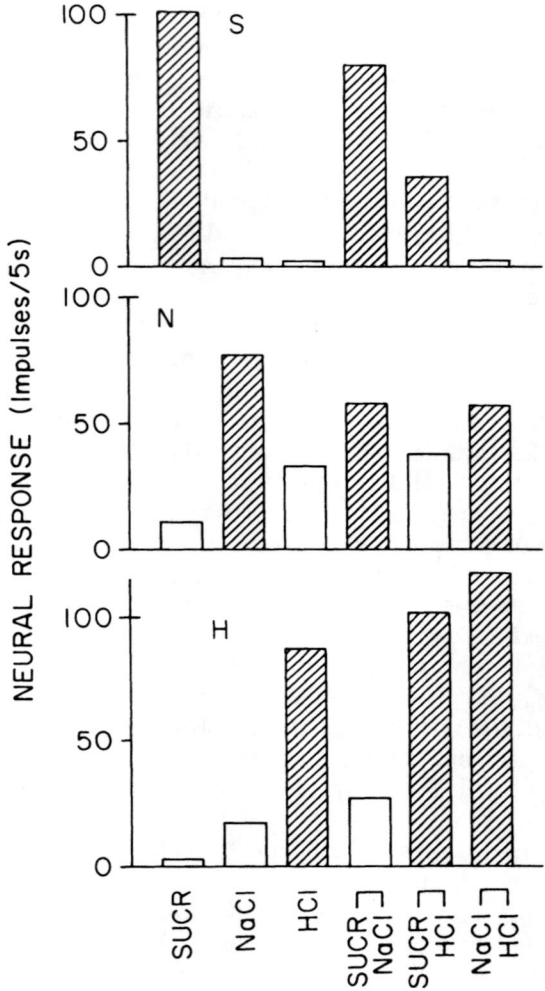

Fig. 8.3 Neural responses of hamster chorda tympani fibers with S, N and H response profiles to binary mixtures and mixture components. Plotted are mean responses (neural impulses/5 s) from 6 S, 4 N, and 6 H profiles, calculated from data presented in Table 1. Bars for responses to stimuli containing sucrose for S profiles, to stimuli containing NaCl for N profiles, and to stimuli containing HCl for H profiles are striped. In fibers with S profiles, the response to the sucrose–HCl or sucrose–NaCl mixture is smaller than the responses to sucrose alone ($p < .05$), and the response to the sucrose–HCl mixture is smaller than the response to the sucrose–NaCl mixture ($p < .05$). In contrast, in fibers with H profiles, the responses to the HCl–sucrose and HCl–NaCl mixtures are larger than the sum of the responses to the two mixture components ($p < .05$). Probabilities are based upon two-tailed binomial tests.

Table 8.2. Neural stimulus activation patterns to mixtures

	Obtained response (i/5s)			Response predicted if stimuli substitute			Response predicted if responses sum		
Stimulus	S	N	H	S	N	H	S	N	H
Sucr + NaCl	**79.8**	58.0	27.0	**101.3**	79.4	17.2	104.6	88.8	20.0
Sucr + HCl	35.5	37.5	**101.7**	**101.3**	36.4	87.5	103.0	44.5	**90.3**
NaCl + HCl	2.5	56.8	**117.7**	3.4	**88.3**	87.8	5.0	110.8	**104.7**

Stimulus activation patterns (mean number of nerve impulses/5 s) across chorda tympani fibers with three different response profiles (S, N, H) are given. Also presented are the patterns predicted if stimuli were to be exact substitutes for one another and the patterns predicted if the responses to component stimuli were to sum. Values for stimulus substitution were obtained from lines fitted to the relationship between log stimulus concentration and log mean response to the most effective stimulus. The lines fit the points quite well; for S profiles, $r = .9977$, for N profiles, $r = .9923$, and for H profiles, $r = .9935$. Values for mixtures containing the most effective stimulus are shown in bold type for each profile-type and for the relevant predictions. Those mixtures elicit significantly smaller responses than predicted for stimulus substitution in S profiles ($p < .001$) and N profiles ($p < .01$) and larger than responses predicted for response summation in H profiles ($p < .05$). Probabilities are based upon two-tailed binomial tests.

heightening of the signal sent by the generalist neurons when a mixture containing stimuli with different tastes is present in the mouth. In fact, the responses to the mixtures are typically even larger than the sum of the responses to the components. The mean effects are seen graphically in the lower pattern in Figure 8.3. When NH_4Cl, which is about three-quarters as effective as HCl (at the center concentration and throughout the stimulus range tested, Fig. 8.2), is mixed with a stimulus with a different distinct taste, neurons with H profiles are activated to about the same level as they are to NH_4Cl alone (Table 8.1). Mixtures of stimuli with different tastes that include the less effective NH_4Cl do not elicit heightened activity in neurons with H profiles as do mixtures that include the most effective HCl. It is interesting to note the similar nonlinear log concentration/response functions seen for both HCl and NH_4Cl in neurons with H response profiles (Fig. 8.2). Response increases at a faster rate above the 'center' concentration than from below it.

Table 8.2 presents the obtained 'stimulus activation patterns' (Frank *et al.*, 1988) for binary mixtures of sucrose, sodium chloride and hydrochloric acid across neurons with S, N and H response profiles in the three columns at the left. The table also presents activation patterns across these functional types of neurons that are predicted from the known effects of the component stimuli according to the 'stimulus substitution' or 'response summation' model. These mixture models have been discussed extensively by psychophysicists (Bartoshuk & Gent, 1985) and have been applied to data recorded from single taste neurons as well (Hyman & Frank, 1980*b*; Travers & Smith, 1984).

The stimulus substitution model assumes that binary mixtures of different chemicals have effects identical to adding two concentrations of a single chemical. In other words, the two chemicals can substitute for one another. This model is often tested in studies of mixtures of taste stimuli of similar taste, assuming that a single receptive process may account for reactions to mixtures with similar perceptual effect. The stimulus substitution model may not be appropriate in this 'perceptual' sense in the present context because stimuli with different distinct tastes are the components of the mixtures. The stimuli cannot substitute if the perceptions do not. Yet it is possible to conceive of the single neuron as the single receptive process in the model. The stimulus substitution predictions in Table 8.2 are based upon straight lines fitted to log concentration versus log response to the most effective stimulus for each functional type of neuron (Hyman & Frank, 1980*b*). Predictions are too large for mixtures containing the most effective stimulus for neurons with S and N profiles, and too small for neurons with H profiles.

The response summation model assumes that the components in binary mixtures have the same effect as they do when presented alone. It might be expected to hold for mixtures of stimuli with different distinct tastes, if the tastes are independently processed. But this 'perceptual' concept may not hold for individual taste neurons, particularly those that may be specialized for identification of chemicals with similar taste (Frank, 1985*b*). The response summation and stimulus substitution models make identical predictions when response and concentration grow at the same rate. In the neural data considered here, however, the response grows more slowly than concentration. A 10-fold increase in concentration elicits a 1.9-fold increase in response in neurons with S profiles, a 2.3-fold increase in response in neurons with H profiles, and a 4.0-fold increase in response in neurons with N profiles (Fig. 8.2). The response summation predictions in Table 8.2 are, expectedly, too large for responses to mixtures that contain the most effective stimulus for neurons with S and N profiles, but they are too small for neurons with H profiles.

In summary, the obtained mixture activation patterns fall outside the range of predictions for the two models. Neurons with S and N profiles show responses to mixtures that are smaller than predictions based on stimulus substitution (compare values in bold print in S and N columns in the left and central triad of columns in Table 8.2) and neurons with H profiles show responses larger than predictions based upon response summation (compare values in bold print in H columns in left and right triad of columns in Table 8.2). The effect of mixing stimuli with distinct tastes on neural activity cannot be predicted by the two common models of mixture interactions developed to deal with sensory perception. Rather, it appears as if processes are at work within the taste bud that muffle the reactions of afferent neurons specialized in detecting nutrient sources but heighten the reactions of generalist afferent neurons when taste stimuli with different distinct tastes are presented together.

IV.   Mixture interactions and identification of components

The mixture interactions seen in responses of single primary afferent taste neurons indicate that the gustatory system of the hamster does not process all mixtures of stimuli with distinct tastes identically. Two of the components within mixtures, sucrose and NaCl, had lessened effects in the sets of neurons most sensitive to them. But mixtures had heightened effects in the set of generalist neurons, which are most sensitive to HCl. Behavioral studies addressed whether these differences in mixture processing are seen in the reaction of hamsters to mixtures.

The ability of hamsters to identify a taste stimulus when mixed with another taste stimulus with a different taste was assessed by establishing a 'conditioned aversion' to a stimulus and measuring its generalization to mixtures. In Figure 8.4, upper, central and lower patterns of generalization are for hamsters that had learned aversions to 0.1M sucrose, 0.1M NaCl, and 0.01M HCl, respectively. The six test solutions were the three solutions to which aversions were learned and their three binary mixtures. The aversions were expressed by a suppression of drinking. In general, the hamsters detected a similarity between a taste stimulus presented alone or within a mixture containing the stimulus and a stimulus with another taste. They did not find any similarity between a stimulus and mixtures that did not contain that stimulus.

However, sucrose and NaCl are less easily identified in mixtures than is HCl. An average 44% loss in expression of the learned aversion to sucrose, an average 71% loss in expression of the aversion to NaCl and an average 11% loss in expression of the aversion to HCl are seen when the conditional stimulus is embedded within either of two mixtures (Fig. 8.4). This result can be compared with the neural data recorded from the chorda tympani (Fig. 8.3), which shows both sucrose (upper panel) and NaCl (central panel) had lessened effects on neurons most sensitive to them when they were embedded in mixtures, but HCl (lower panel) did not. These neural data suggest that sucrose and NaCl were less intense when presented in mixtures than when presented alone, but HCl was not. Since the expression of an aversion increases with the strength of a test stimulus (Nowlis, 1974), a simple intensity shift in the 'target' stimulus

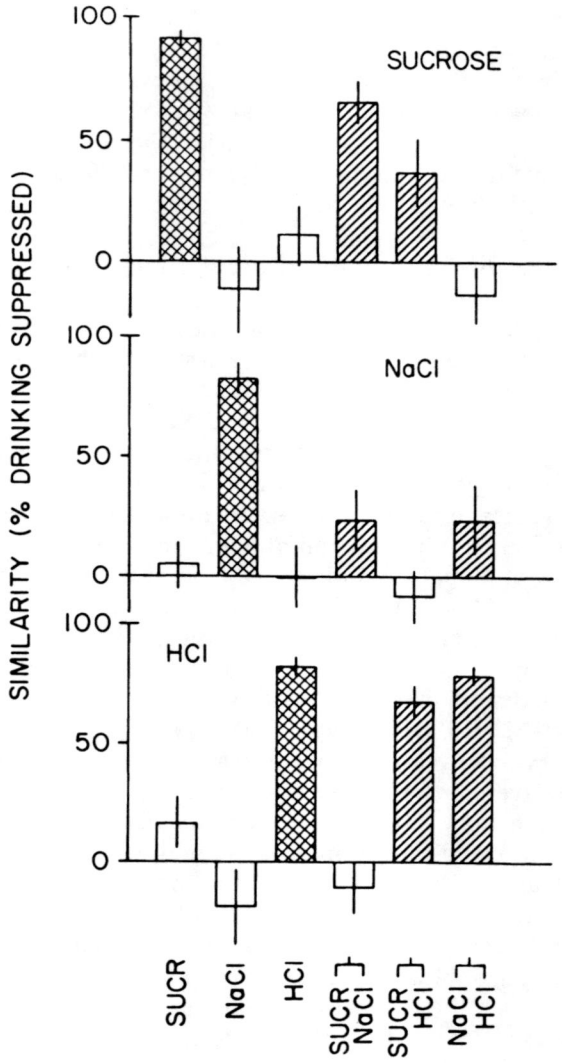

Fig. 8.4 Generalization of aversions learned to components with dissimilar tastes to binary mixtures in hamsters. Twelve hamsters learned an aversion to sucrose, 12 learned an aversion to NaCl, and 12 learned an aversion to HCl. Each of the three patterns shows the 'mean per cent suppression' to one conditional stimulus (cross-hatched bar) and five other test stimuli (listed along the abscissa). The percentage that drinking was reduced relative to controls is plotted. Striped bars are for mixtures containing the conditional stimulus. Line segments at the centers of bar indicate $+/-$ one standard error of the mean per cent suppression (Nowlis & Frank, 1981 *b*). Hamsters detect a similarity ($p < .01-.05$) between a binary mixture containing 0.1 M sucrose and sucrose itself; they detect a similarity ($p < .01$) between a binary mixture containing 0.01 M HCl and HCl itself; but they hardly detect a similarity ($p < .05-.10$) between a binary mixture containing 0.1 M NaCl and NaCl itself.

(Rabin, unpublished doctoral dissertation, Yale University, New Haven, Connecticut, 1986) may be reflected in the behavior. However, the effect of sucrose was most severely affected neurally, but NaCl seems most affected behaviorally. There are many possible reasons the anterior lingual neural mixture effects do not exactly match the behavioral discriminations. The NaCl used in the neural study was one-tenth of the molarity of the NaCl used in the behavioral study. The neurons with N profiles sampled were all highly reactive (see above). Mixture interactions may be occurring in gustatory receptive fields other than the anterior lingual field or in the gustatory central nervous system.

Pure sucrose (Fig. 8.3, upper pattern) or pure NaCl (Figure 8.3, central pattern) are more effective than mixtures in stimulating neurons specialized for their detection. Aversions learned to mixtures should not show decrements when tested against components if shifts in neural efficacy (intensity) of 'target' sucrose or NaCl were to account for expression of an aversion. Figure 8.5 shows the patterns of generalization of aversions learned to a 0.1M sucrose and 0.1M NaCl mixture (upper pattern), a sucrose and 0.01M HCl mixture (central pattern), and a NaCl and HCl mixture (lower pattern). The aversions were expressed as suppressions of drinking of the three component stimuli (Fig. 8.5, three bars to the left in each panel). Overall, the aversion to the binary mixture (i.e. the conditional stimulus) generalizes nearly unabated to the two component stimuli. There are no siginificant losses in the expressions of individual mixture aversions to the components. These results (Fig. 8.5) are consistent with the idea that the hamsters are showing some decrement in the expression of a learned aversion to NaCl or sucrose (Fig. 8.4) because their neural responses are lessened in mixtures (Fig. 8.3).

However, average expression of an aversion to a component is 86% of the expression of an aversion to the conditional stimulus mixture (Fig. 8.5). This is a significant loss ($p < .05$, two-tail binomial test). The neural effect of sucrose or NaCl would increase upon extraction from a mixture (Fig. 8.3), and an increase in stimulus effect (intensity) ought to lead to an increase in the expression of the aversion (Nowlis, 1974).

Effects other than intensity changes are also suggested by inter-mixture generalizations of aversions (Fig. 8.5, three bars to the right in each panel). Aversions learned to binary mixtures of stimuli with different distinct tastes generalize to other binary mixtures that contain one component in common with the conditional stimulus mixture. But, an average 69% loss in expression of the aversion for the sucrose-NaCl mixture (Figure 8.5, upper pattern), 59% loss for the NaCl-HCl mixture (Fig. 8.5, lower pattern), and a 35% loss for the sucrose-HCl mixture (Fig. 8.5, central pattern) is seen for the two inter-mixture generalizations. Neural activity (Fig. 8.3) to the common inter-mixture component

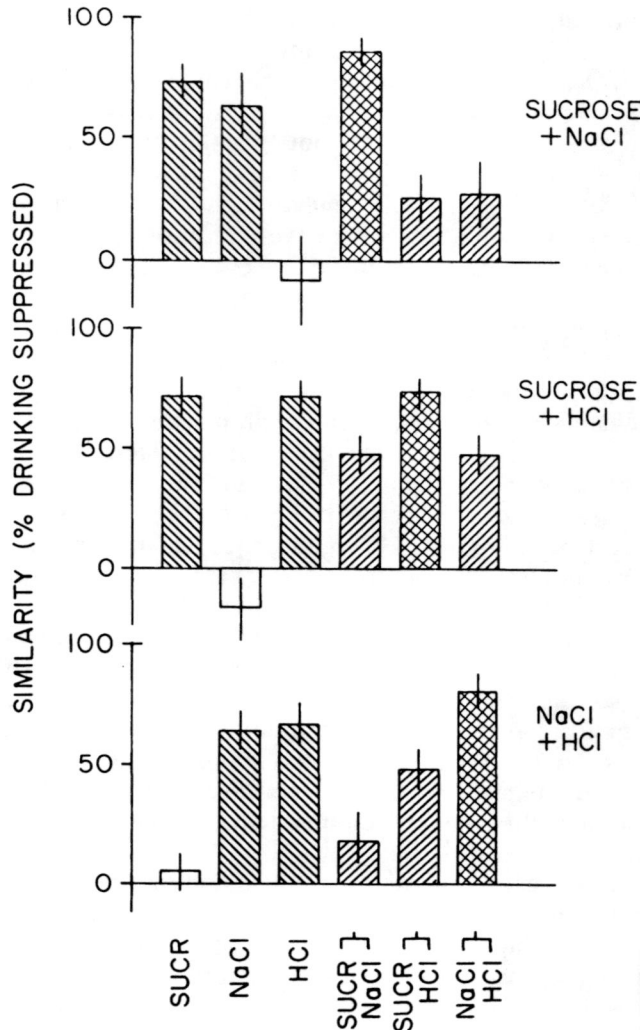

Fig. 8.5 Generalization of aversions learned to binary mixtures of stimuli with dissimilar tastes to mixture components and other binary mixtures. Twelve hamsters learned an aversion to sucrose–NaCl mixture, 12 learned an aversion to a sucrose–HCl mixture, and 12 learned an aversion to a NaCl–HCl mixture. Each pattern shows the 'mean per cent suppression' to one conditional stimulus (cross-hatched bar) and five other test stimuli (listed along the abscissa). The percentage that drinking was reduced relative to controls is plotted. Striped bars are for test stimuli having a component in common with the conditional stimulus. Line segments at the center of bars indicate $+/-$ one standard error of the mean per cent suppression (Nowlis & Frank, 1981b). Hamsters readily detect a similarity ($p < .01$) between component stimuli (0.1 M sucrose, 0.1 M NaCl, 0.01 M HCl) and binary mixtures of these stimuli with dissimilar tastes, and also detect a similarity ($p < .01-.05$) between mixtures that have one component in common.

Table 8.3. Per cent learned aversions generalize to mixture components

Stimulus configuration		Component A			
Tested	Learned	NaCl	Sucrose	HCl	Mean
A	(A + X)	76.7	91.6	90.0	86.1
(A + X)	A	28.9[a]	56.0[a]	88.5	57.8
(A + X)	(A + Y)	27.4[a]	47.5[a]	62.1	45.7
Mean		44.3	65.0	80.2	63.2

Percentages are (per cent suppression to test stimulus)/(per cent suppression to stimulus to which aversion was learned) × 100. A is the 'target' (the mixture component that is found in both conditional (learned) and test stimuli); X and Y are other components. A value of 100 per cent would indicate no loss in expression of the aversion. A value of 0 per cent would indicate that the aversion did not generalize.

[a] The animals show less ($p < .01$) generalization to the test stimulus than to the stimulus to which the aversion was learned and there is a greater ($p < .05$) decrement in generalization when A is NaCl than when A is sucrose. Probabilities are based on two-tailed binomial tests.

decreases an average 29% for the sucrose–NaCl mixture, decreases 6% for the NaCl-HCl mixture, but increases 70% for the sucrose-HCl mixture. These numbers are averages based on the assumption that identification of sucrose, NaCl or HCl is solely dependent upon activity of neurons with S, N or H response profiles, respectively. Neurons with S profiles (Fig. 8.3, upper panel) are much less active to *sucrose* + HCl than to *sucrose* + NaCl, and neurons with N profiles (Fig. 8.3, central panel) are about equally active to *NaCl* + HCl and sucrose + *NaCl*. Neurons with S profiles, (Fig. 8.3, upper panel) are much more active to *sucrose* + NaCl than to *sucrose* + HCl, and neurons with H profiles (Fig. 8.3, lower panel) are somewhat more active to NaCl + *HCl* than to sucrose + *HCl*. Neurons with N profiles (Fig. 8.3, central panel) are equally active to sucrose + *NaCl* and *NaCl* + HCl, but neurons with H profiles (Fig. 8.3, lower panel) are somewhat less active to sucrose + *HCl* than to NaCl + *HCl*. Although the order of the quantitative losses in expression of aversions (sucrose-NaCl > NaCl-HCl > sucrose-HCl) is correlated with the order of the effects of mixture interactions seen in the nerve, there is a general loss ($p < .05$, two-tailed binomial test) in inter-mixture generalizations of aversions that cannot be accounted for by intensity shifts suggested by activity changes occurring at the gustatory periphery. From a perceptual point of view, it would be astonishing if a full-blown aversion would be seen to half of the complex stimulus to which it was learned.

Table 8.3 presents a summary of results of the studies on cross-generalizations between binary mixtures and component stimuli with distinctly different tastes. HCl is most easily identified in a test stimulus, regardless of the learned configuration (compare column means for NaCl, sucrose and HCl). Sucrose and NaCl are less readily identified. NaCl is the most difficult to identify within a stimulus configuration that differs from the one in which the aversion to it was learned (note values in first data column). Also, identification is easiest when a stimulus is presented in pure form (compare means for rows). There is a general decline in expression of aversions to stimuli in new configurations. However, the behavioral data are partially accounted for by the reduced neural effects of sucrose or NaCl and heightened neural effects of HCl in mixtures of stimuli with dissimilar tastes. This suggests that chorda tympani neural activity plays a role in identification of stimuli with distinctly different tastes in hamsters.

V.   Mixture interactions and intake of taste solutions

The gustatory system is commonly thought to play a role in the recognition of chemicals with nutritional or harmful effects when ingested. It may identify an energy source (e.g. a sucrose solution) as well as stimulate ingestion of nutrients when an animal is hungry. Replete hamsters, for example, prefer to drink sucrose solutions above 10 mM in strength rather than water. Similarly, the gustatory system may identify a source of sodium (e.g. NaCl solution) as well as stimulate ingestion at a time an animal is in need. Desert animals such as hamsters probably monitor their sodium balance very closely. Replete hamsters prefer to drink water to NaCl solutions of 10 mM and stronger. There are specialist gustatory channels for these two classes of nutrient that are influenced by the purity of the nutrient source (Fig. 8.3). If sodium and sugar are mixed, both specialist channels show reduced activity. The activity of the specialist channels are also reduced by mixing other contaminants such as non-sodium salts or acids with the nutrients. In fact, stimuli that activate a third generalist gustatory channel, such as hydrochloric acid or ammonium chloride, influence neural activity in the specialist channels quite significantly. In addition, these generalist channels, which may play a role in the detection of harmful contaminants within potential nutrient sources, show heightened activity to mixtures (Fig. 8.3). In contrast to the specialist neurons, they are more active to a complex stimulus. Replete hamsters prefer to drink water to citric acid solutions, which also activate the generalist channel (Frank *et al.*, 1988).

In Table 8.4, preferences (relative to water) of hamsters for solutions of 0.5M sucrose, 0.1M NaCl, 0.01M citric acid, and the binary mixtures

Table 8.4. Two-bottle preference for components and mixtures

Solution A	N	Preference (%)		Amount consumed (mL) per animal in 48 hours	
		Mean	SE	Solution A	Water
Water	85	49.8	1.4	12.5	12.6
.5 M sucrose	20	72.9	3.2	17.3	6.7
.1 M NaCl	44	28.2	2.5	8.2	18.4
.01 M citric acid	20	16.5	1.8	3.5	17.0
Sucr + NaCl	10	67.9	5.5	15.7	7.8
Sucr + H-Citr	10	62.6	5.7	13.0	8.2
NaCl + H-Citr	10	9.5	2.5	1.7	15.8

The preference values are mean and standard error (SE) of the per cent of fluid that the N animals drank from the bottle containing solution A. Fifty per cent would ind cate that the animals drank equally from the two bottles. (From Hettinger & Frank, 1985.)

of these three stimuli with different distinct tastes are given. The hamsters were in a replete, fluid-balanced state. Taste solutions preferred to water are sucrose (2.7 to 1), the sucrose-NaCl mixture (2.1 to 1) and the sucrose-citric acid mixture (1.7 to 1). Animals adjusted their intake of sucrose when it was mixed with NaCl or citric acid. Thus, 17 mL of sucrose, 16 mL of the sucrose-NaCl mixture, and 13 mL of the sucrose-citric acid mixture were ingested in 48 hours. Preference for 0.25M sucrose to water (70.8 +/− 4.5%) differs slightly from preference for 0.5M sucrose (72.9 +/− 3.2). Intake of the two concentrations of sucrose was identical (17 mL/48 hours). With drastically reduced activity in afferents with S profiles due to a lower concentration (Fig. 8.2) or mixing (Fig. 8.3, upper panel), intake of sucrose is slightly affected. Intake of sucrose cannot be controlled simply by afferents of the chorda tympani branch of the VIIth cranial nerve in hamsters. Afferent gustatory channels of other cranial nerves (IX and X) have been directly implicated in the control of ingestion in catfish (Atema, 1971; Finger & Morita, 1985; Finger, 1987). The VIIth cranial nerve appears to be more involved in locating and sampling food in these highly gustatory animals.

Sodium chloride intake is strongly affected by adding chemicals of different distinct tastes to the solution (Table 8.4). Hamsters prefer the NaCl-sucrose mixture to water but prefer water to solutions of NaCl (2.5 to 1) and the NaCl-citric acid mixture (9.5 to 1). The preference for drinking water over the NaCl-citric acid mixture is greater than the preference for water over NaCl ($p < .01$). The hamsters ingested 16 mL of the NaCl-sucrose mixture, 2 mL of the NaCl-acid mixture, compared with 8 mL of NaCl alone. NaCl intake was modified in two directions, upward by sucrose and downward by citric acid. Chorda tympani neurons with N profiles show a reduction of activity with the addition of either sucrose or acid (Fig. 8.3, central panel). It is likely that NaCl specialist neurons identify and monitor the nutrient's purity, but sodium intake is complexly regulated. The level of activity of chorda tympani neurons with N response profiles in rats is very sensitive to sodium deprivation (Contreras & Frank, 1979). In this case, the intake of NaCl is increased when the activity of this specialist gustatory channel is reduced.

Intake of citric acid is also strongly affected by adding chemicals with other tastes (Table 8.4). Hamsters prefer to drink the acid-sucrose mixture to water but prefer water to solutions of citric acid (5.1 to 1) and the NaCl-acid mixture. The preference for water over citric acid is greater than the preference for water over NaCl ($p < .05$). The hamsters drank 13 mL of the acid-sucrose mixture, 2 mL of the acid-NaCl mixture, compared to 3.5 mL of citric acid alone. Intakes of citric acid and NaCl, two 'unpalatable' stimuli, are readily influenced by the addition of stimuli with different tastes in replete laboratory hamsters. The salt-acid mixture is less 'palatable' than either component. The negative 'palatability' is strengthened, not 'averaged' in the mixture. The generalist chorda tympani neurons show increased activity to mixtures (Fig. 8.3, lower panel) and may identify potentially harmful individual chemicals and chemical mixtures.

VI.   Review and conclusions

Peripheral gustatory mixture interactions occur within the series of events that begin with stimuli interacting with taste receptor cell membranes and end with the generation of neural action potentials in primary afferent neurons. These mixture interactions were observed in the activities of single chorda tympani taste neurons in hamsters and their effect on discrimination and ingestion was assessed. Mixtures of chemicals with distinctly different tastes for hamsters are processed in two different ways. Reactions to chemicals with specific nutrient value are lessened in neurons specialized for their identification when specific activators (sucrose or sodium chloride) are embedded in mixtures. In contrast, mixing stimuli with distinctly different tastes augments reactions of generalist neurons that respond to many 'unpalatable' chemicals. Hamsters identify components in binary mixtures of taste stimuli with distinctly different tastes, suggesting that the mixture interactions affect stimulus intensity rather than stimulus quality. The decreased activity in specialist and increased activity in generalist primary afferent gustatory channels is reflected in the hamsters' ability to identify particular target stimuli. The peripheral gustatory mixture interactions are not clearly reflected in intake. Even so, ingestion of simple mixtures by hamsters is systematically related to ingestion of mixture components. The information for identification of nutrient and harmful chemicals in mixtures may be provided by the neurons in the chorda tympani nerve. Ingestion, a more complex phenomenon than sampling and identification, must require integration of a variety of sensory inputs.

Acknowledgments

This work was supported by the US National Science Foundation (grant BNS-8519638). The author is grateful for the excellent experimental work of T. P. Hettinger, A. M. Hyman and G. H. Nowlis, and B. G. Rehnberg's critical reading of an earlier version of this manuscript.

References

Atema, J. (1971). Structures and functions of the sense of taste in the catfish (*Ietalurus natalis*). *Brain, Behavior, & Evolution* **4**, 273–294.

Bartoshuk, L. M. (1975). Taste mixtures: is mixture suppression related to compression? *Physiology & Behavior* **14**, 643–649.

Bartoshuk, L. M., & Cleveland, C. T. (1977). Mixtures of substances with similar tastes: a test of a psychophysical model of taste mixture interactions. *Sensory Processes* **1**, 177–186.

Bartoshuk, L. M., & Gent, J. F. (1985). Taste mixtures: an analysis of synthesis. In D. W. Pfaff, ed., *Taste, olfaction, and the central nervous system*, pp. 210–232. The Rockefeller University Press, New York.

Boudreau, J. C., Anderson, W., & Oravec, J. (1975). Chemical stimulus determinants of cat geniculate ganglion chemoresponsive group II unit discharge. *Chemical Senses & Flavour* **1**, 495–517.

Boudreau, J. C., Oravec, J. J., & Hoang, N. K. (1982). Taste systems of goat geniculate ganglion. *Journal of Neurophysiology* **48**, 1226–1242.

Braveman, N. S., & Bronstein, P., eds (1985). Experimental assessments and clinical applications of conditioned food aversions. *Annals of the New York Academy of Sciences*, vol. 43. New York Academy of Sciences, New York.

Cameron, A. T. (1947). *The taste sense and the relative sweetness of sugars and other substances (Report 9)*. Sugar Research Foundation, New York.

Camhi, J. M. (1984). *Neuroethology: nerve cells and the natural behavior of animals*. Sinauer Associates, Sunderland, Massachusetts.

Contreras, R. J., & Frank, M. E. (1979). Sodium deprivation alters neural responses to gustatory stimuli. *Journal of General Physiology* **73**, 569–594.

Finger, T. E. (1978). Gustatory nuclei and pathways in the central nervous system. In T. E. Finger & W. L. Silver, eds, *Neurobiology of taste and smell*, pp. 331–354. John Wiley, New York.

Finger, T. E., & Morita, Y. (1985). Two gustatory systems: facial and vagal gustatory nuclei have different brainstem connections. *Science* **227**, 776–778.

Frank, M. E. (1973). An analysis of hamster afferent taste nerve response functions. *Journal of General Physiology* **61**, 588–618.

Frank, M.E. (1985*a*). Sensory physiology of taste and smell discriminations using conditioned food aversion methodology. In N. S. Braveman & P. Bronstein, eds., Experimental assessments and clinical applications of conditioned food aversions, *Annals of the New York Academy of Sciences* vol. 43, pp. 89–99. New York Academy of Sciences, New York.

Frank, M. E. (1985*b*). On the neural code for sweet and salty tastes. In D. W. Pfaff, ed., *Taste, olfaction, and the central nervous system*, pp. 107–128. The Rockefeller University Press, New York.

Frank, M. E. (1985*c*). Quantitative co-variance of taste responses: Empirical criteria for 'natural' types of mammalian peripheral gustatory neurons. *Chemical Senses* **10**, 430.

Frank, M. E., Contreras, R. J., & Hettinger, T. P. (1983). Nerve fibers sensitive to ionic taste stimuli in chorda tympani of the rat. *Journal of Neurophysiology* **50**, 941–960.

Frank, M. E., Bieber, S. L., & Smith, D. V. (1988). The organization of taste sensibilities in hamster chorda tympani nerve fibers. *Journal of General Physiology* **91**, 861–896.

Gent, J. F., Goodspeed, R. B., Zagraniski, R. T., & Catalanotto, F. A. (1987). Taste and smell problems: validation of questions for the clinical history. *Yale Journal of Biology & Medicine* **60**, 27–35.

Halpern, B. P. (1987). Human judgements of MSG taste: quality and reaction times. In Y. Kawamura & M. R. Kare, eds, *Umami: a basic taste*, pp. 327–354. Marcel Dekker, New York.

Hettinger, T. P., & Frank, M. E. (1985). Preferences of hamsters for solutions of chemicals with sweet, salty, sour, bitter, sulfurous, soapy, alkaline, or combined flavors: analytic hedonic processing. *Chemical Senses* **10**, 444.

Hyman, A. M., & Frank, M. E. (1980*a*). Effects of binary taste stimuli on the neural activity of the hamster chorda tympani. *Journal of General Physiology* **76**, 125–142.

Hyman, A. M., & Frank, M. E. (1980*b*). Sensitivities of single nerve fibers in the hamster chorda tympani to mixtures of taste stimuli. *Journal of General Physiology* **76**, 143–173.

Kelling, S. T., & Halpern, B. P. (1983). Taste flashes: reaction times, intensity, and quality. *Science* **219**, 412–414.

Moskowitz, H. R. (1973). Models of sweetness additivity. *Journal of Experimental Psychology* **99**, 88–98.

Murphy, C., & Cain, W. S. (1980). Taste and olfaction: independence versus interaction. *Physiology & Behavior* **24**, 601–605.

Murphy, C., Cain, W. S., & Bartoshuk, L. M. (1977). Mutual action of taste and olfaction. *Sensory Processes* **1**, 204–211.

Nowlis, G. H. (1974). Conditioned stimulus intensity in acquired alimentary aversions in rats. *Journal of Comparative & Physiological Psychology* **86**, 1173–1184.

Nowlis, G. H., & Frank, M. E. (1977). Qualities in hamster taste: behavioral and neural evidence. In J. Le Magnen & P. MacLeod, eds, *Olfaction and Taste VI*, pp. 241–248. Information Retrieval, London.

Nowlis, G. H., & Frank, M. E. (1981*a*). Quality coding in gustatory systems of rats and hamsters. In D. M. Norris, ed., *Perception of Behavioral Chemicals*, pp. 58–80. Elsevier, Amsterdam.

Nowlis, G. H., & Frank, M. E. (1981*b*). On the hamster's response to taste mixtures. Paper presented at the 4th Annual Meeting of the Association for Chemoreception Sciences, Sarasota, Florida.

Nowlis, G. H., Frank, M. E., & Pfaffmann, C. (1980). Specificity of acquired aversions to taste qualities in hamsters and rats. *Journal of Comparative & Physiological Psychology* **94**, 932–942.

O'Mahony, M., & Ishii, R. (1987). The umami taste concept: implications for the dogma of four basic tastes. In Y. Kawamura & M. R. Kare, eds, *Umami: a basic taste*, pp. 75–93. Marcel Dekker, New York.

Rolls, E. T., Yaxley, S., Sienkiewicz, Z. J., & Scott, T. R. (1985). Gustatory responses of single neurons in the orbitofrontal cortex of the macaque monkey. *Chemical Senses*, **10**, 443.

Scott, T. R., Yaxley, S., Sienkiewicz, Z. J., & Rolls, E. T. (1986). Gustatory responses in the frontal opercular cortex of the alert cynomologus monkey. *Journal of Neurophysiology* **56**, 876–890.

Travers, S. P., & Smith, D. V. (1984). Responsiveness of neurons in the hamster parabrachial nuclei to taste mixtures. *Journal of General Physiology* **84**, 221–250.

9

Neural and behavioral mechanisms of taste mixture perception in mammals

David V. Smith

Department of Otolaryngology and Maxillofacial Surgery,
University of Cincinnati College of Medicine
Cincinnati, Ohio, USA

I. Introduction

Many of the chapters in this volume which are about the neurophysiology of mixture perception have involved invertebrate organisms and olfactory receptors. The inherently complex nature of odorants and our inability to agree on what classes of stimuli might comprise primary olfactory qualities have, almost by default, guided the neurophysiological investigation of the olfactory system toward the study of complex stimuli, i.e. olfactory mixtures. On the other hand, those of us working in gustatory physiology are all too familiar with the four 'basic' qualities of sweet, salty, sour and bitter and this perspective has largely guided our approach to the physiological and behavioral analysis of taste (Frank, 1973; McBurney & Gent, 1979; Nowlis & Frank, 1977, 1981 Nowlis, Frank & Pfaffmann, 1980; Perrotto & Scott, 1976; Pfaffmann, 1974; Pfaffmann, Frank, Bartoshuk & Snell, 1976; Smith & Theodore, 1984; Smith, Travers & Van Buskirk, 1979; Smith, Van Buskirk, Travers & Bieber, 1983a, 1983b; Travers & Smith, 1979, 1984; Van Buskirk & Smith, 1981). Thus, the neurophysiological analysis of mammalian gustatory systems has been dominated by the assumption of four basic taste qualities (see Erickson, 1982, 1985), with little attention to the effects of complex stimuli.

The effects of simple taste mixtures on peripheral gustatory neural responses have been noted in cats (Kruger & Boudreau, 1972) and rats (Beidler, 1953; Miller, 1971; Sato, Ogawa & Yamashita, 1971; Wang, 1973), and investigated in some detail in hamsters (Hyman & Frank,

Copyright © 1989 by Academic Press Australia.
All rights of reproduction in any form reserved.

1980*a*, 1980*b*). The latter investigators noted that the effects of binary mixtures on both the integrated response of the whole chorda tympani nerve and the responses of individual fibers were complex and difficult to predict from the responses to the components. Generally, mixtures of two stimuli with different taste qualities (e.g. sucrose and NaCl) resulted in an integrated chorda tympani response that was smaller that the sum of the responses to the two components (Hyman & Frank, 1980*a*). This suppression was strongest with mixtures of electrolytes and non-electrolytes (e.g. NaCl and sucrose). With more homogeneous mixtures (i.e. two non-electrolytes such as sucrose and d-phenylalanine or two electrolytes such as NaCl and HCl), the response of the chorda tympani to the mixture was similar to that predicted by the concentration-response function of either stimulus. Individual fibers in the hamster chorda tympani nerve responded differentially to these mixtures, depending upon the fiber's classification (Hyman & Frank, 1980*b*). Sucrose-best fibers, for example, responded to a mixture of d-phenylalanine and sucrose as if one of the stimuli had been increased in concentration. However, these same fibers responded to a mixture of sucrose or d-phenylalanine and an electrolyte as if the concentration of the sweet stimulus had been reduced, i.e. the mixture suppressed the response to the fiber's best stimulus. The responses of HCl-best fibers were quite different, with the responses to mixtures generally approaching the sum of the responses to the individual components (Hyman & Frank, 1980*b*).

In contrast to the neurophysiological analysis of taste mixtures, the psychophysical effects of taste mixtures have been studied rather extensively (Bartoshuk, 1975; Bartoshuk & Cleveland, 1977; Beebe-Center, Rodgers, Atkinson & O'Connell, 1959; Fabian & Blum, 1943; Frank & Archambo, 1986; Gillan, 1982; Lawless, 1979, 1982; Kamen, Pilgrim, Gutman & Kroll, 1961; Kroeze, 1982; Kroeze & Bartoshuk, 1985; Kunznicki & Ashbaugh, 1982; Moskowitz, 1971, 1972; Pangborn, 1960; Rifkin & Bartoshuk, 1980). In mixtures of two stimuli with different taste qualities, human subjects have little or no difficulty recognizing the components in the mixture (Bartoshuk, 1975; Indow, 1969; Lawless, 1979; Pangborn, 1961, 1962; Pangborn & Trabue, 1967). Thus, the quality of a stimulus is not changed appreciably when it is mixed with another. On the other hand, it is common to see a change in the perceived intensity of a stimulus when it is a component of a mixture. Most frequently, when stimuli with different taste qualities are mixed, in what may be referred to as a heterogeneous mixture, the intensities of the individual qualities are suppressed below what they are when unmixed (Bartoshuk, 1975; Lawless, 1979; Moskowitz, 1972; Pangborn, 1961). With mixtures of stimuli of similar quality (i.e. homogeneous mixtures), the perceived intensity of the mixture is greater than that of either component (Bartoshuk & Cleveland, 1977; Frijters & DeGraaf, this volume; Moskowitz, 1974a, 1974b; Stone & Oliver, 1969; Yamaguchi, Yoshikawa, Ikeda & Ninomiya, 1970), its intensity being predictable from the slope of the psychophysical function relating perceived intensity to stimulus concentration (Bartoshuk, 1975). For those substances with compressed psychophysical functions (slopes < 1.0), their mixtures will show suppression, i.e. their intensities will be less than the simple additive sum of the intensities of the components. For those substances with expanded psychophysical functions (slopes > 1.0), on the other hand, their mixtures will show synergism, i.e. their intensities will be greater than the simple additive sum of the component intensities (Bartoshuk, 1975; Bartoshuk & Cleveland, 1977). Manipulating the experiment to change the psychophysical function from a compressed to an expanded one has been shown to change the mixture interaction from apparent suppression to apparent synergism (Bartoshuk & Cleveland, 1977). True synergism in taste mixtures, in which the perceived intensity of the mixture is greater than can be predicted by any additive model, is seen only rarely, and seems to involve particular kinds of stimuli (Rifkin & Bartoshuk, 1980).

The similarities and differences among gustatory stimuli, including mixtures, have been studied in several mammalian species by measuring the degree of stimulus generalization following conditioned taste aversions (Nachman, 1963; Nowlis, 1974; Nowlis & Frank, 1977, 1981; Nowlis et al., 1980; Smith & Theodore, 1984; Smith et al., 1979; Tapper & Halpern, 1968; Wiggins, Smith & Frank, 1987). The patterns of generalization to stimuli representing the four basic taste qualities following conditioned aversion to one of several taste stimuli suggest that rats and hamsters classify a variety of stimuli into four groups, corresponding to the human qualities of sweet, salty, sour, and bitter (Nowlis & Frank, 1977, 1981; Nowlis et al., 1980). Hamsters made ill after tasting a two-component mixture, one of which was familiar and one of which was novel, did not show a subsequent aversion to the familiar substance but reduced their intake of the novel component and the mixture to a similar degree (Nowlis & Frank, 1981). This finding suggests that these animals responded to the novel component in the mixture as though it had the same taste as when it was unmixed, i.e. that there was not a unique quality to the mixture itself. In addition, for stimuli that have multiple tastes to humans, like strong Na-saccharin, the pattern of generalization reflects these multiple qualities (Nowlis & Frank, 1977, 1981; Wiggins et al., 1987). Thus, the conditioned aversion procedure has proven to be an effective tool for examining the way in which experimental animals respond to and classify gustatory stimuli, including mixtures.

The remainder of this chapter will describe work that we have done on the neural and behavioral processing of taste mixtures, using mammalian preparations. In a behavioral experiment with rats, we used the conditioned aversion procedure to address the question of whether animals can respond quantitatively to the component qualities in a mixture (Smith & Theodore, 1984). The electrophysiological responses of third-order cells in the parabrachial nuclei of the hamster provided us with information about how taste mixtures are processed in the central nervous system (Travers & Smith, 1984). Although these studies, and those described in this volume by Frank, represent much of what has been done in the behavioral and neurophysiological analysis of taste mixture processing by mammals, much more work is needed in this area.

II.   Analysis of taste mixtures by the rat

Rats were trained to take their daily water intake by licking a drinking spout located just outside a testing chamber. Water was presented at 10 s intervals, separated by intertrial intervals of 20 s. After this initial training, the animal was either conditioned to avoid 0.3 M sucrose, 0.1 M NaCl, or 0.003 M HCl, or served as a control animal (with no conditioning). Conditioning was induced by intraperitoneal injection of cyclophosphamide (75 mg/kg IP) 15 minutes after the animal's first experience with the conditioned stimulus. Controls were injected with comparable volumes of isotonic saline. On the test day, each animal was randomly presented with distilled water, the conditioned stimulus (CS), and binary mixtures of the CS with the other stimuli and with 0.001 M quinine hydrochloride (QHCl). These mixtures were made by combining these four stimuli in the volumetric ratios: 100% CS to 0% other, 75% CS to 25% other, 50% CS to 50% other, 25% CS to 75% other, and distilled water (0% CS). Table 9.1 shows the concentrations of each of these four stimuli when mixed in these proportions. Responses were the average numbers of licks given by control and experimental rats to each of the stimuli over five 10 s trials of each stimulus (see Smith & Theodore, 1984, for a more detailed description of these methods).

The mean number of licks ($\pm$ SEM) to sucrose and its mixtures (with NaCl, HCl and QHCl) by control rats and by those with conditioned aversions to 0.3 M sucrose are shown in Figure 9.1. Mixtures of sucrose with NaCl or HCl at all concentrations were licked freely by control animals (>60 licks/10 s), as shown by the solid lines in Figure 9.1. However, adding QHCl to sucrose resulted in a considerable suppression

Table 9.1 Concentrations (M) of each stimulus for each proportion used in the stimulus mixtures for study of stimulus generalization

Stimulus	100%[a]	75%	50%	25%	0%
NaCl	0.1	0.075	0.05	0.025	H_2O
Sucrose	0.3	0.225	0.15	0.075	H_2O
HCl	0.003	0.00225	0.0015	0.00075	H_2O
QHCl	0.001	0.00075	0.0005	0.00025	H_2O

[a] Conditioning stimuli (except 0.001 M QHCl).

of licking (to <31 licks/10 s). The dashed lines in Figure 9.1 show the mean number of licks by animals with aversions to 0.3 M sucrose. The greater the sucrose concentration with mixtures of NaCl or HCl, the greater the decrement in licking rate below that of control animals. Mixtures of sucrose with QHCl were licked at very low rates (10 licks/ 10 s) by these conditioned animals, regardless of sucrose concentration. The CS (0.3 M sucrose) was licked by these conditioned rats at a very low rate (mean = 9.9 licks/10 s). A 2 × 4 factorial analysis of variance revealed the group, treatment and interaction effects shown in Figure 9.1 all to be significant ($p < .05$), except for the group effect for the sucrose/QHCl mixtures ($p = .054$). The interaction effects in this

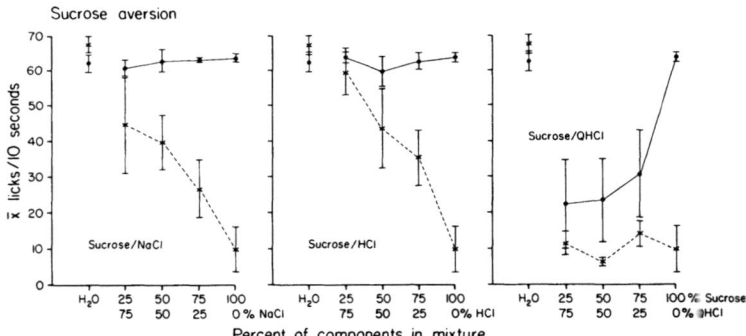

Fig. 9.1. Responses (mean licks/10 s) of control (solid lines) and sucrose-aversion (dashed lines) rats to mixtures of sucrose with the other basic taste stimuli (NaCl, HCl and QHCl). The percentages of the components in the mixtures are given along the abscissa (see Table 9.1 for concentrations in these mixtures). Error bars represent ± 1 SEM (From Smith & Theodore, 1984.)

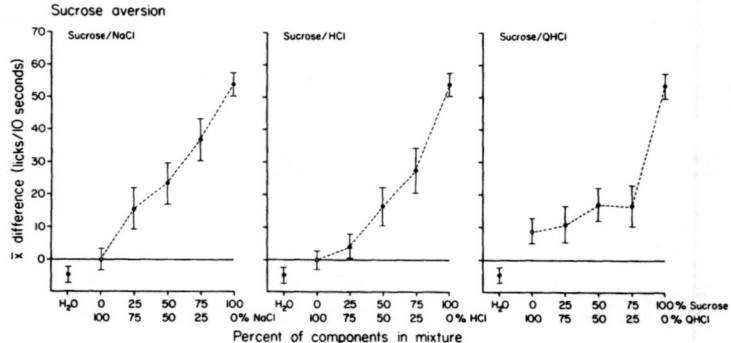

Fig. 9.2.   Mean difference in licks/10 s ($\pm$ SE$_{diff}$) between control rats and those with sucrose aversions to mixtures of sucrose with the other basic taste stimuli. Percentages of the
components in the mixtures are given along the abscissa. Horizontal lines represent zero
difference between the two groups. (From Smith & Theodore, 1984.)

analysis reflect an increasing difference between control and conditioned
animals as the sucrose concentration increases. Similar effects occur for
the mixtures tested with the NaCl- and HCl-conditioned rats.

   The results shown in Figure 9.1 are replotted in Figure 9.2 as the difference between control animals and those with a conditioned aversion
to 0.3 M sucrose. The solid circles depict the mean differences in the data
shown in Figure 9.1 ($\pm$ SE$_{diff}$). The open circles show the differences between control animals and those with sucrose aversions when tested with
pure solutions of 0.1 M NaCl, 0.003 M HCl and 0.001 M QHCl. The
latter data were obtained from other groups of rats under conditions
similar to those used in this mixture experiment and are shown here for
comparative purposes. These difference-score functions depict the
degree of generalization of animals with sucrose aversions to mixtures
containing sucrose (solid circles) and to the other basic taste compounds
(open circles). These animals did not generalize at all from a sucrose
aversion to pure NaCl (two-tailed t test, $p > .10$), although there was a
tendency for sucrose-averse animals to be slightly more hesitant about
licking pure QHCl than control rats (two-tailed t test, $p < .05$). Rats with
sucrose aversions reduced their licking rate to mixtures of sucrose with
NaCl in a monotonic fashion as the concentration of sucrose in the
mixture was increased. This reduction of licking rate is depicted as an increase in the mean difference shown in Figure 9.2. The generalization

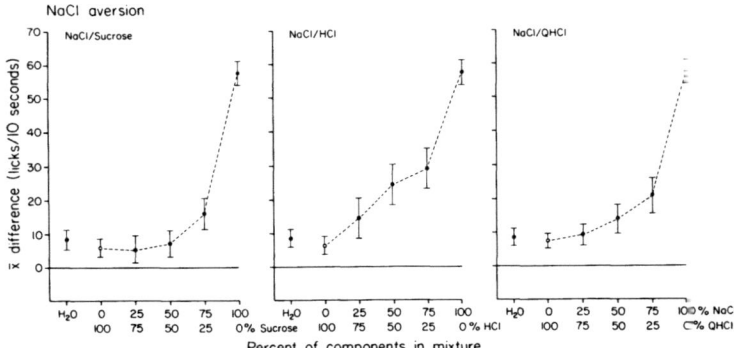

Fig. 9.3. Mean difference in licks/10 s ($\pm$ SE$_{diff}$) between control rats and those with NaCl aversions to mixtures of NaCl with the other basic taste stimuli. Percentages of the components in the mixtures are given along the abscissa. Horizontal lines represent zero difference between the two groups. (From Smith & Theodore, 1984.)

function for sucrose-HCl mixtures was also monotonic with respect to sucrose concentration, although the licking rate to a 25% sucrose – 75% HCl mixture was not significantly different between the control and aversion animals (two-tailed *t* test, $p < .05$), suggesting no generalization at all to this concentration of sucrose. Because of the strong suppression of licking by control rats of the sucrose–QHCl mixtures, there was no relation to sucrose concentration for these mixtures.

The differences in licking rates for control and NaCl-aversion rats are shown in Figure 9.3. The relationship of the difference in licking rate to NaCl concentration was monotonic, whether NaCl was mixed with sucrose, HCl or QHCl. However, both the NaCl–sucrose and NaCl–QHCl mixtures showed a weaker relationship to NaCl concentration than the NaCl-HCl mixtures. In fact, the differences between the control and NaCl-aversion rats for the 25% NaCl – 75% sucrose and the 50% NaCl – 50% sucrose mixtures were not significant (two-tailed *t* test, $p > .05$). This lack of difference in the NaCl–sucrose mixtures was due to only a slight reduction in licking rate by the NaCl-aversion animals, whereas the very small difference in the NaCl–QHCl mixtures was due to depressed licking of these QHCl-containing stimuli by the control rats. Thus, the natural preference for sucrose and aversion to QHCl appear to interact with the differences shown by control and conditioned animals (see Smith & Theodore, 1984).

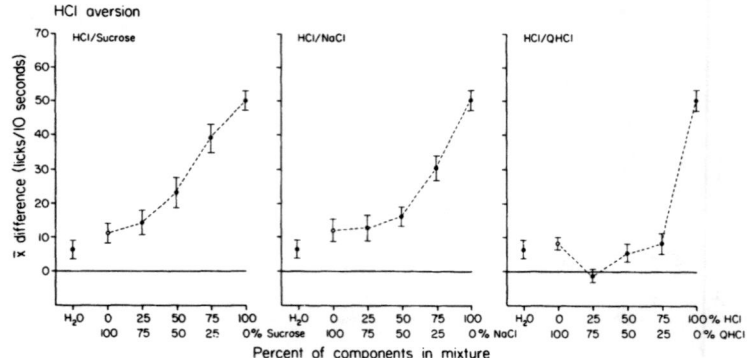

Fig. 9.4. Mean difference in licks/10 s ($\pm$ SE$_{\text{diff}}$) between control rats and those with HCl aversions to mixtures of HCl with the other basic taste stimuli. Percentages of the components in the mixtures are given along the abscissa. Horizontal lines represent zero difference between the two groups. (From Smith & Theodore, 1984.)

The differences between control animals and those with HCl aversions are shown in Figure 9.4. For the HCl–sucrose and HCl–NaCl mixtures, the generalization functions are clearly monotonic with respect to HCl concentration. However, for the HCl–QHCl mixtures, the licking rate for the 25% HCl – 75% QHCl mixture was not significantly different following HCl aversion (two-tailed t test, $p > .05$) and those for the remaining mixtures were unrelated to HCl concentration. As with the other aversions (Fig. 9.2 and Fig. 9.3), the natural aversion to QHCl by control animals serves to preclude a measurable relationship of these mixtures to the concentration of the CS.

The results of this experiment demonstrate that rats are capable of responding to the similarity between a taste mixture and its components. Further, within the limits imposed by this behavioral task, their responses reflect the strength of the conditioned aversive stimulus in these binary mixtures. Stimulus generalization along a concentration dimension has been shown with unmixed concentrations of NaCl (Nowlis, 1974). In those studies, the stronger the NaCl concentration used in training, the stronger the aversion, and the stronger the concentration used in testing, the stronger the aversion manifested during the generalization test. In the present experiments, only a single, relatively strong concentration of each CS was used in training, with several weaker concentrations used as mixture components during testing. In most instances, except for the limits imposed by natural gustatory preferences and aversions, the rats in the present experiments generalized to these weaker stimuli along a concentration gradient, showing less aversion as the strength of the CS was reduced in the test mixtures. There were limits on this generalization related to the natural preferences and aversions of these animals to gustatory stimuli. Most obvious in this regard was the suppression of the licking rate of control animals by the addition of QHCl to any of the other three stimuli (see Fig. 9.1; also evident in difference scores in Fig. 9.2 – Fig. 9.4). Of these four stimuli, only QHCl elicits aversive reflexes in the rat (Grill & Norgren, 1978) and only QHCl suppressed the licking rate of control animals in the present experiment. The other three stimuli (sucrose, NaCl and HCl) were all licked avidly by these animals under these experimental conditions. Because of the suppression of licking rate by QHCl, none of the generalization functions for mixtures of these other compounds with QHCl were very clearly related to CS concentration (see Fig. 9.2 – Fig. 9.4). In these instances, the potential differences between control and aversion animals were severely limited by the already low licking rates of the control rats. Thus, these differences could not reflect the conditioned animals' ability to respond to the quality of the CS in mixtures with QHCl.

Just as natural aversions may limit this technique as a means of assessing taste similarity, natural preferences may also interact with the generalization of conditioned aversions. For example, animals with conditioned aversions to sucrose generalized to mixtures of sucrose with NaCl in an almost linear fashion with respect to the sucrose concentration in the mixture (Fig. 9.2). However, animals with NaCl aversion did not respond to the two weaker concentrations of NaCl when they were mixed with sucrose (Fig. 9.3), although they responded readily to NaCl if it was mixed with HCl (Fig. 9.3). One explanation for this lack of reciprocity may lie in the naturally strong preference of rats for sucrose (Pfaffmann, 1957), which may have overridden the conditioned aversion to NaCl. Because rats can lick no faster than 6 to 7 licks/s (Halpern, 1975), the control animals could not reflect any stronger preference for NaCl–sucrose mixtures, since they were already licking at their maximum rate. If they could, then the differences between control and NaCl-aversion animals might have reflected the rat's ability to respond to the concentration of NaCl in mixtures with sucrose. Note that sucrose preference did not override an HCl aversion (Fig. 9.4).

The compression of these generalization functions by unlearned preferences and aversions is, in a way, analogous to the effects of a category scale on human judgments of sensory magnitude, which tends to produce skewed response distributions at its extremes (Engen, 1971). Whereas human observers cannot assign numbers higher or lower than the end points of a category scale, these rats cannot lick faster than their maximum rate nor slower than that necessary to obtain sufficient gustatory input upon which to base a decision about acceptance or rejection. Thus, these factors result in some non-linearities in the generalization functions obtained using this conditioned aversion procedure. However, except for these limitations, rats are able to respond to a conditioned aversive stimulus when it is mixed with other compounds, even at relatively weak concentrations (see Fig. 9.2 – Fig. 9.4). Investigators using the conditioned aversion procedure as a means of assessing taste similarity should, however, be cautious of the potential influence of natural preferences and aversions on this measure.

Another explanation for the non-reciprocity in sucrose and NaCl aversions could lie in the known effects of mixture interactions on perceived taste quality. It has been shown in human psychophysical studies that perceived saltiness is suppressed in mixtures of NaCl and sucrose (Pangborn, 1962). Further, recent work with adrenalectomized rats has shown that, although these animals are capable of recognizing NaCl when it is mixed with sucrose, they drink considerably less NaCl in this mixture than when the same concentration is presented alone (McCutcheon & Brown, 1983), suggesting that sucrose might be suppressing the taste of NaCl, as it is known to do in human studies. In the present experiment, the apparent inability of animals with NaCl aversions to respond to NaCl in mixtures with sucrose could be a reflection of a similar mixture suppression in rats.

Previous behavioral work with hamsters has shown that if a taste aversion is conditioned to a two-component mixture, this aversion generalizes to the two components presented unmixed (Nowlis & Frank, 1977, 1981). Further, if one component is rendered ineffective for taste-aversion learning by making it thoroughly familiar to the animal prior to conditioning, the aversion generalizes to the novel component and to the mixture, but not to the familiar component. This suggests that the animals are responsive to the components of the mixed stimulus. In addition, the fact that adrenalectomized rats, which show an increased preference for NaCl, will consume NaCl when it is mixed with sucrose in preference to a pure sucrose solution (McCutcheon & Brown, 1983) suggests that rodents are able to respond appropriately to the separate components of a taste mixture. The results presented in Figures 9.1–9.4 further demonstrate that animals respond to the similarity between the components in taste mixtures and the same components presented unmixed. In this case, the conditioning was to a pure taste stimulus and the animals were required to respond to the presence of this taste quality in the mixed test stimuli. Within the limits of the techniques employed, which were discussed above, rats appear to be quite capable of responding to the components in a taste mixture in a manner that is related to their concentration.

III. Electrophysiological responses of central gustatory neurons

Cells in the brain stem of the hamster are slightly more broadly responsive to the four basic taste stimuli than fibers in the chorda tympani nerve (Smith & Travers, 1979; Travers & Smith, 1979; Van Buskirk & Smith, 1981). This breadth of tuning is even greater in pontine cells of the parabrachial nuclei (PbN) than in those of the solitary nucleus (NTS) in the medulla, particularly for cells responding best to sucrose (Van Buskirk & Smith, 1981). Thus, the responses of these broadly tuned PbN cells were examined following stimulation of the anterior portion of the tongue with the four basic stimuli and their binary mixtures. The stimuli used were 0.03 M NaCl, 0.1 M sucrose, 0.003 M HCl, and 0.001 M QHCl, and the six possible two-component, undiluted mixtures of these compounds. These mixtures were prepared so that the concentrations of each of the components in the mixtures were the same as their concentrations in the single-component stimuli. The responses of each of 34 PbN neurons were obtained to these 10 stimuli. The four basic stimuli were presented in random order for each cell, followed by a random presentation of the six mixtures. When possible, the entire series was presented three times. Twenty-three cells were successfully stimulated at least twice with all 10 stimuli. Responses were recorded from the PbN of anesthetized hamsters using glass-insulated tungsten microelectrodes. Activity of the cells was amplified by conventional means and stored on magnetic tape for off-line analysis. Action potentials arising from single cells were identified by consistency of spike amplitude, duration and waveform. The numbers of action potentials occurring in successive 500 ms bins were accumulated for a 30 s period for each stimulus, comprised of 5 s of the pre-stimulus period without a distilled water rinse flowing, 5 s with the rinse flowing, 10 s with the stimulus flowing, and 10 s of post-stimulus rinse. The mean spontaneous discharge rate (during the pre-stimulus period) was routinely subtracted from the rate of impulse discharge during the initial 5 s of the stimulus period to provide a measure of the change in impulse frequency due to the stimulus presentation. The criterion for a response to a particular stimulus was an increase in firing rate by more than one standard deviation above the

Table 9.2   Frequency of mixture responses greater than, less than, or equal to the response to the more effective mixture component

Type of component		Frequency (%) of mixture responses		
1st component	2nd component	Greater	Less	Equal
+	+	14/62 (22.6)	7/62 (11.3)	41/62 (66.1)
+	0	3/48 (6.3)	12/48 (25.0)	33/48 (68.8)
0	0	3/15 (20.0)		12/15 (75.0)
+	–	1/11 (9.1)	2/11 (18.2)	8/11 (72.7)
–	–			1/1
All component pairs		21/137 (15.3)	21/137 (15.3)	95/137 (69.3)

+ = excitatory response, – = inhibitory response, 0 = no response.

mean spontaneous rate (impulses/s) for that cell on at least two trials. Two responses were considered to be reliably different from one another if the differences between the two mean responses exceeded the sum of their standard deviations. For a more detailed description of these methods, see Travers and Smith (1984).

For most PbN cells, the response to a mixture was the same as the response to the more effective component of the mixture presented alone. Across the 23 cells that were stimulated more than once with the 10 mixtures, the distribution of the types of responses to the components and the mixtures is shown in Table 9.2. The types of responses evoked by the individual components (excitatory, inhibitory, or no response) are shown on the left of the table as +, –, or 0 respectively. The frequencies of the responses to the mixtures of these components that were greater than, less than or equal to the response to the more effective component are shown to the right. In 95 out of 137 cases (69%), there was no reliable difference between a mixture response and the response of a cell to the more effective component of that mixture. However, the remaining 31% of the mixture responses (42/137) were reliably different from the responses evoked by their more effective component. Twenty-one mixture responses (15%) were greater and 21 (15%) were smaller than the response to their more effective component. There was a significant relationship between the type of mixture response (i.e. greater than, less than or equal to its more effective component) and the types of responses (+, –, or 0) evoked by the mixture's components ($\chi^2 = 9.73$, $p < .05$, $df = 4$). Mixture responses greater than the response to the more effective component occurred more frequently (14/62 greater versus 7/62 less) when both components were excitatory. However, when one component was excitatory and the other elicited no response, mixture responses smaller than the response to the more effective component occurred more frequently (12/48 less versus 3/48 greater). However, the category of a mixture response was not tightly coupled to the type of responses evoked by the mixture's components. A neuron that increased its firing rate to two different stimuli did not necessarily respond better to the mixture of those stimuli than it did to the more effective component of that mixture presented alone (i.e. the response was greater only 22.6% of the time). Further, a neuron that gave an excitatory response to one stimulus and did not respond to the other did not always respond to the mixture of these stimuli as if only the excitatory stimulus was present. In fact, the response was often reduced (25% of the time). Overall however, the most consistent result (69.3% of the responses) was a response to the mixture that was no different from the response to its more effective component.

The 34 cells recorded from the PbN were classified into best-stimulus classes on the basis of their responses to the four basic stimuli (Frank, 1973). Among these cells, 13 were sucrose-best, 13 were NaCl-best, 4 were HCl-best, and 4 were QHCl-best. Mean responses to the 10 stimuli were calculated for each of these best-stimulus classes and are shown in Figure 9.5. The responses to the best stimulus and to mixtures containing the best stimulus for each class of neurons are shaded in Figure 9.5. Even though some of the individual neurons responded better to mixtures than to single-component stimuli (see Table 9.2), their effects were not apparent in the mean responses for a group of neurons. Mixtures of the four basic stimuli did not elicit mean responses that were appreciably greater than the response evoked by the best stimulus for any given class of cells. In HCl- and QHCl-best cells, none of the mean responses to the mixtures were significantly greater than the response to the best stimulus (i.e. to HCl or QHCl, respectively). However, there was a very slight but significant increase in the response to the sucrose/NaCl mixture in both sucrose-best (Sandler's A test, $A = 0.243$, $p < .05$) and NaCl-best ($A = 0.273$, $p < .05$) cells. Sucrose plus NaCl, which was more effective for sucrose- and NaCl-best neurons than the best stimulus alone, contained the best stimulus as a component of the mixture (shaded bars, Fig. 9.5). However, some mixtures that contained the best stimulus for a class of cells actually evoked less vigorous responses than the best stimulus alone. In NaCl-best neurons, for example, a mixture of NaCl plus HCl elicited a response that was 0.79 times as effective as NaCl alone ($A = 0.261$, $p < .05$). For three of the four QHCl-best cells, a mixture of NaCl and QHCl elicited a smaller response than that evoked by QHCl alone. This mixture was only 0.75 times as effective as QHCl in QHCl-best neurons. Mixtures not containing the best stimulus for a class of cells (unshaded bars, Fig. 9.5) always evoked mean responses that were smaller than the response to the best stimulus. For both sucrose- and NaCl-best neurons, the mean responses evoked by these sideband mixtures were significantly smaller

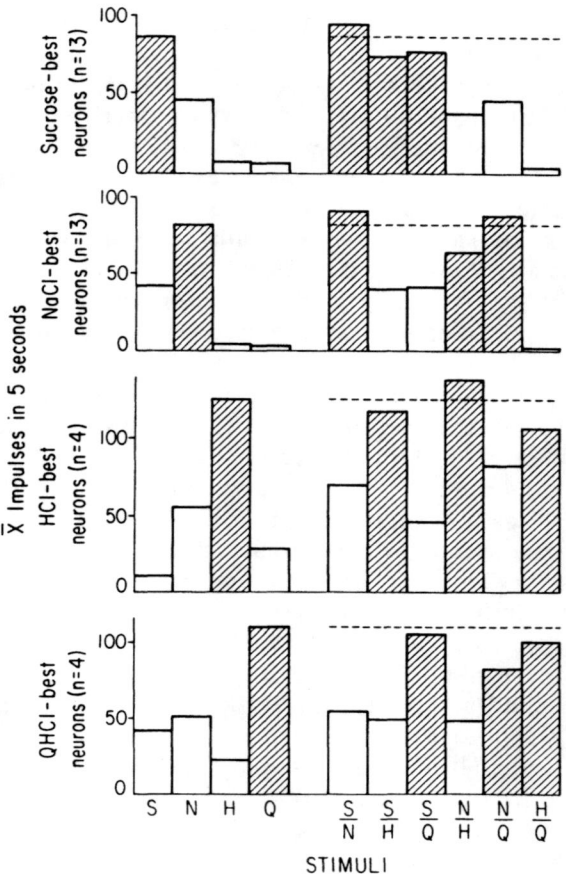

Fig. 9.5. Mean firing rates to all 10 stimuli for each of the four best-stimulus classes of cells in the hamster PbN. The shaded bars indicate the best stimulus and the mixtures containing the best stimulus for each class of cells. The dashed lines denote the mean firing rate to the best stimulus. Abbreviations: S, sucrose; N, NaCl; H, HCl; Q, QHCl; S/N, sucrose/NaCl mixture, etc. (From Travers & Smith, 1984.)

than the mean responses evoked by the best stimuli ($p < .05$, all comparisons). In general, the response to mixtures containing the best stimulus were very nearly the same as the response to the best stimulus (shaded bars, Fig. 9.5) and those not containing the best stimulus were smaller (unshaded bars).

Table 9.3. Across-neuron correlations (Pearson r^a) among the four basic stimuli and their mixtures

	S	N	H	Q	S/N	S/H	S/Q	N/H	N/Q
H/Q	−.05	+.07	+.77	+.79	+.03	+.49	+.39	+.55	−.34
N/Q	+.48	+.91	+.23	+.30	+.75	+.58	+.59	+.70	
N/H	+.16	+.59	+.73	+.11	+.50	+.72	+.30		
S/Q	+.85	+.56	+.11	+.52	+.77	+.75			
S/H	+.63	+.54	+.59	+.21	+.75				
S/H	+.87	+.85	+.02	−.02					
Q	+.10	+.06	+.30						
H	−.17	+.04							
N	+.61								

[a] For $n = 34$, $r = +0.44$, $p < .01$, and $r = +0.34$, $p < .05$.

Across-neuron correlations have been used to quantify the similarity between the patterns of activity evoked by two stimuli across a population of cells (Erickson, 1963, 1968). If two stimuli evoke similar patterns, the across-neuron correlation between them will be high, but the across-neuron correlation will be low if the stimuli generate dissimilar patterns. The across-neuron correlations among the four stimuli and their six mixtures are shown in Table 9.3. Stimulus pairs having a common component (e.g. sucrose–NaCl and sucrose, $r = +0.87$, or sucrose–NaCl and NaCl–HCl, $r = +0.50$) tended to correlate significantly and more highly than those pairs without a common component (e.g. sucrose–NaCl and HCl–QHCl, $r = +0.03$). The mean correlation for all the stimulus pairs having a common component ($+0.63$) was significantly greater ($t = 6.716$, $p < .001$) than the mean for the pairs without a common component ($+0.21$). In general, the across-neuron pattern for a mixture correlated significantly with each of the across-neuron patterns generated by the individual components, but not with those evoked by stimuli not in the mixture. For example, the across-neuron correlation between sucrose plus NaCl and sucrose alone was $+0.87$ and between this mixture and NaCl alone was $+0.85$, whereas this mixture correlated only $+0.02$ and -0.02 with HCl and QHCl, respectively.

The across-neuron patterns of activity for sucrose, NaCl, and the mixture of these two stimuli, are shown in Figure 9.6. In this figure, the cells are ordered along the abscissa according to their response to sucrose, which is shown by the solid circles. The responses of these cells to NaCl are depicted by the triangles and dashed lines and to the sucrose plus

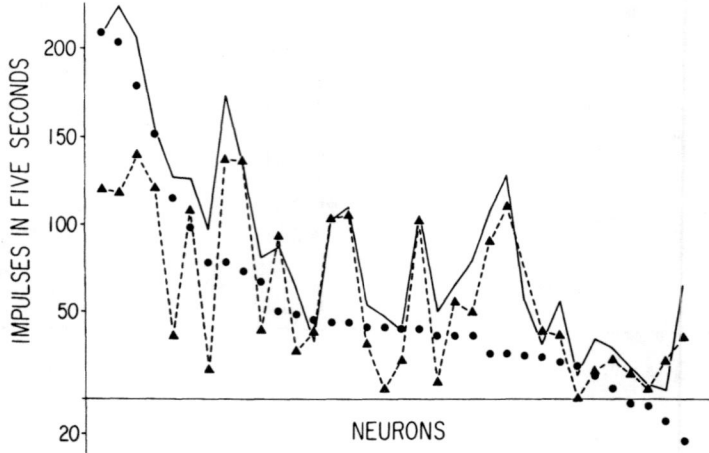

Fig. 9.6. Across-neuron patterns for sucrose, NaCl, and for the mixture of these stimuli. The neurons in each pattern are arranged along the abscissa according to their responsiveness to sucrose. The filled circles with no connecting line indicate the sucrose pattern, the filled triangles and dashed line designate the pattern for NaCl, and the solid line shows the pattern produced by the sucrose plus NaCl mixture. (From Travers & Smith, 1984.)

NaCl mixture by the solid line. The pattern to the mixture bears some similarity to the patterns to both components. More importantly, this pattern follows the response to the more effective component in each cell. That is, when the largest response is to sucrose (left side of the figure), the mixture pattern is similar to sucrose, and when the largest response is to NaCl (several peaks in the middle of the figure), the mixture pattern is similar to NaCl. The across-neuron patterns to all the mixtures were well predicted from responses to the more effective components of the mixtures. For example, the pattern for the sucrose plus NaCl mixture (Fig. 9.6) correlated very highly (+0.97) with a hypothetical pattern composed of each cell's response to this mixture's more effective component, either sucrose or NaCl. The mixture patterns for the other five mixtures also correlated highly with hypothetical patterns derived in this fashion, with these correlations for all six mixtures being: S + N, +0.97; S + H, +0.93; S + Q, +0.94; N + H, +0.87, N + Q, +0.91; and H +Q, +0.98. The mean correlation between the observed mixture patterns and these hypothetical mixture patterns was +0.93. This suggests that the response to a mixture is best predicted by the response to the mixture's more effective component (see also the best-stimulus analysis in Fig. 9.5)

The relationships among all of the across-neuron patterns to the four basic stimuli and their mixtures is best appreciated through multi-dimensional scaling (Bieber & Smith, 1986). The results of such an analysis (KYST) of the across-neuron correlations among these stimuli is depicted in Figure 9.7, where the similarities and differences among the stimuli are represented in three-dimensional space. In this figure, closer proximity in the space represents greater similarity in the across-neuron patterns. The four basic stimuli (S, N, H and Q) are dispersed to the corners of the space and five of the six mixtures are in positions

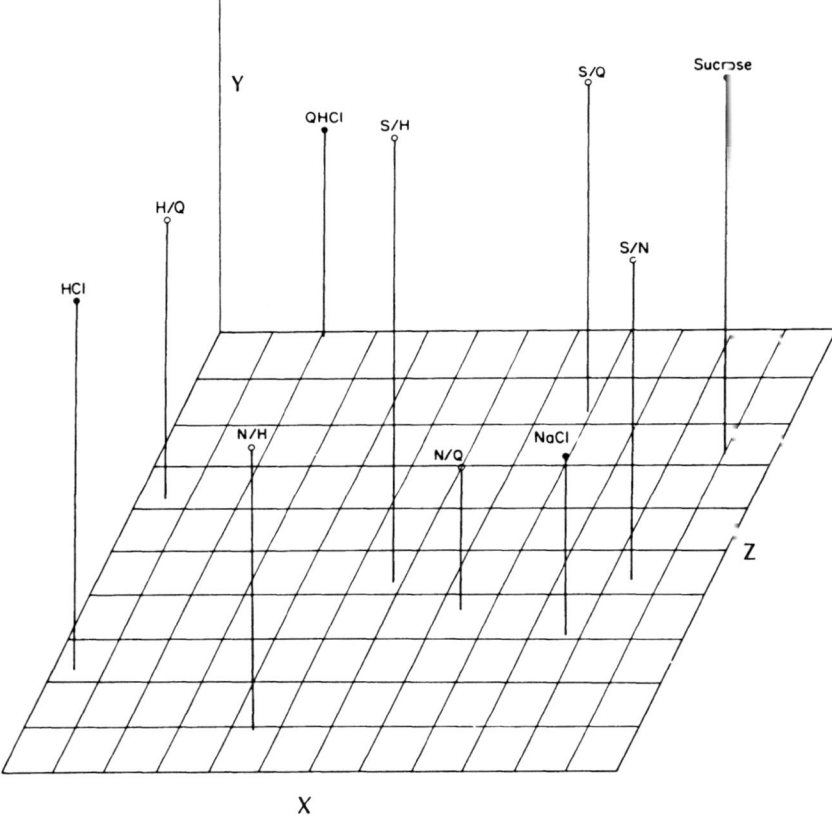

Fig. 9.7. Three-dimensional representation of the relationships among the 10 stimuli generated by a multidimensional scaling analysis (KYST) of the across-neuron correlations shown in Table 9.3. Filled circles designate the positions of the single component stimuli and the open circles indicate the positions of the mixtures. Abbreviations are the same as in Figure 9.5. (From Travers & Smith, 1984.)

intermediate between their components. These five mixtures all correlated significantly with both of their single components. A single mixture, NaCl plus QHCl, correlated significantly with only one of its components (NaCl). This mixture lies much closer to NaCl than to QHCl in Figure 9.7, a placement that is consistent with the across-neuron correlations among these stimuli ($+0.91$ between NaCl and the mixture, and $+0.30$ between QHCl and the mixture).

A major finding of this study was that usually the response of a cell in the hamster PbN to a binary mixture was not different from the response to the more effective component of the mixture. The best predictor of the magnitude of a cell's response to a two-component mixture is its response to the mixture's more effective component. Thus, it is not surprising that the across-neuron patterns evoked by the mixtures are well predicted by hypothetical patterns derived from the responses to the more effective components of the mixtures. This finding is also evident when examining the mean responses to mixtures and their single components in the four best-stimulus classes of cells (Fig. 9.5). Thus, the neural response to the dominant component of a mixture tends to predict the response to the mixture. Although not based on the kind of careful factorial experiment necessary to test a mixture model of perceived intensity, this result is compatible with the 'dominant component' model discussed (McBride, this volume) in the context of information integration theory (Anderson, 1981; see also Frank & Archambo, 1986).

Fifteen % of the responses to mixtures in these PbN cells were greater and 15% were smaller than the response to the more effective component (Table 9.2). Enhancement occurred most often when both component stimuli evoked excitatory responses, although the magnitude of the response increase was relatively small. Suppression of the response to the more effective component of the mixtures occurred most often when one component elicited an excitatory response and the other no response. Suppression was evident not only in the responses of certain individual neurons, but also in the mean responses of NaCl- and QHCl-best neurons. This decrease in neural response could be an analogue of psychophysical mixture suppression (Bartoshuk, 1975; Lawless, 1979; Moskowitz, 1972; Pangborn, 1961), although it occurs in only a small percentage of cases. In general, the amount of mixture suppression or enhancement seen in these PbN cells was less than has been reported in hamster chorda tympani fibers (Frank, this volume; Hyman & Frank, 1980b).

A clear case of mixture suppression in these neural data occurred for NaCl–QHCl mixtures, in which the response to the mixture was smaller than that to QHCl in QHCl-best cells (Fig. 9.5) and in which the response pattern to the mixture was clearly dominated by the response to NaCl (Table 9.3, Fig. 9.7). In human psychophysical studies, the bitterness of QHCl is suppressed more than the saltiness of NaCl in a QHCl–NaCl mixture (Bartoshuk, 1979). Further, although hamsters usually generalize an aversion conditioned to a two-component mixture to both stimuli tested individually, an aversion to a mixture of 0.1 M NaCl and 0.001 M QHCl generalizes strongly to NaCl but not to QHCl (Nowlis & Frank, 1977). Thus, for this NaCl–QHCl mixture, we see a suppression of the response to QHCl in hamster brainstem neurons, in hamster behavioral responses, and in human psychophysical judgments.

IV. Conclusions

There has been relatively little work on the behavioral or neural responses to taste mixtures by mammals, and much of what has been done is reported in this volume. It is quite clear from behavioral studies that mammals are quite capable of recognizing and responding appropriately to the individual components of binary mixtures (McCutcheon & Brown, 1983; Nowlis & Frank, 1981; Nowlis *et al.*, 1980; Smith & Theodore, 1984). The work reported here demonstrates that rats can recognize and respond quantitatively to the strength of a mixture component. The sum of this kind of behavioral work (McCutcheon & Brown, 1983; Nowlis & Frank, 1981; Nowlis *et al.*, 1980; Smith & Theodore, 1984) suggests strongly that the components of binary mixtures do not take on a new quality, but remain perceptible in the mixture, just as they do to humans (Bartoshuk, 1975; Indow, 1969; Lawless, 1979; Pangborn, 1961, 1962; Pangborn & Trabue, 1967). The responses to binary mixtures of individual gustatory nerve fibers (Hyman & Frank, 1980b) or single cells in the parabrachial nuclei (Travers & Smith, 1984) are complex. The overwhelming tendency of PbN cells is a response to a binary mixture similar to the response to the mixture's more effective component. A 'dominant component' model has been proposed for the psychophysical effects of heterogeneous binary taste mixtures (McBride, this volume). Further work on the neurophysiology of mixtures should address the quantitative interaction between both homogeneous and heterogeneous mixture components in an attempt to demonstrate a neural substrate for some of the psychophysical models of taste mixture interaction, such as the equiratio model (Frijters & De Graaf, this volume) and the information integration approach (Anderson, 1981; Frank & Archambo, 1986; McBride, this volume).

Acknowledgments

This work was supported by the US National Institute of Neurological and Communicative Disorders and Stroke Grant NS-10211, Research Career Development Award NS-00168, and Jacob K. Javits Neuroscience Investigator Award NS-23524. Some of the results reported here have been published previously (Smith & Theodore, 1984; Travers & Smith, 1984).

References

Anderson, N. H. (1981). *Foundations of information integration theory.* Academic Press, New York.

Bartoshuk, L. M. (1975). Taste mixtures: is mixture suppression related to compression? *Physiology & Behavior* **14**, 643–649.

Bartoshuk, L. M. (1979). Taste interactions in mixtures of sucrose with NaCl and sucrose with QHCl (abstract). *Society for Neuroscience* **5**, 125.

Bartoshuk, L. M., & Cleveland, C. T. (1977). Mixtures of substances with similar tastes: a test of a psychophysical model of taste mixture interactions. *Sensory Processes* **1**, 177–186.

Beebe-Center, J. G., Rodgers, M. S., Atkinson, W. H., & O'Connell, D. N. (1959). Sweetness and saltiness of component solutions of sucrose and NaCl as a function of concentration of solutes. *Journal of Experimental Psychology* **4**, 231–234.

Beidler, L. M. (1953). Properties of chemoreceptors of tongue of rat. *Journal of Neurophysiology* **16**, 595–607.

Bieber, S. L., & Smith, D. V. (1986). Multivariate analysis of sensory data: a comparison of methods. *Chemical Senses* **11**, 19–47.

Engen, T. (1971). Psychophysics: II. Scaling methods. In J. W. Kling & L. A. Riggs, eds, *Woodworth and Schlosberg's experimental psychology,* pp. 47–86. Holt, Rinehart and Winston, New York.

Erickson, R. P. (1963). Sensory neural patterns and gustation. In Y. Zotterman ed., *Olfaction and taste I,* pp. 205–213. Pergamon Press, New York.

Erickson, R. P. (1968). Stimulus coding in the topographic and non-topographic afferent modalities: on the significance of activity in individual sensory neurons. *Psychological Review* **75**, 447–465.

Erickson, R. P. (1982). Studies on the perception of taste: do primaries exist? *Physiology & Behavior* **28**, 57–62.

Erickson, R. P. (1985). Definitions: a matter of taste. In D. W. Pfaff, ed., *Taste, olfaction, and the central nervous system,* pp. 129–150. Rockefeller University Press, New York.

Fabian, F. W., & Blum, H. B. (1943). Relative taste potency of some basic food constituents and their competitive and compensatory action. *Food Research* **8**, 179–193.

Frank, M. (1973). An analysis of hamster afferent taste nerve response functions. *Journal of General Physiology* **61**, 588–618.

Frank, R. A., & Archambo, G. (1986). Intensity and hedonic judgments of taste mixtures: an information integration analysis. *Chemical Senses* **11**, 427–438.

Gillan, D. (1982). Mixture suppression: the effect of spatial separation between sucrose and NaCl. *Perception & Psychophysics* **32**, 504–510.

Grill, H. J., & Norgren, R. (1978). The taste reactivity test. I. Mimetic responses to gustatory stimuli in neurologically normal rats. *Brain Research* **143**, 263–279.

Halpern, B. P. (1975). Temporal patterns of liquid intake and gustatory neural response. In D. Denton & J. Coghlan, eds, *Olfaction and taste V*, pp. 47–52. Academc Press, New York.

Hyman, A. M., & Frank, M. E. (1980a). Effects of binary taste stimuli on the neural activity of the hamster chorda tympani. *Journal of General Physiology* 76, 125–142.

Hyman, A. M., & Frank, M. E. (1980b). Sensitivities of single nerve fibers in the hamster chorda tympani nerve to mixtures of taste stimuli. *Journal of General Physiology* 76, 143–173.

Indow, T. (1969). An application of the τ scale of taste: interaction among the four qualities of taste. *Perception & Psychophysics* 5, 347–351.

Kamen, J. M., Pilgrim, F. L., Gutman, N. J., & Kroll, B. J. (1961). Interaction of suprathreshold taste stimuli. *Journal of Experimental Psychology* 62, 343–356.

Kroeze, J. H. A. (1982). After repetitive sucrose stimulation saltiness suppression in NaCl-sucrose mixtures is diminished: implications for a central mixture suppression mechanism. *Chemical Senses* 7, 81–92.

Kroeze, J. H. A., & Bartoshuk, L. M. (1985). Bitterness suppression as revealed by split-tongue taste stimulation in humans. *Physiology & Behavior* 35, 779–783.

Kruger, S., & Boudreau, J. C. (1972). Responses of cat geniculate ganglion tongue units to some salts and physiological buffer solutions. *Brain Research* 47, 127–145.

Kuznicki, J. T., & Ashbaugh, N. (1982). Space and time separation of taste mixture components. *Chemical Senses* 7, 39–62.

Lawless, H. T. (1979). Evidence for neural inhibition in bittersweet taste mixtures. *Journal of Comparative & Physiological Psychology* 93, 538–547.

Lawless, H. T. (1982). Paradoxical adaptation to taste mixtures. *Physiology & Behavior* 25, 149–152.

McBurney, D. H., & Gent, J. F. (1979). On the nature of taste qualities. *Psychological Bulletin* 86, 151–167.

McCutcheon, B., & Brown, L. (1983). Response to NaCl taste in mixtures with sucrose by sodium deficient rats. *Physiology & Behavior* 30, 405–408.

Miller, Jr, I. J. (1971). Peripheral interactions among single papilla inputs to gustatory nerve fibers. *Journal of General Physiology* 57, 1–25.

Moskowitz, H. R. (1971). Intensity scales for pure tastes and for taste mixtures. *Perception & Psychophysics* 9, 51–56.

Moskowitz, H. R. (1972). Perceptual changes in taste mixtures. *Perception & Psychophysics* 11, 257–262.

Moskowitz, H. R. (1974a). Models of additivity for sugar sweetness. In H. R. Moskowitz, B. Scharf & J. C. Stevens, eds, *Sensation and measurement: papers in honor of S. S. Stevens*, pp. 379–388. Reidel, Dordrecht, The Netherlands.

Moskowitz, H. R. (1974b) Sourness of acid mixtures. *Journal of Experimental Psychology* 4, 640–647.

Nachman, M. (1963). Learned aversion to the taste of lithium chloride and generalization to other salts. *Journal of Comparative & Physiological Psychology* 56, 343–349.

Nowlis, G. H. (1974). Conditioned stimulus intensity and acquired alimentary aversions in the rat. *Journal of Comparative & Physiological Psychology* 86, 1173–1184.

Nowlis, G. H., & Frank, M. (1977). Qualities in hamster taste: Behavioral and neural evidence. In J. LeMagnen & P. MacLeod, eds, *Olfaction and taste VI*, pp. 241–248. Information Retrieval, London.

Nowlis, G. H., & Frank, M. E. (1981). Quality coding in gustatory systems of rats and hamsters. In D. M. Norris, ed., *Perception of behavioral chemicals*, pp. 59–30. Elsevier North Holland, Amsterdam.

Nowlis, G. H., Frank, M. E., & Pfaffmann, C. (1980). Specificity of acquired aversions to taste qualities in hamsters and rats. *Journal of Comparative & Physiological Psychology* **94**, 932–942.

Pangborn, R. M. (1960). Taste interrelationships. *Food Research* **5**, 245–256.

Pangborn, R. M. (1961). Taste interrelationships. II. Suprathreshold solutions of sucrose and citric acid. *Journal of Food Science* **26**, 648–655.

Pangborn, R. M. (1962). Taste interrelationships. III. Suprathreshold solutions of sucrose and NaCl. *Journal of Food Science* **27**, 495–500.

Pangborn, R. M., & Trabue, I. M. (1967). Detection and apparent taste intensity of salt-acid mixtures in two media. *Perception & Psychophysics* **2**, 503–509.

Perrotto, R. S., & Scott, T. R. (1976). Gustatory neural coding in the pons. *Brain Research* **110**, 283–300.

Pfaffmann, C. (1957). Taste, mechanisms in preference behavior. *American Journal of Clinical Nutrition* **5**, 142–147.

Pfaffmann, C. (1974). Specificity of the sweet receptors of the squirrel monkey. *Chemical Senses & Flavour* **1**, 61–67.

Pfaffmann, C., Frank, M., Bartoshuk, L. M., & Snell, T. C. (1976). Coding gustatory information in the squirrel monkey chorda tympani. In J. M. Sprague & A. N. Epstein, eds, *Progress in psychobiology and physiological psychology*, vol. 6, pp. 1–27. Academic Press, New York.

Rifkin, B., & Bartoshuk, L. M. (1980). Taste synergism between monosodium glutamate and disodium 5'-guanylate. *Physiology & Behavior* **24**, 1169–1172.

Sato, M., Ogawa, H., & Yamashita, S. (1971). Comparison of potentiating effect on gustatory response by disodium 2-methyl mercapto-5'-inosinate with that by 5'-IMP. *Japanese Journal of Physiology* **21**, 669–679.

Smith, D. V., & Theodore, R. M. (1984). Conditioned taste aversions: generalization to taste mixtures. *Physiology & Behavior* **32**, 983–989.

Smith, D. V., & Travers, J. B. (1979). A metric for the breadth of tuning of gustatory neurons. *Chemical Senses* **4**, 215–229.

Smith, D. V., Travers, J. B., & Van Buskirk, R. L. (1979). Brainstem correlates of gustatory similarity in the hamster. *Brain Research Bulletin* **4**, 359–372.

Smith, D. V., Van Buskirk, R. L., Travers, J. B., & Bieber, S. L. (1983a). Gustatory neuron types in hamster brain stem. *Journal of Neurophysiology* **50**, 522–540.

Smith, D. V., Van Buskirk, R. L., Travers, J. B., & Bieber, S. L. (1983b). Coding of taste stimuli by hamster brainstem neurons. *Journal of Neurophysiology* **50**, 541–558.

Stone, H., & Oliver, S. M. (1969). Measurement of the relative sweetness of selected sweeteners and sweetener mixtures. *Journal of Food Science* **34**, 215–222.

Tapper, D. N., & Halpern, B. P. (1968). Taste stimuli: a behavioral categorization. *Science* **161**, 708–710.

Travers, J. B., & Smith, D. V. (1979). Gustatory sensitivities in neurons of the hamster nucleus tractus solitarius. *Sensory Processes* **3**, 1–26.

Travers, S. P., & Smith, D. V. (1984). Responsiveness of neurons in the hamster parabrachial nuclei to taste mixtures. *Journal of General Physiology* **84**, 221–250.

Van Buskirk, R. L., & Smith, D. V. (1981). Taste sensitivity of hamster parabrachial pontine neurons. *Journal of Neurophysiology* **45**, 144–171.

Wang, M. B. (1973). Analysis of taste receptor properties derived from chorda tympani nerve firing patterns. *Brain Research* **54**, 314–317.

Wiggins, L. L., Smith, D. V., & Frank, R. A. (1987). Taste processing in the rabbit: generalization of learned taste aversions. *Chemical Senses* **12**, 708.

Yamaguchi, S., Yoshikawa, T., Ikeda, S., & Ninomiya, T. (1970). Studies on the taste of some sweet substances. II. Interrelationships among them. *Agricultural & Biological Chemistry* **34**, 187–197.

PART THREE
HUMAN PERCEPTION OF MIXTURES

10

Attention and learning in the perception of odor mixtures

Michael D. Rabin

International Flavors and Fragrances, Research and Development
Union Beach, New Jersey, USA

William S. Cain

John B. Pierce Foundation Laboratory and Yale University,
New Haven, Connecticut, USA

I. Introduction

It is straightforward, using standard psychophysical methods to determine how well an observer can discriminate among odors. But if such an
investigation is performed, does the measure of discrimination provide
an accurate reflection of the olfactory system's discriminative prowess?
The literature on 'educating the senses', or perceptual learning, implies
that the simple discrimination experiment, wherein participants have
little prior familiarity with the stimuli, might provide only a conservative
answer regarding olfactory prowess. Indeed, particular kinds of experience with odors appear to improve ability to tell two odors apart (Rabin,
in press). This report extends the investigation of the role of experience in
successive discrimination to the simultaneous discrimination of components in mixtures.

PERCEPTION OF COMPLEX SMELLS
AND TASTES ISBN 0 12 042990 X

173

Copyright © 1989 by Academic Press Australia.
All rights of reproduction in any form reserved.

A. *Perceptual learning and discrimination*

Perceptual learning refers to the phenomenon whereby an organism improves its ability to extract information from the environment (Gibson, 1969). Improvement can come from relatively short-term deliberate experience or practice with the stimuli of interest, or it can arise in the natural course of an organism's development. The term discrimination refers to the process whereby an organism resolves differences between stimuli. These may be quantitative as in resolving two different weights, or qualitative as in resolving different colors. For example, in a now classic example of learning to discriminate, Lunn (1948) described how Japanese chick sexing experts were brought to the United States to teach their trade to poultry farmers. To an untrained observer, the genital regions of newly hatched male and female chicks appear the same. Trained sexers, however, could discriminate at up to 99.5% accuracy.

We all experience examples of perceptual learning in daily life. An unfamiliar video game may initially seem like a 'blooming, buzzing confusion' of lights and sounds. After some practice, though, we learn the proper discriminative responses to the stimuli and learn to play the game competently. For perceptual learning to be demonstrated in a discrimination experiment, participants should show improvement in responding to differences between stimuli.

B. *Improving discrimination*

Training participants by giving them labels to apply to stimuli has emerged as an important functional procedure for improving certain types of discrimination. In the early stages of research on perceptual learning, labels received attention because of the theory of 'acquired distinctiveness of cues' (Miller & Dollard, 1941). Briefly, the theory — a reflection of the prevailing Zeitgeist — proposed that stimuli become more distinct when paired with distinct responses, the labels. The belief was that new responses add to the distinctiveness of the cues eliciting them. When, for instance, a child learns early on that there is such a thing as milk chocolate and dark chocolate, the distinction between these two products might become exaggerated over any previous distinction between the products. Distinct verbal labels can be seen to add to the distinctiveness of the stimuli giving rise to them and lead to better discrimination (Goss, 1953; Jeffrey, 1953; Segal, 1964).

Later, Gibson (1969) argued that verbal labels, rather than making stimuli more discriminable by providing discriminative responses, might focus attention on the discriminative features or invariant patterns already present in the stimuli. Dark chocolate is not only darker than milk chocolate, it is also harder, shinier and more bitter. Gibson's discriminative features hypothesis argued that verbal labels act to draw attention to the inherent properties of the stimulus. Gibson also pointed out that verbal labels might increase the efficiency of remembering, as might be required when stimuli are presented successively. So, labels may draw attention to definitive, encodable features of a stimulus and may aid discrimination particularly in non-simultaneous presentation of a standard and a comparison stimulus (Murray & Lee, 1977; Pick, 1965).

C. Odor mixtures

The evidence indicates that the mixture of a small number of chemically pure odorants rarely results in qualities not apparent in the components. For binary mixtures, the more intense component dominates the overall quality, although the perceived intensities of both components are subject to mutual weakening. Thus, mixture quality perception is describable largely in terms of masking and counteraction rather than the emergence of new qualities.

Laing and Willcox (1983) performed a major investigation of both odor quality and intensity for several binary odor mixtures. Participants in their study profiled the odor quality of mixtures and individual components using an adjective list profiling technique. Laing and Willcox reported that:

(a) mixtures composed of two odors of equal perceived (unmixed) intensity are profiled as a combination of the separate odor profiles, although some features are missing in the mixture;
(b) very small changes in the ratio of the perceived intensities of the individual components shifted the perception of the mixture largely in the direction of the stronger component.

However, even though the participants reported the dominant profile of the mixture as that of the more intense component, the mixture might have been discriminable from the dominant odor alone. As the authors pointed out, participants may or may not have perceived the mixture as unique among the stimuli. A discrimination procedure could offer some resulution on this issue.

D. Perceptual learning in olfaction

Rabin (in press) has shown that odor quality discrimination depends upon a participant's familiarity with the stimulus odors. For reasons that are not entirely clear in terms of a theoretical explanation, a person can tell two potentially confusable odors apart better if he or she is familiar with them. If we extrapolate this result, obtained with successive presentation of odors, to mixtures, then we would expect that increasing familiarity with components of a mixture might enhance ability to attend to them selectively. We can, however, achieve the same goal of increased discrimination by choosing stimuli with which a person is already familiar and for which he or she already possesses a good label. Previous research (Rabin & Cain, 1984) has shown that familiar odors are encoded in memory better than unfamiliar odors. In that case, the labels attached to the odors seem critical. A familiar odor generally can be labeled consistently over time and the label may then mediate the memory.

Although labels might improve successive discrimination between two odors by virtue of increasing the efficiency of remembering, it is not yet clear that they can aid in the simultaneous task of resolving mixtures. In that case, memory is less of an issue.

We are told that perfumers and flavorists are able to identify even very subtle components of mixtures. Are such persons naturally better able to do so, or are they able to do so because they possess a richer set of terms with which to resolve the mixtures? If the latter is true, then performance on a discrimination task involving mixtures may improve if a person has a better olfactory vocabulary or if allowed to work with those stimuli for which their personal lexicon is already developed.

Although familiar odors and readily labeled odors are better encoded in memory, unfamiliar odors that are not easily labeled can also be encoded in some way. Hence, odor memory and successive odor discrimination are not determined by denotative labels entirely. Other dimensions of olfactory experience may play a role. A prominent candidate would be pleasantness. In everyday life, people seem quite adept at detecting off-notes in foods even when the off-notes may be of unknown quality. Perhaps affective responses, then, are also important in determining whether an odorant is pure or is mixed with some trace contaminant.

In the experiment reported below, we explored whether familiarity with the components of an odor mixture would influence their salience in a mixture. We also explored whether pleasantness might influence salience. The study takes perceptual learning into account by attempting to relate whether the odor knowledge that a participant brings into an experiment determines discrimination of components in mixtures. In this task, a target odor comprising a single major component was paired either with itself (a 'same' response), or with a transformation comprising the major component and an accompanying minor component ('different').

The major component could vary in familiarity and the minor component in familiarity or pleasantness. The minor component, as the name suggests, was weak with respect to the target, so that the task became one of detecting the minor component's presence against the background of the major component, rather than a simple intensity discrimination. Because both familiarity and pleasantness appear to vary greatly from person to person, this experiment used a stimulus set individually selected for each participant.

Issues addressed were:

(a) whether familiarity with an odor might enhance one's ability to detect whether it contains what might be thought of as a contaminant (minor component);
(b) whether familiarity with the minor component assists in discriminating it from a background consisting of a stronger odor (major component);
(c) whether the pleasantness of the minor component would influence its detection against the background of the major component. The design, in addition, sought to relate discrimination performance to the intensity of the minor component.

II. Method

Five males and five females participated. All were between the ages of 20 and 30 (mean age = 24) and were paid for their time.

Stimuli were selected to represent bimodal values of familiarity and pleasantness (i.e. familiar versus unfamiliar, pleasant versus unpleasant) for each individual participant. Three familiar and three unfamiliar stimuli of equal pleasantness were chosen (the familiarity dimension), and two pleasant and two unpleasant stimuli of equal familiarity were chosen (pleasantness dimension). Therefore, stimuli varying along the familiarity dimension were chosen to be of approximately equal pleasantness, and stimuli varying in pleasantness were chosen to be of approximately equal familiarity.

Participants first smelled 30 odors in an odor library during an evaluation session. The 30 were previously matched for perceived intensity. The entire set was presented twice. During the first pass through the set, participants rated pleasantness on a scale ranging from −5 to +5, rated familiarity in a 0 to 10 scale, and finally generated a label for the odor. Participants were instructed to provide the most appropriate label possible and were told that they would be asked to label the odors again on a second pass through the set. During the second pass, participants labeled the odors again trying to use the labels they had provided during the first pass. Twenty seconds elapsed between presentations of successive odorants and approximately two minutes elapsed between the two passes through the library.

Participants' ratings were later evaluated to choose the major and minor components for the discrimination task. Four types of stimuli were defined for each participant as follows:

1. *Familiar (F+).* These stimuli were given high familiarity ratings and labeled consistently, i.e. with the same label, during both runs through the library. Consistent labeling is a measure of encodability.
2. *Unfamiliar (F−).* These stimuli were chosen based on low familiarity ratings and inconsistent labeling.
3. *Pleasant (P+).* Pleasant stimuli were chosen from those odors given high pleasantness ratings.
4. *Unpleasant (P−).* These odorants were chosen from among those given low pleasantness ratings.

The set of major components contained three F+ stimuli and three F− stimuli. These six were of approximately equal, moderate perceived intensity, i.e. equal to 5 on a 10 point intensity scale. The set of minor components included the same F+ and F− stimuli but at an intensity equal to 2. The set of minor components also included two P+ and P− stimuli also at an intensity equal to 2.

It was not always possible to select odorants from the extremes of either the familiarity or pleasantness scales, since pleasantness and familiarity tend to be correlated (Engen & Ross, 1973; Lawless & Cain, 1975). In this experiment too, familiarity and pleasantness showed a positive correlation (average $r = 0.43$, $p < .05$). Compromises were necessary to ensure that the pleasantness of the F+ and F− stimuli was about equal and that the familiarity of the P+ and P− stimuli was about equal.

The F+ and F− stimuli differed significantly in rated familiarity ($t = 7.42$, $p < .0001$), and the P+ and P− stimuli differed in rated pleasantness ($t = 16.81$, $p < .0001$). However, the P+ and P− stimuli also differed in their rated familiarity ($t = 3.36$, $p < .01$), thereby confounding familiarity and pleasantness for these stimuli as the correlation noted above would suggest. This confounding, though, would work against the experimental prediction that high familiarity would aid discrimination, whereas low pleasantness would also aid discrimination.

Table 10.1. Odorant distribution into categories

Odorant	Per cent of each category accounted for			
	F+	F−	P+	P−
Amyl acetate	7	0	10	0
Amyl butyrate	3	3	5	5
p-anisaldehyde	3	0	15	5
Anisole	0	7	0	0
Benzaldehyde	10	0	0	0
Butyl alcohol	0	7	0	5
n-butyric acid	0	0	0	15
l-carvone	13	0	0	0
Cineole	3	3	10	5
Cinnamyl n-butyrate	3	3	0	0
Ethyl butyrate	7	0	15	0
Ethyl propionate	0	7	0	0
Ethyl valerate	3	3	0	0
tr-2-hexenal	3	10	0	0
cis-3-hexen-1-ol	7	7	5	0
n-hexyl salicylate	0	3	0	0
Hydroxycitronellal	0	10	5	0
Isovaleric acid	0	0	0	15
Lavandin abrialis	7	0	15	0
Linalool	3	10	0	0
Linalyl acetate	3	10	5	0
Methyl disulfide	0	0	0	20
Methyl propionate	0	0	0	15
Octanal	7	13	0	5
b-phenethyl alcohol	3	0	0	5
sec-phenethyl alcohol	0	0	0	0
Pinene	7	3	10	0
Propyl butyrate	7	0	5	5
Pyridine	0	0	0	0
trans-2-decen-1-al	0	0	0	0

The F+ and F− stimuli did not differ in their rated pleasantness (t = 1.85, $p > .05$). Table 10.1 shows the percentage of cases where each stimulus was determined to be F+, F−, P+ and P−. Amyl acetate, for instance, fell into the F+ set 7% of the time and into the P+ set 10% of the time. No participant found it unfamiliar enough or unpleasant enough for it to fall into the F− or P− categories. Amyl butyrate, on the other hand, fell into the F+ and F− categories an equal number of times. Hence, some participants found it familiar and some did not. Some found it pleasant, and some found it unpleasant. Such a pattern, which occurred not infrequently, highlights the individuality of odor experience and preference.

Stimuli were prepared by injecting odorants via a syringe into Interflo™ pellets (Chromex Corp., #p–375) made of compressed filaments of polypropylene. Interflo acts as an excellent adsorbent, capable of holding a relatively large amount of liquid odorant and releasing its vapors slowly over time. Each cylindrical pellet measures approximately 1 cm in length by 1cm in diameter and can retain up to approximately 0.2mL of fluid. For these experiments 0.15mL was injected into each pellet. All odorants were colorless, so there was no need to hide pellets from participants' view.

Pellets were presented to participants in white glass jars (60 mL capacity) with plastic screw-on lids. Jars were washed and reused as necessary, but lids were disposed of after use with one odorant as were the pellets. Four jars of each odorant were prepared so that no jar was presented twice consecutively. Only one sniff was allowed per jar for each trial. Two pellets were used in each jar. One contained the major component, while the other contained the minor component or the mineral oil diluent alone. Hence, mixing occurred in the vapor phase rather than in the liquid phase.

A two-interval, same-different task was used as the test of discrimination. The first odor of a pair could be the major component or the major/minor complex. If the first odor was the major component, then the second could be the major component ('same') or the complex ('different'). If the first odor was the complex, then the second odor was the major component ('different').

Table 10.2 shows how the major component transformations were prepared for each participant. The numerals 1 to 6 represent the major component items for each participant. The numerals 21 to 24 represent the P+ and P− stimuli (1 + 2 and 6 + 24, for example, represent mixtures of stimuli 1 and 2, and 6 and 24 respectively). Discrimination pairings can be determined by reading across the table (1 versus 1 + 2, 1 versus 1 + 3, ... 6 versus 6 + 24). Note that the stimuli were not the same for each participant because they were chosen from the library to represent a range of familiarity and pleasantness for each participant. In actual testing, trials were run in both possible orders, e.g. 3 versus 3 + 1 and 3 + 1 versus 3. In addition, an equal number of 'same' trials were randomly interspersed among the 'different' trials. 'Same' trials, as implied above, consisted of two identical major components.

The influence of minor component intensity was studied by adding a second intensity level to eight of those components. These eight major/minor complexes contained the minor component at an intensity equal to 4 on the intensity scale (see asterisks in Table 10.2). A total of 248 trials completed the entire matrix of discrimination pairings.

Table 10.2. Set of stimuli for discrimination task. Entries show the major/minor complex

	Familiarity of minor component						Pleasantness of minor component			
	1(F+)	2(F+)	3(F+)	4(F−)	5(F−)	6(F−)	21(P+)	22(P+)	23(P−)	24(P−)
MAJOR COMPONENT										
1 (F+): —		$1+2^a$	$1+3$	$1+4$	$1+5^a$	$1+6$	$1+21^a$	$1+22$	$1+23^a$	$1+24$
2 (F+): $2+1$		—	$2+3$	$2+4$	$2+5$	$2+6$	$2+21$	$2+22$	$2+23$	$2+24$
3 (F+): $3+1$		$3+2$	—	$3+4$	$3+5$	$3+6$	$3+21$	$3+22$	$3+23$	$3+24$
4 (F−): $4+1$		$4+2^a$	$4+3$	—	$4+5^a$	$4+6$	$4+21^a$	$4+22$	$4+23^a$	$4+24$
5 (F−): $5+1$		$5+2$	$5+3$	$5+4$	—	$5+6$	$5+21$	$5+22$	$5+23$	$5+24$
6 (F−): $6+1$		$6+2$	$6+3$	$6+4$	$6+5$	—	$6+21$	$6+22$	$6+23$	$6+24$

F+: familiar   F−: unfamiliar   P+: pleasant   P−: unpleasant

[a] These major/minor complexes appeared with the minor component present at two intensities, one approximately 50% stronger than the other. The presence of two different intensities added eight stimuli to the set.

Participants in the discrimination experiment were told that they would be presented with stimuli that were either identical or different in that one had an extra odorant added. Two such sessions were run per participant. The first session was comprised of all of the F+ and F− transformations (left side of Table 10.2), consisted of 136 trials, and lasted approximately 1.5 hours. The second session was comprised of all of the P+ and P− transformations (right side of Table 10.2), consisted of 112 trials, and lasted approximately 1 hour. Sessions were always run in this order so that familiarity of the minor component items would not be increased merely by virtue of exposure to the odorants during the experiment. The two sessions took place within one week of the evaluation session and either within one day of each other or, if held on the same day, 20 minutes apart.

As regards analysis of responses, a hit was scored if a participant correctly called identical stimuli the same. A false alarm (FA) was scored if a pairing of a major component versus major/minor complex was called same. The FA rate, therefore, represents the rate at which participants failed to detect the minor component against the background of the major component.

The parameter A' was used as the measure of discrimination (Pollack & Norman, 1964; Gescheider, 1985). One advantage of A' is that it corresponds to per cent correct in a two-alternative forced-choice task and thus provides an intuitive and readily graspable interpretation. Eight separate A's were calculated for each participant, one for each possible combination of major component and minor component:

1. F+F+	5. F−F+
2. F+F−	6. F−F−
3. F+P+	7. F−P+
4. F+P−	8. F−P−

Four separate hit rates were used in order to calculate the appropriate A'. One hit rate was calculated for each of the two types of major components, F+ and F−, by pooling the total hits for the three F+ odors together, and the three F− odors together, and dividing by the appropriate number of 'same' trials. This was done for each of the two test periods to adjust for changes that might occur due to the type of minor components used (during session 1 the major components varied in familiarity and during session 2 they varied in pleasantness). Therefore, four hit rates used were:

1. one for F+ major components used during session 1;
2. one for F+ major components used during session 2;
3. one for F− major components used during session 1;
4. one for F− major components used during session 2.

Together with these four hit rates, eight FA rates were calculated corresponding to each of the eight cells of the experiment listed above. For example, for the F+F− cell, FA rate was based on all trials in which an F+ major component was paired with an F− minor component. For the F−P− cell, as another example, FA rate was based on all 'same' responses to pairings of all F− stimuli with P− minor components. The corresponding hit rate for the first example was case 1 above, while for the second example it was case 4.

III. Results

Figure 10.1 depicts discrimination performance for the eight pairings of major and minor components for the low concentrations of minor components only. Both major component type and minor component type influenced the difficulty of the discrimination.

A three way analysis of variance was performed on the data with sex of participant as a between-group factor, and major component type and minor component type as repeated measures. Sex of the participant played no role ($F(1,8) = 0.06$, ns). Both major component type ($F(1,8)$

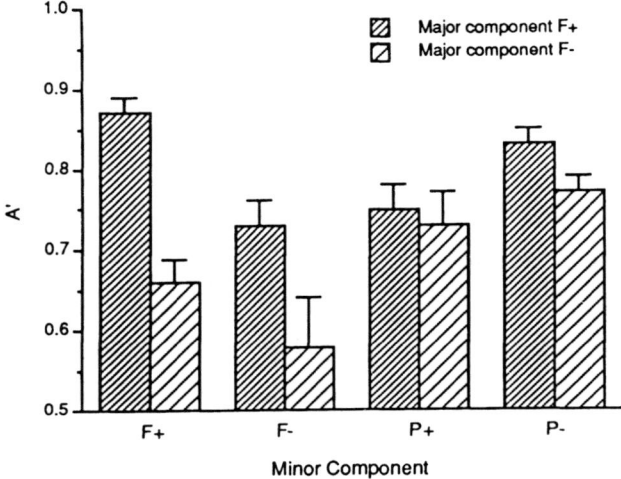

Fig. 10.1. Discrimination performance for the eight combinations of major and minor component stimuli for the low concentrations of minor component only.

$= 29.36, p < .001)$ and minor component type $(F(3,24) = 7.85, p < .001)$ did influence discrimination performance. The interaction between major and minor component types did not quite reach significance $(F(3,24) = 2.60, p = .08)$.

The results, then, implied that a familiar major component is more likely to be discriminated from an adulterated version of itself than is an unfamiliar one. Furthermore, the data suggest that a familiar minor component is more likely to be apparent than an unfamiliar one, whereas an unpleasant minor component is more likely to be apparent than a pleasant one. A contrast was performed with the following weights assigned to the minor components: $F+ = 1, F- = -1, P+ = -1$, and $P- = 1$. The contrast proved significant $(F(1,8) = 9.90, p = .01)$. Hence, the data support the idea that both major component familiarity and minor component familiarity or pleasantness can influence the ability to discriminate between an odor and a transformation of itself.

A. Intensity of minor component

Only one higher intensity pairing was used for each section of the discrimination matrix (Table 10.2). Calculation of a metric such as A' for each cell per individual participant becomes inappropriate owing to the

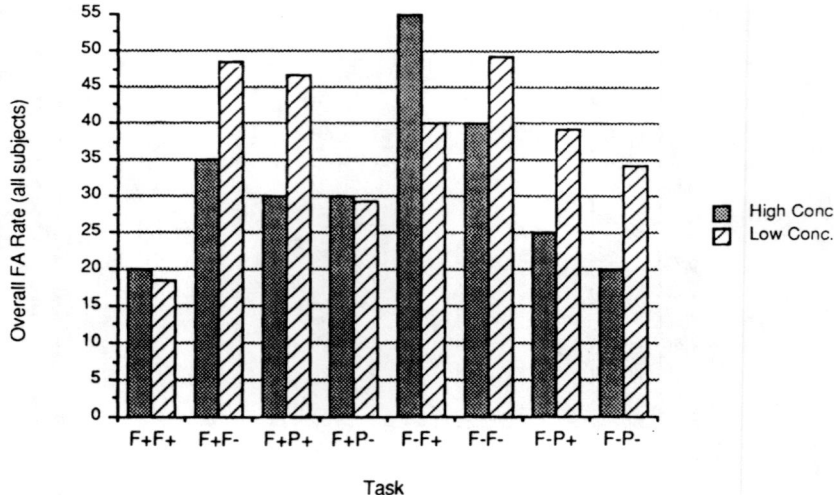

Fig. 10.2. False alarm rates (per cent) for the eight combinations of major and minor component stimuli for both low and high concentrations of minor components.

small number of trials per condition, nevertheless, FA rates can be used for the comparison. A false alarm occurred when a participant responded 'same' when presented with pairs that were in fact different. FA rates were calculated by pooling the total number of FAs across all participants for the higher intensity minor components in order to compare them to FA rates for the lower intensity minor components. This comparison appears in Figure 10.2.

For all but three of the possible pairings of major and minor components, the FA rate for the higher intensity minor components was lower than that of the lower intensity minor component. This indicates that when a minor component was stronger, it was generally easier to detect against the background of the major component. This generalization was not true for the two situations where the minor component was a familiar one, suggesting that a familiar odor can be a salient stimulus independent of its concentration for the two intensities tested. This also appears to hold when the minor component was an unpleasant one mixed with a familiar major component, although this did not apply to the case where an unpleasant minor component was mixed with an unfamiliar major component.

Although the data could not be analyzed statistically because of the relatively small number of observations per cell, the following two consistencies appeared:

1. A more intense minor component is easier to detect than a weaker one.
2. The familiarity of the minor component can make it a more salient stimulus, so that intensity can be a less important cue than familiarity.

IV. Discussion

Three questions were posed in the introduction:

1. Does familiarity with an odor enhance one's ability to detect whether it contains a minor component?
2. Does familiarity with the minor component assist in discriminating it from a background consisting of a stronger odor?
3. Does the pleasantness of the minor component influence its detection against the background of a stronger odor?

A. Familiarity of major components

High familiarity of a major component, as indexed by numerical rating of familiarity and the ability to label it consistently, facilitates the detection of a minor component's presence in the odor. One possible explanation is that participants possess a more finely tuned perceptual representation of a familiar stimulus and are better able to judge whether or not another stimulus, a transformation of the familiar one, violates the perceptual boundaries of the first. In this interpretation, accurate labeling indicates the presence of a unique perceptual category congruous with discriminative capacity. The finding parallels speech research, suggesting that perceptual categories are subject to the effects of training and experience (Carney, Widen, & Viemeister, 1977; Pisoni, Aslin, Perey, & Hennessy, 1982).

B. Familiarity of minor components

The familiarity of the minor component also influenced ease of discrimination. This result implies that participants could not detect the minor component as readily if it were unfamiliar. In fact, Figure 10.2 shows that for the low concentration F− minor components, the FA rate was above

45%, suggesting a serious detection problem, i.e. participants could not detect its presence when smelled against the major component. The FA rate for F+ minor components, though, averaged about 30%. It seems, then, that the familiar minor components were *good* stimuli in the sense that Garner (1974) uses the term, implying that they exist in unique classes and share few overlapping features with other stimuli. *Bad* stimuli, such as unfamiliar ones, may have very few unique perceptual features for a participant and therefore blend in more readily against a stronger background.

C. *Pleasantness of minor components*

Although pleasantness did appear to influence detection of the minor component, with unpleasant stimuli more detectable than pleasant ones, this effect was not as large as the effect of familiarity. The finding that pleasantness can influence discrimination is not surprising, given the common experience, mentioned earlier, of detecting impurities or spoilage in food products.

   One factor working against finding a larger effect of pleasantness was the confounding of pleasantness with familiarity. P+ stimuli averaged higher familiarity than P− stimuli, thereby adding somewhat to the detectability of the P+ stimuli. In spite of this advantage, however, P− stimuli were still more readily detected against the background of a major component.

   Why did average A′ for unfamiliar major components appear to improve in the part of the experiment where pleasantness was a variable? As noted earlier, P+ and P− minor components were always tested during session 2 of the experiment after the F+ and F− minor components. Therefore, order of testing was tied to the type of minor component used. It is possible that the type of minor component interacted with type of major component to raise the overall detectability of the F− major components. However, a hypothesis more consistent with the other results reported in this chapter is that the average familiarity of the F− major components may have been raised during the first session, thereby leading to improved discrimination during session 2, although it was still below that of F+ major components.

V.   Conclusions

The results demonstrate that the ability to detect the individual components of odor mixtures is not a simple function of stimulus intensity. If stimulus intensity were the only determinant of the ability to analyze an odor mixture, then the various types of minor component used would have affected ease of discrimination equally.

Furthermore, the results suggest that even when intensity of a minor component is increased, detectability may still interact with cognitive aspects of the stimulus. One implication of Figure 10.2 was that familiarity of the minor component may be more important than its intensity in determining its salience, at least for the intensities tested.

Laing and Willcox (1983) noted in their study of mixture perception that their profiling technique was not necessarily sensitive enough to detect whether participants could actually discriminate between the binary mixtures even though the profiles may have been similar. Their results implied that when perceived intensity of one of the mixture components increased over the other, profiling of the stimulus shifted categorically in the direction of the more intense. The results of the present experiment suggest that discrimination may indeed provide a picture of perception different from that provided by profiling alone. The results supplement those of Laing and Willcox in suggesting that cognitive attributes of odor stimuli will influence detection of mixture components.

The present finding helps to explain why perfumers and odor experts may be able to treat odor mixtures analytically, whereas the layperson may not. Endowing a layperson with a perfumer's broad experience would make subtle mixture components more salient stimuli.

The conclusion that experience assists discrimination between two unmixed odors therefore extends to odor mixtures as well. Hence, cognitive processing of odor mixtures operates according to principles like those that apply to simple stimuli. Furthermore, the results reveal that psychophysical measures, without attention to cognitive variables, may fail to predict the quality of odor mixtures, since perceived quality will probably vary from individual to individual, depending on that person's past olfactory history.

Acknowledgments

This work was supported by a grant from the Fragrance Research Fund and US National Institutes of Health Grant NS-21644.

References

Carney, A. E., Widen, G. P., & Viemeister, N. F. (1977). Noncategorical perception of stop consonants differing in VOT. *Journal of the Acoustical Society of America* **2**, 961-970.

Engen, T., & Ross, B. M. (1973). Long-term memory of odors with and without verbal descriptors. *Journal of Experimental Psychology* **100**, 221–227.

Garner, W. R. (1974). *The processing of information and structure.* Lawrence Earlbaum Associates, Potomac.

Gescheider, G. A. (1985). *Psychophysics, method, theory and application.* Lawrence Earlbaum Associates, Hillsdale.

Gibson, E. J. (1969). *Principles of perceptual learning and development.* Prentice Hall, Englewood Cliffs, New Jersey.

Goss, A. E. (1953). Transfer as a function of type and amount of preliminary experience with task stimuli. *Journal of Experimental Psychology* **46**, 419–428.

Jeffrey, W. E. (1953). The effects of verbal and nonverbal responses in mediating an instrumental act. *Journal of Experimental Psychology* **45**, 327–333.

Laing, D. G., & Willcox, M. E. (1983). Perception of components in binary odor mixtures. *Chemical Senses* **7**, 249–264.

Lawless, H. T., & Cain, W. S. (1975). Recognition memory for odors. *Chemical Senses and Flavour* **1**, 331–337.

Lunn, J. H. (1948). Chick sexing. *American Scientist* **36**, 280–287.

Miller, N. E., & Dollard, J. (1941). *Social learning and imitation.* Yale University Press, New Haven, Connecticut.

Murray, F. S., & Lee, T. S. (1977). The effects of attention-directing training on recognition memory task performance of three-year-old children. *Journal of Experimental Child Psychology* **23**, 430–441.

Pick, A. D. (1965). Improvement of visual and tactual form discrimination. *Journal of Experimental Psychology* **69**, 331–339.

Pisoni, D. B., Aslin, R. N., Perey, A. J., & Hennessy, B. L. (1982). Some effects of laboratory training on identification and discrimination of voicing contrasts in stop consonants. *Journal of Experimental Psychology: Human Perception and Performance* **8**, 297–314.

Pollack, I., & Norman, D. A. (1964). A non-parametric analysis of recognition experiments. *Psychonomic Science* **1**, 125–126.

Rabin, M. D. (in press). Experience facilitates olfactory quality discrimination. *Perception and Psychophysics.*

Rabin, M. D., & Cain, W. S. (1984). Odor recognition: familiarity, identifiability, and encoding consistency. *Journal of Experimental Psychology: Learning, Memory, and Cognition* **10**, 316–325.

Segal, E. M. (1964). Demonstration of acquired distinctiveness of cues using a paired-associate learning task. *Journal of Experimental Psychology* **67**, 587–590.

11

The role of physicochemical and neural factors in the perception of odor mixtures

David G. Laing

CSIRO Division of Food Research,
Sydney, Australia

I. Introduction

Little is known about how much and where mixtures are processed by the olfactory system, what factors determine how many odorants can be perceived, and what influence the physicochemical features of the stimuli and the receptor system have on the perception of mixtures. Nevertheless, each of these unknowns must be resolved if a scientific basis for the perception of components of odor mixtures is to be established.

Perhaps the most important problem to resolve initially concerns the role of peripheral and central components of the olfactory system in the processing of mixtures. If substantial processing occurs at the odor receptors, through competitive or non-competitive binding, an understanding of additivity, masking and synergism, the most common consequences of mixing odors, is likely to be gained through structure-activity studies. On the other hand, if processing is primarily a central phenomenon, an intimate knowledge of neural circuitry and response properties of cells in different regions of the olfactory system, for example, the olfactory bulb, will be required.

Our initial studies with mixtures (Laing & Willcox, 1983; Laing, Panhuber, Willcox & Pittman, 1984) had two aims:

1. to determine whether odor quality, odor intensity, or both, determined which components were perceived in a mixture.
2. to find odorants that strongly suppressed (masked), or enhanced the perception of other odorants, so that physiological studies of where and how these effects occurred could be pursued.

PERCEPTION OF COMPLEX SMELLS
AND TASTES ISBN 0 12 042990 X

Copyright © 1989 by Academic Press Australia.
All rights of reproduction in any form reserved.

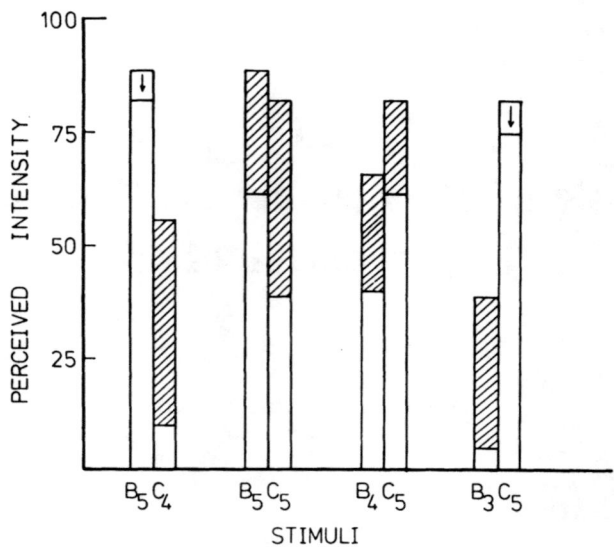

Fig. 11.1. Mixtures of benzaldehyde (B) and carvone (C). Estimates of the perceived intensity of some of the test stimuli. The height of each bar is the arithmetic mean of the scores of all subjects recorded for each stimulus when the stimulus was a single odorant. For example, with the mixture B_5C_4, the height of the left-hand bar indicates the mean perceived intensity of B_5 when presented alone, and the height of the right-hand bar indicates the intensity of C_4 presented alone. Mean estimates of B_5 and C_4 in the mixture $B_5 C_4$ are indicated by the height of the bar up to where cross-hatching commences (C_4) or, where there is no cross-hatching, the lower mark within the bar (B_5). Cross-hatching indicates a component was suppressed in a mixture and the effect is significant at the $p < .01$ level. Perceived intensity is measured in millimetres on a 130 mm line scale. Subscripts indicate the concentration level of the stimulus. There were five concentrations of each substance, with level 5 being the highest concentration and level 1 the lowest.

Briefly, two types of odor interactions were observed. Odorants either suppressed each other in a mutual or reciprocal manner, so that the odorant with the highest perceived intensity before mixing was dominant (Fig. 11.1), or there was a one-sided non-reciprocal interaction where only one odor was suppressed over a wide range of concentrations (Fig. 11.2). Of special significance in the former case was the observation that an odorant need only be of slightly higher intensity than the other before mixing, to completely or almost completely suppress the lower intensity odorant. In contrast, with odor pairs where only one odorant was suppressed (Laing *et al.*, 1984), suppression occurred even when the suppressing odorant had the lower intensity before mixing. Clearly, perceived intensity was important when reciprocal suppression occurred, and quality and intensity were important when non-reciprocal suppression was found.

Although two very clear effects were observed in these studies, there was no indication that the effects were of peripheral or central origin, and the molecular structures of the odorants offered no clue as to the importance of physicochemical factors. In the meantime, two independent physiological studies (Mozell & Jagodowicz, 1973; Mackay-Sim, Shaman & Moulton, 1982) provided the basis for the development of a hypothesis (Laing, 1987) that promised to link physicochemical features of odorants and the characteristic response of receptor cells with the perception of components of odor mixtures. The principle aim of the present chapter is to describe this hypothesis and the series of experiments that investigated its validity.

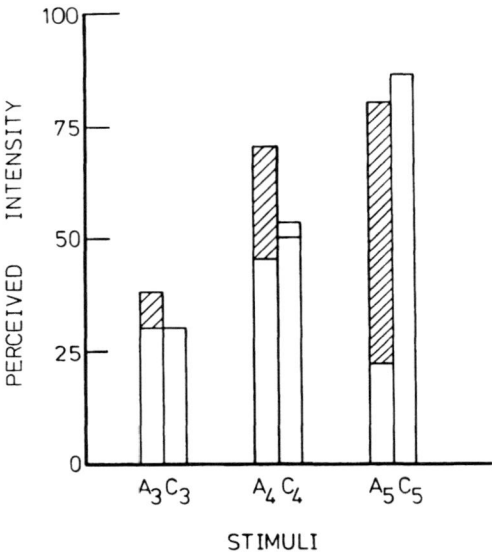

Fig. 11.2. Mixtures of propionic acid (A) and carvone (C). Estimates of the perceived intensity of some of the test stimuli. The height of each bar is the arithmetic mean of the scores of all subjects recorded for each stimulus when that stimulus was a single odorant. For example, with A_4C_4, the height of the left-hand bar indicates the mean perceived intensity of A_4 when it was presented alone, and the height of the right-hand bar indicates the intensity of C_4 when presented alone. Mean intensity estimates of A_4 and C_4 when they were presented in the mixture A_4C_4 are indicated by the height of the bar up to where cross-hatching commences (A_4) or, where there is no cross hatching, the lower mark within the bar (C_4). Cross-hatching indicates a component was suppressed in a mixture and is significant at the $\underline{p} < .01$ level. The subscripts indicate the concentration level where 5 is the highest level and 1 the lowest.

II.   The differential adsorption hypothesis

First, Mozell and Jagodowicz have demonstrated with frogs that inhaled odorants are differentially adsorbed, in a chromatographic-like process, by the aqueous mucus which covers the olfactory receptor epithelium, according to their air-water (mucus) partition coefficients. Also, during a simulated sniff, Hornung, Lansing and Mozell (1975) showed that odorants with high coefficients (non-polar odorants), such as hydro-carbons, are adsorbed poorly, but at a fairly constant rate, as they pass over frog olfactory mucus and become evenly distributed, whilst odor-ants with low air-to-water partition coefficients (polar odorants), such as acids and alcohols, are rapidly adsorbed and become mainly confined to the anterior epithelium (Fig. 11.3). Since the olfactory epithelium of all terrestrial animals, including humans, is covered by a layer of aqueous mucus, it is possible that the differential adsorption seen in the frog may be common to all species. Thus, the major point here is that differential adsorption provides a means for separating mixture components before they reach receptor cells.

In the study by Mackay-Sim *et al.* (1982), electrophysiological meas-urements (electro-olfactograms) on the olfactory receptor epithelium of the salamander showed that the regions that contain receptor cells with the highest responsivity to hydrocarbons are sited posteriorly near the internal nares, whilst those for alcohols are located anteriorly (Fig. 11.4). Although the adsorption of odorants has not been studied in the salamander, the flattish arrangement of dorsal and ventral epithelia suggests differential adsorption is also likely to occur in a similar manner in this animal.

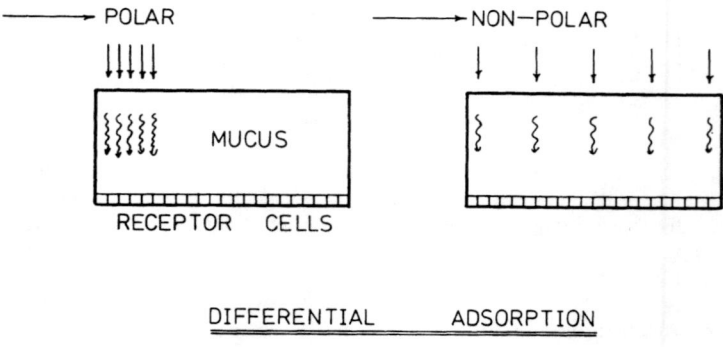

Fig. 11.3. Diagrammatic representation of the differential adsorption of polar and non-polar odorants by the olfactory mucus.

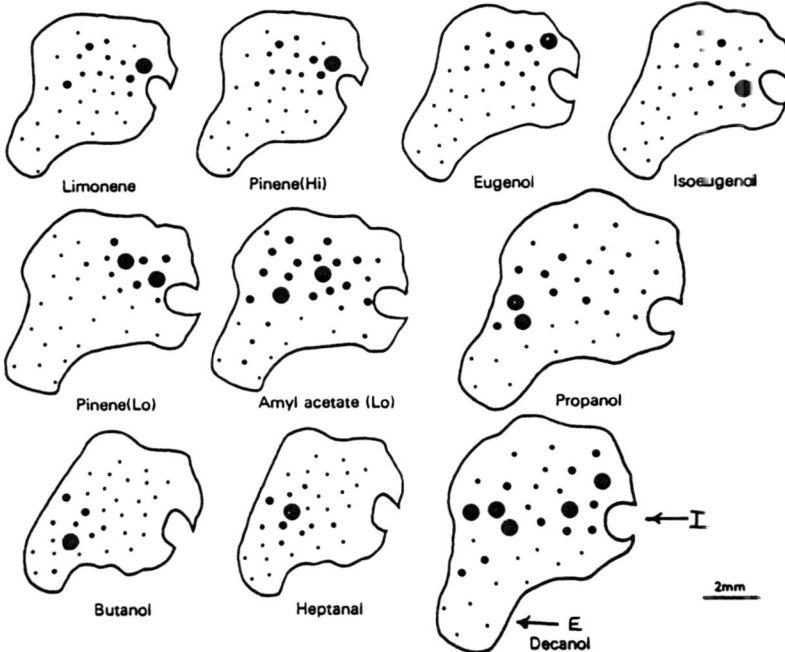

Fig. 11.4. Schematic representation of left ventral olfactory epithelium of the salamander. The region nearest the bottom of the page of each outline is the anterior epithelium, nearest the top is the posterior epithelium. Medial and lateral regions are at the left and right sides respectively. The external (E) and internal (I) nares are indicated by arrows. Dots show recording sites. The size of a dot indicates the magnitude of electrical activity recorded. Large dots indicate sites where the highest activity was found. (Adapted from Mackay-Sim, Shaman and Moulton (1982), with permission.

Together, the findings of Mozell & Jagodowicz (1973), Hornung *et al.* (1975) and Mackay-Sim *et al.* (1982) indicated that odorants are differentially adsorbed by the olfactory mucus, and receptor cells appear to be arranged in a manner that is related to the polarity of the odor molecules to which they show their greatest response. From this the following hypothesis was developed. During (a) sniff(s) odorants are selectively adsorbed by the olfactory mucus according to their polarity, which may enhance the amount of odorant reaching a particular region at which receptor cells are most responsive to that substance. Hence, it was predicted that when a mixture is sniffed there will be minimal interaction between non-polar and polar odorants, because their regions of responsivity are widely separated. In particular, a polar odorant will not interfere with the reception of a non-polar odorant, because very little if any of the former will reach regions containing receptors for the non-polar odorant during a sniffing episode. Similarly, because a small number of molecules of non-polar odorants are adsorbed by the anterior epithelium and the sensitivity of anterior cells for hydrocarbons is low (Mackay-Sim *et al.*, 1982), non-polar odorants may have little effect on the reception of polar odorants. Conversely, interaction between two non-polar, or two polar odorants would be more likely to occur, because their regions of responsivity and adsorption overlap significantly and increase the likelihood that the odorants will compete for sites on the membranes of receptor cells.

A.   Methods and results

To determine whether this hypothesis was applicable to mammals, Bell, Laing and Panhuber (1987) conducted two experiments. In the first experiment, humans were asked to identify and estimate the perceived intensity of components of binary mixtures. The stimuli were five concentrations of each of the non-polar odorants (+)-limonene (orange), α-pinene (camphoraceous) and the polar odorant propionic acid (vinegar), and all possible binary mixtures of these. Stimuli were presented to 20 subjects via an air dilution olfactometer (Laing, Panhuber & Baxter, 1978). In the second experiment, a mixture of limonene and propionic acid was presented to rats injected with tritiated 2-deoxyglucose (2-DG). The odorants were presented in an air stream to the rats, which were restrained in a 'wind tunnel'. In this experiment we hoped to determine the physiological basis for the human responses to the mixture. Characteristically, odor stimulation increases 2-DG uptake in specific glomeruli in the olfactory bulb, and the glomeruli that exhibit uptake vary with the type of odorant used. Odor-induced metabolic activity, detected with 2-DG at the glomeruli, primarily represents responses of the interface synapses between nose and brain. More specifically, 2-DG uptake represents a predominant metabolic response of the olfactory receptor cell axon's presynaptic terminals in the glomerulus (Benson, Burd, Landis & Shepherd, 1985). By mapping the patterns of glomeruli that respond to limonene and propionic acid, presented separately and as mixtures, it could be tested whether masking is a peripheral event: if it is, there should be a change in the glomerular activity pattern for the masked odor. Absence of disturbance to the glomerular activity patterns that characterized the single odors would imply that other, more central mechanisms are responsible for mixture suppression.

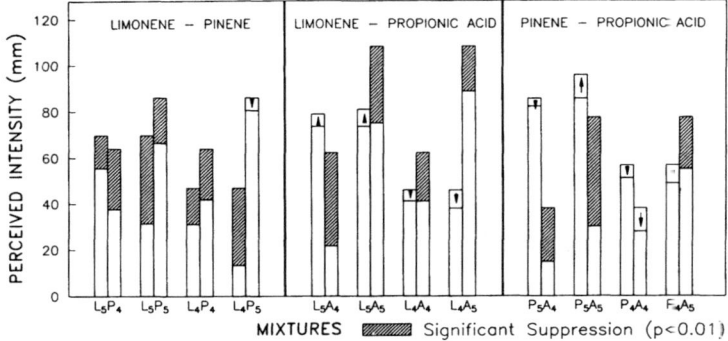

Fig. 11.5. Estimates of the perceived intensity of some of the test stimuli of each odor pair. The height of each bar is the arithmetic mean of the scores of all subjects recorded for each stimulus when that stimulus was a single odorant. For example, with the mixture of limonene (L) and pinene (P), L_5P_5, the height of the left-hand bar, indicates the mean perceived intensity of L_5 when it was presented alone, and the height of the right-hand bar indicates the intensity of P_5 when presented alone. Mean estimates of the components of the mixture L_5P_5 are indicated by the height of the bar up to where cross-hatching commences. Cross-hatching indicates a component was suppressed in a mixture and is significant at the $p < .01$ level.

In the experiment with humans, the lack of suppression of the hydrocarbons by the acid and the occurrence of reciprocal suppression between the hydrocarbons (Fig. 11.5) supported the proposal that suppression in mixtures is a peripheral phenomenon and that the hypothesis may be correct. However, what was not predicted was the finding that the hydrocarbons strongly suppressed perception of the acid (Fig. 11.5), implying that these molecules reached and interfered with the reception process of cells in the anterior epithelium that are stimulated by the acid.

This view is supported by the reduced uptake of 2-DG in response to the mixture, in glomeruli that always showed high uptake when propionic acid was the sole stimulus (Fig. 11.6). The reduced uptake of 2-DG with the mixture suggests the hydrocarbon was inhibiting receptor cells that were normally activated by the acid. This was an exciting finding and suggested that electrophysiologists, in addition to identifying regions of the epithelium that are excited by specific odors, should also seek out the areas where these odors inhibit cell responses. Clearly, if an understanding of odor interactions is to be achieved the areas of excitation and inhibition for different odorants will need to be defined.

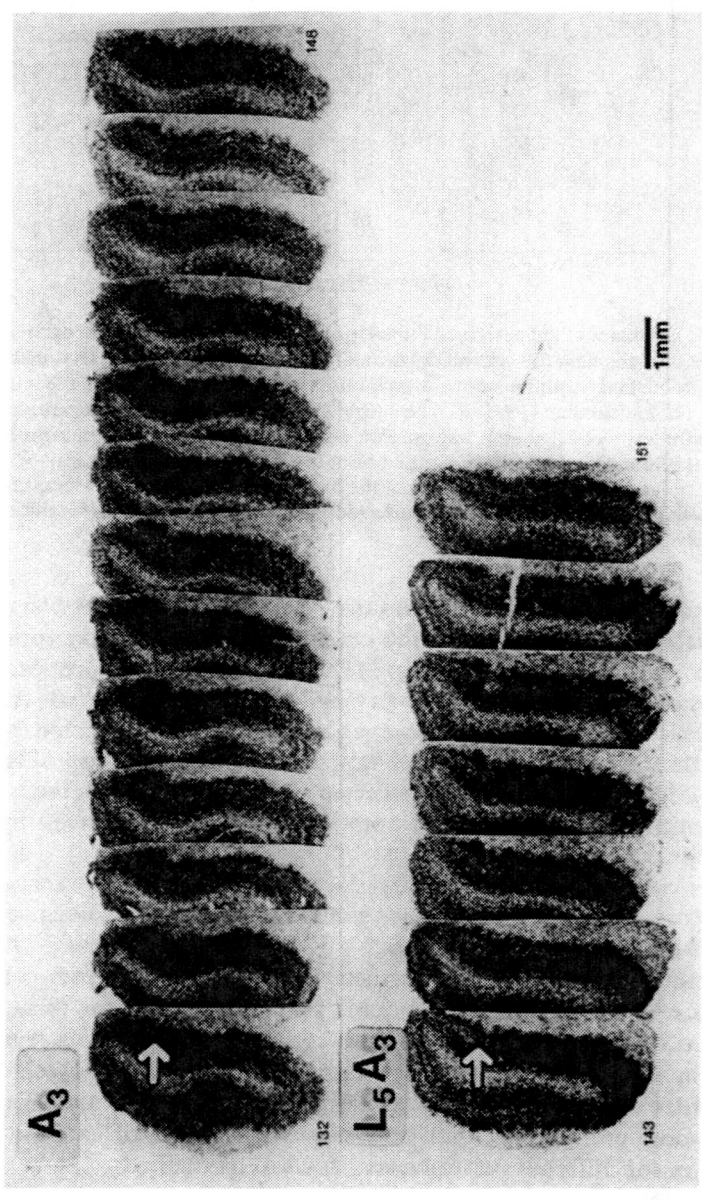

B. Discussion

The results of the above study suggested that some hydrocarbon molecules reached and inhibited acid-sensitive cells, whereas the acid molecules appeared to have failed to reach or inhibit regions where cells show their highest excitability for the hydrocarbon. The polarity of the odorant, therefore, appeared to determine whether one component of a mixture will reach and hence be available to interfere with the action of another component. The human data, therefore, supported the view that odorants of similar polarity compete to excite cells in similar epithelial regions, whilst odorants of different polarity will compete only when a component gains access to the zone normally excited by the other. The rat data showed that it is not necessary for the competing odor on its own to have an excitatory effect on the regions excited by the other odor if it has a different polarity. If an odorant of different polarity has access to the excitatory epithelial zones of another odorant, without exciting them, it is possible for it to exert inhibitory effects on cells excited by the other odorant.

III.   The hypothesis — a second test

Recently, a more searching study of the differential adsorption hypothesis (Laing, in press) indicated that the mechanisms underlying odor interactions are more complex than indicated above. In that study the interactions of limonene and propionic acid with three additional odorants, octane, (−)-carvone and n-butanol, were investigated. All the odorants were chosen because their adsorption by frog receptor epithelium and/or their ability to stimulate nerves that project from the anterior or posterior regions of the frog epithelium has been determined (Hornung *et al.*, 1975; Mozell, 1964; Mozell, 1970; Mozell & Jagodowicz, 1973). Carvone and butanol, which have low air-to-water partition coefficients, are largely adsorbed by the anterior epithelium, and, during a simulated sniff, stimulate nerves that innervate the anterior epithelium (Fig. 11.7). On the other hand octane, like limonene, is poorly adsorbed by frog epithelium but, stimulates nerves from the anterior and posterior

---

◁ **Fig. 11.6.**   Autoradiographs from serial sections cut at around 3 mm from the anterior pole of the olfactory bulb are shown for representative rats stimulated with propionic acid (A₃), and the mixture (L₅A₃). The numbers represent the position (20 μm sections) of the first and last sections that mark the beginning and end of 2-deoxyglucose uptake in the region around the 3 mm mark. A much reduced uptake of 2-deoxyglucose was found with the mixture. (Adapted from Bell et al., 1987).

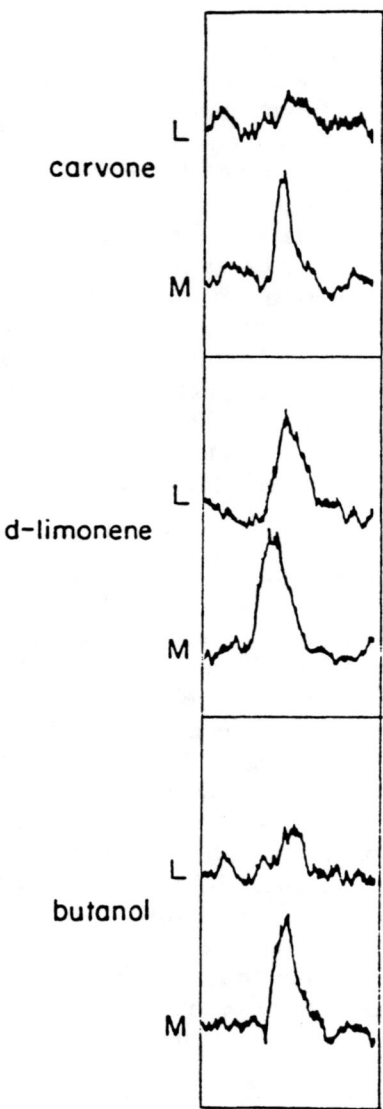

Fig. 11.7. Summated neural discharges in the lateral (upper) and medial (lower) branches of the frog olfactory nerve, as a result of input from receptor cells sited at the internal (lateral) and external (medial) nares respectively. The odorants used to stimulate the receptor cells are indicated alongside each pair of traces. (Adapted from Kurtz and Mozell, 1985, with permission.)

regions (Fig. 11.7). The similar magnitude of stimulation of the two regions by the hydrocarbons is different from that found by Mackay-Sim *et al.* (1982) with the salamander, where a fairly clear separation of receptor cells that respond best to hydrocarbons or alcohols was observed. However, the frog is the only animal in which both adsorption by the mucus and receptor responses have been measured, so selection of odorants on the basis of this information provided a more thorough test of the differential adsorption hypothesis. Accordingly, in the second investigation of the hypothesis, the predictions were that, in mixtures containing only anterior or anterior-posterior stimulators, there is a greater chance of reciprocal effects occurring than with mixtures of anterior and anterior-posterior stimulators, because the former types are adsorbed in similar regions of the epithelium. In contrast, with the latter type of mixtures, competition for receptor sites or cells is only likely to occur in the anterior epithelium where overlap of their adsorption and stimulation zones occurs. Thus, in the latter case the cells in the posterior epithelium that respond to the anterior-posterior stimulator should be unaffected by an anterior stimulator. The prediction in this case, therefore, was that an anterior-posterior stimulator will affect the reception of an anterior stimulator to a greater degree than the latter will affect an anterior-posterior stimulator.

A. Methods and results

In the study, 20 humans were presented with all possible pairings of the five odorants, each of which was prepared and presented at three concentrations via a 15-channel air dilution olfactometer. Each stimulus, whether a single odorant or a mixture, was evaluated in triplicate by each subject. A summary of the results from each of the eight odor pairs is given in Tables 11.1 to 11.3. The results from the earlier studies of the odor pairs limonene-pinene (Bell *et al.*, 1987) and carvone–propionic acid (Laing *et al.*, 1984) are also included.

Examination of the results from mixtures containing only anterior and anterior-posterior stimulators (Table 11.1) shows that the hypothesis failed to predict the outcome in the majority of cases. Although the dominance of limonene in mixtures with butanol was as predicted, the opposite result was obtained with mixtures of octane and propionic acid, octane–butanol and limonene–carvone, where the polar anterior stimulators all suppressed the anterior-posterior stimulators in a one-sided manner. With the odor pair octane–carvone, although carvone dominated in most mixtures, octane reduced the perceived intensity of carvone on a number of occasions.

Table 11.1. Frog model: predictions and effects with mixtures containing components that stimulate different regions of the olfactory epithelium

Anterior	Mixture versus	anterior-posterior stimulators	Result[b]
Propionic acid	←	Limonene[a]	+
Butanol	←	Limonene[a]	+
Carvone[a]	←	Limonene	−
Propionic acid[a]	←	Octane	−
Butanol[a]	←	Octane	−
Carvone[a]	←	Octane	−

Arrow is directed at the component the hypothesis predicted would be suppressed.
[a] Indicates the dominant odorant.
[b] (+) or (−) indicates the prediction was correct or incorrect respectively.

Table 11.2. Frog model: predictions and effects with mixtures containing components that stimulate similar regions of the olfactory epithelium

Anterior	Mixture versus	anterior stimulators	Result[b]
Propionic acid	⇄	Carvone[a]	−
Propionic acid	⇄	Butanol[a]	−
Butanol	⇄	Carvone[a]	−

⇄ indicates reciprocal suppression was predicted.

[a] Indicates the dominant odorant
[b] (+) or (−) indicates the prediction was correct or incorrect respectively.

As regards the effects of mixing odors that have similar adsorption and stimulating characteristics, the hypothesis failed to predict the outcome with more than half the odor pairs listed in Tables 11.2 and 11.3. For example (Table 11.2), butanol and carvone strongly suppressed propionic acid in a one-sided manner and carvone was perceived as the dominant odorant in mixtures with butanol. Only with mixtures of limonene and pinene, and limonene and octane (Table 11.3) — all anterior-posterior stimulators — was reciprocal suppression observed as predicted. Overall, the adsorption and stimulating characteristics of the odorants did not reliably predict whether reciprocal or non-reciprocal effects would occur.

Table 11.3. Frog model: predictions and effects with mixtures containing components that stimulate similar regions of the olfactory epithelium

Anterior-posterior	Mixture versus	anterior-posterior stimulators	Result[b]
Limonene[a]	⟵⟶	Octane[a]	+
Limonene	⟵⟶	Pinene	+

⟵⟶ indicates reciprocal suppression was predicted.

[a] Indicates the dominant odorant
[b] (+) or (−) indicates the prediction was correct or incorrect respectively.

B. Discussion

Although the results do not rule out the possibility that differential adsorption is the first step in the processing of components of odor mixtures, they indicate that differential adsorption and responses to odorants in the frog do not provide a reliable indicator of the odorants that will be perceived in a mixture by humans.

A similar conclusion is reached if the salamander is used as the model. That is, if limonene and octane are viewed as posterior stimulators, and carvone, butanol and propionic acid as anterior stimulators, the predictions again generally fail (Table 11.4), the most glaring failures being the suppression of limonene by carvone, octane by propionic acid, and octane by butanol. These three examples clearly indicate that, contrary to the hypothesis, polar odorants can reach and interfere with the reception of non-polar odorants during a sniff or sniffing episode.

C. General discussion

The two features of the receptor system that were used to predict the outcome of mixing odors in the studies described above were the differential adsorption of odorants by the mucus, and the arrangement and responsivity of the receptor cells. Despite the outcome of these studies, it is still possible that these two features play a significant role in the process of identifying mixture components. It is almost certain, for example, from our knowledge of the adsorption of odor molecules by aqueous media, that odorants will be differentially adsorbed by the olfactory mucus. Water-soluble odorants like alcohols and acids will always be adsorbed in substantially greater quantities than the water-insoluble hydrocarbons.

Table 11.4. Salamander model: predictions and effects with mixtures containing components that stimulate different regions of the olfactory epithelium

Anterior	Mixture versus	posterior stimulators	Result[b]
Propionic acid	← / →	Limonene[a]	+
Butanol	← / →	Limonene[a]	+
Carvone[a]	← / →	Limonene	−
Propionic acid	← / →	Octane	−
Butanol[a]	← / →	Octane	−
Carvone[a]	← / →	Octane	−

Length of arrows indicates predicted direction of suppression. Short arrows indicate less suppression was predicted.
[a] Indicates the dominant odorant.
[b] (+) or (−) indicates the prediction was correct or incorrect respectively.

However, what we do not know in the case of humans is whether the inspired odor stream flows across, or is directed perpendicularly at the mucus. Although in either case differential adsorption will determine how much of each odorant is adsorbed, it is the direction of flow that will determine the distribution of odorants across the epithelium. If flow is across the mucus, then the resultant separation of odorants by this means may yet prove to be the initial step and an integral part of the process of identifying mixture components. Similarly, whether the receptors are arranged randomly, as in the frog, or ordered, as in the salamander, is likely to influence how odors will affect each other.

Although interpretation of the results at this stage is necessarily speculative, one particularly attractive possibility is that odor interactions are dependent on the availability of receptor sites for different odors on receptor cells. For example, propionic acid suppressed perception of octane but had little effect on the perception of the other hydrocarbon, limonene. The acid, therefore, affected octane sensitive cells far more than cells sensitive to limonene. This suggests that the receptor molecule or receptor cell membrane of limonene-sensitive cells

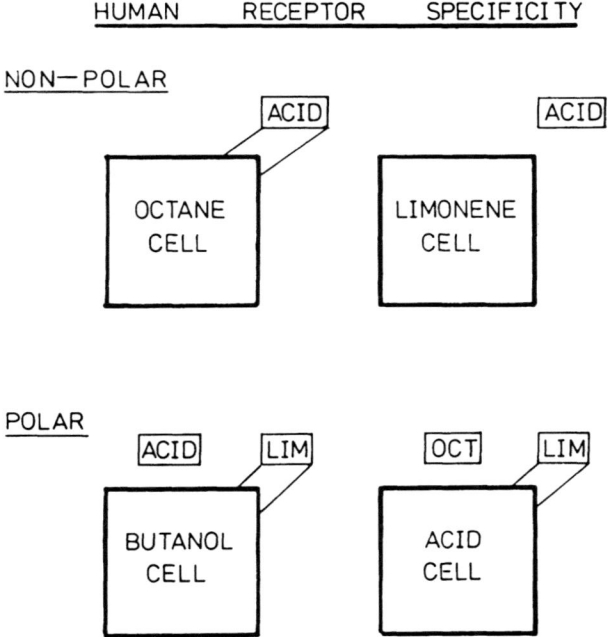

Fig. 11.8. Specificity of human olfactory receptor cells. Each large box represents a cell type, each small box an odorant. ACID, LIM and OCT are the odorants propionic acid, limonene and octane respectively. An odorant linked by a pair of lines to a cell indicates the cell has sites for that odorant that permit the odor to inhibit or excite the cell. Odorants not linked to a cell do not affect the cell. For example, propionic acid scarcely affects the perception of limonene when the two are presented as a mixture, whereas limonene strongly suppresses the perception of propionic acid.

have few sites that allow the acid to interact with the cell, whereas with octane-sensitive cells there may be sites that allow the acid to stimulate the cell or to bind to sites that allow it to interfere and inhibit stimulation of the cell by octane. Figure 11.8 provides a summary of what the data indicate in terms of the responsivity of human receptor cells.

IV. Conclusions

From the limited amount of data available on the effects of mixing odors, it is still not possible to specify a single physicochemical property of odorants, or one feature of the receptor system, that will help to predict how perception of one odor will be affected by the presence of another.

Clearly, air to water (mucus) partition coefficients of odorants, the differential adsorption properties of the olfactory mucus, and receptor-cell responsivity, as found in the frog and salamander, do not account for the effects reported here. Whether the availability of sites for different odors on receptor cells, as suggested above, is important, remains to be seen. Encouragingly, recent psychophysical studies by Laing and Willcox (1987) suggest that the different times taken by odorants to diffuse across the olfactory mucus and stimulate receptor cells may be another factor that plays a role in determining which odor will dominate in a mixture. This proposal, however, awaits further study with a wider range of odorants.

References

Bell, G. A., Laing, D. G., & Panhuber, H. (1987). Odour mixture suppression: evidence for a peripheral mechanism in human and rat. *Brain Research* **426**, 8–18.

Benson, T. E., Burd, G. D., Landis, D. M. D., & Shepherd, G. M. (1985). High resolution 2-deoxyglucose in quick-frozen slabs of neonatal rat olfactory bulb. *Brain Research* **339**, 67–78.

Gesteland, R. C., Lettvin, J. Y., & Pitts, W.H. (1965). Chemical transmission in the nose of the frog. *Journal of Physiology* **181**, 525–559.

Hornung, D. E., Lansing, R. D., & Mozell, M. M. (1975). Distribution of butanol molecules along bullfrog olfactory mucosa. *Nature* **254**, 617–618.

Kurtz, D. B., & Mozell, M. M. (1985). Olfactory stimulation variables. Which model predicts the olfactory nerve response? *Journal of General Physiology* **86**, 329–352.

Laing, D. G. (1987). Coding of chemosensory stimulus mixtures. *Annals of the New York Academy of Sciences* **510**, 61–66.

Laing, D. G. (in press). Relationship between the differential adsorption of odorants by the olfactory mucus and their perception in mixtures. *Chemical Senses.*

Laing, D. G., & Willcox, M. E. (1983). Perception of components in binary odour mixtures. *Chemical Senses* **7**, 249–264.

Laing, D. G., & Willcox, M. E. (1987). An investigation of the mechanisms of odor suppression using physical and dichorhinic mixtures. *Behavioural Brain Research* **26**, 79–87.

Laing, D. G., Panhuber, H., & Baxter, R. I. (1978). Olfactory properties of amines and n-butanol. *Chemical Senses & Flavour* **3**, 149–166.

Laing, D. G., Panhuber, H., Willcox, M. E., & Pittman, E. A. (1984). Quality and intensity of binary odor mixtures. *Physiology & Behavior* **33**, 309–319.

Mackay-Sim, A., Shaman, P., & Moulton, D. G. (1982). Topographic coding of olfactory quality: odorant-specific patterns of epithelial responsivity in the salamander. *Journal of Neurophysiology* **48**, 584–596.

Mozell, M. M. (1964). Evidence for sorption as a mechanism of the olfactory analysis of vapours. *Nature* **203**, 1181–1182.

Mozell, M. M. (1970). Evidence for a chromatographic model of olfaction. *Journal of General Physiology* **56**, 46–63.

Mozell, M. M., & Jagodowicz, M. (1973). Chromatographic separation of odorants by the nose: retention times measured across *in vivo* olfactory mucosa. *Science* **181**, 1247–1249.

12

Models for describing intensity interactions in odor mixtures: a reappraisal

P. Laffort

Laboratoire de Physiologie de la Chimioréception (UA 1190),
Centre National de la Recherche Scientifique, Gif-sur-Yvette, France

I. Introduction

The first studies on olfactory interactions were carried out at the same time as early experiments into olfaction by Passy (1895), Backman (1918) and Zwaardemaker (1907, 1925). From the beginning, models appeared to be necessary in olfaction, and more generally speaking, in all chemical senses, since the phenomena, unlike those encountered in general pharmacology are not clearly understandable.

One of the many examples of this difficulty is provided by a psychophysical experiment reported by Cain and Drexler (1974) on binary mixtures of pyridine and several pleasant second components. Different situations were simultaneously observed for a given pair of odorants. For some pairs of odorants, the overall perceived intensity was smaller than the perceived intensity of one of the two components: this situation is often called subtraction (Berglund, Berglund & Lindvall, 1976). For other pairs of odorants, the situation was that of the 'strongest component model', that is, an absence of the influence of the smallest component. Finally, on the other cases, the classical situation of hypo-addition was observed: the total perceived intensity was smaller than the sum of the perceived intensities of the components and greater than each one taken separately. The experimental data of Cain and Drexler (1974) are reported in Figure 12.1 (first comment); the other data in this diagram will be discussed below.

PERCEPTION OF COMPLEX SMELLS
AND TASTES ISBN 0 12 042990 X

Copyright © 1989 by Academic Press Australia.
All rights of reproduction in any form reserved.

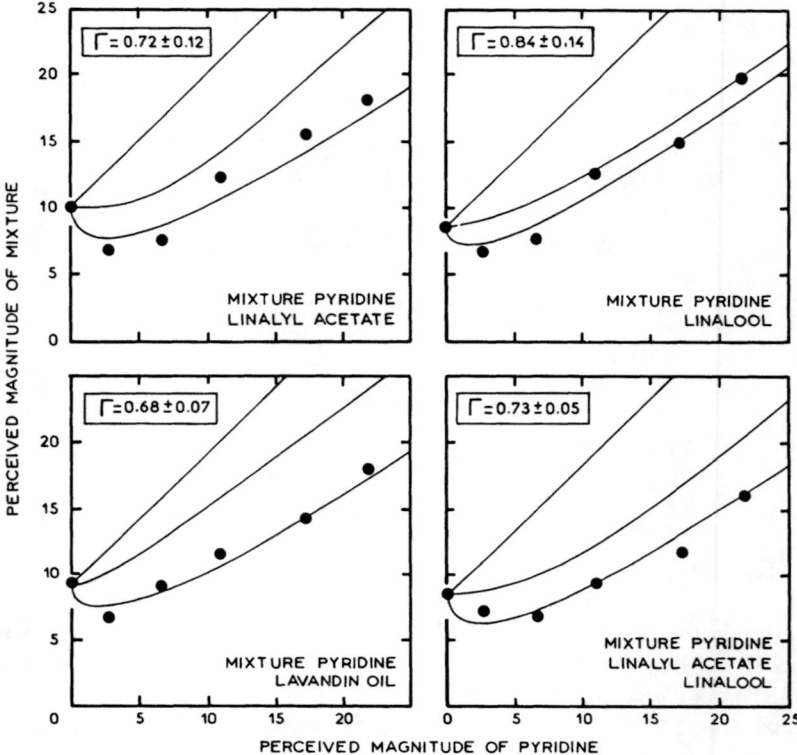

Fig. 12.1. Experimental psychophysical data from Cain and Drexler (1974). Three theoretical curves have been drawn, corresponding to three models: additivity on the top, U at the bottom, and UPL2 in the middle.

Comments

1. Experimental points exhibit a phenomenon of subtraction on the left of each diagram and a phenomenon of hypo-addition on the right.
2. The U model curve (as well as the curve of the vector model, not represented here) is in agreement with the above observation.
3. Γ reflects the distance between the two lower curves and $\cos \alpha_U$ reflects the distance between the two extreme curves.
4. Γ values of less than 1, reflecting a *true* inhibition, are clearly observed in at least three diagrams.

Other similar findings were made in electrophysiological experiments, which are too complex to describe here (Etcheto, Pichon & Laffort, 1982). The phenomenon of hypo-addition is not very informative by itself, and is also observed when a substance is added to itself.

A model must provide a single number for each pair of odorants, whatever the relative concentrations and levels of magnitude. In addition, this number must have a reference value (for example 0, 1 or 100) each time a substance is added to itself. These numbers may then be used for predictive or structure-activity purposes, for central and peripheral comparisons, and their changes along the pathways of the central nervous system can be observed. In this chapter, four historical periods will be considered, showing the progress in reaching this objective. The qualitative aspects of models of mixture interactions, studied by a few authors (Ekman, Engen, Künnapas & Lindman 1964; Laing & Willcox, 1983; Laing, Panhuber, Willcox & Pittman, 1984; Gregson, 1980, 1986), will be left aside.

II. First period: 1895–1970

Practically no progress was made into the understanding of olfactory interactions until 1971. Most of the experiments performed before this date expressed the perceived intensity of the mixtures in terms of the concentrations of the components. The following works may be mentioned in addition to those cited above:

Psychophysical experiments. Rosen, Peter & Middleton (1962); Guadagni, Buttery, Okano & Burr (1963); Baker (1964); Jones & Woskow (1964); Borelli & Angleraud (1965); Kendall & Neilson (1966); Laffort (1968); Koster (1968, 1969).

Electrophysiological experiments. MacLeod (1968). Two additional works published after 1971 may be considered to belong to this period intellectually, Köster & MacLeod (1974) and Muller (1979).

When the results were expressed in this way, a complete lack of similarity was often observed between two iso-intensity curves of binary mixtures, studied at two levels of perception. Moreover, in terms of relative concentrations, successive situations of synergy, inhibition and absence of interaction were often observed along the same iso-intensity curves. This is illustrated in Figure 12.2, from Köster (1969). In fact, for reasons explained below we call these interactions 'apparent'.

It may be concluded that during this period, the need for a synthetic view of olfactory interactions was not reached at all.

III. Second period: 1971–78

In 1971, Berglund, Berglund & Lindvall suggested that the perceived intensity of mixtures (ψ_{AB}) should be expressed as a function of the perceived intensities of the components (ψ_A and ψ_B). A regression

208 *Laffort*

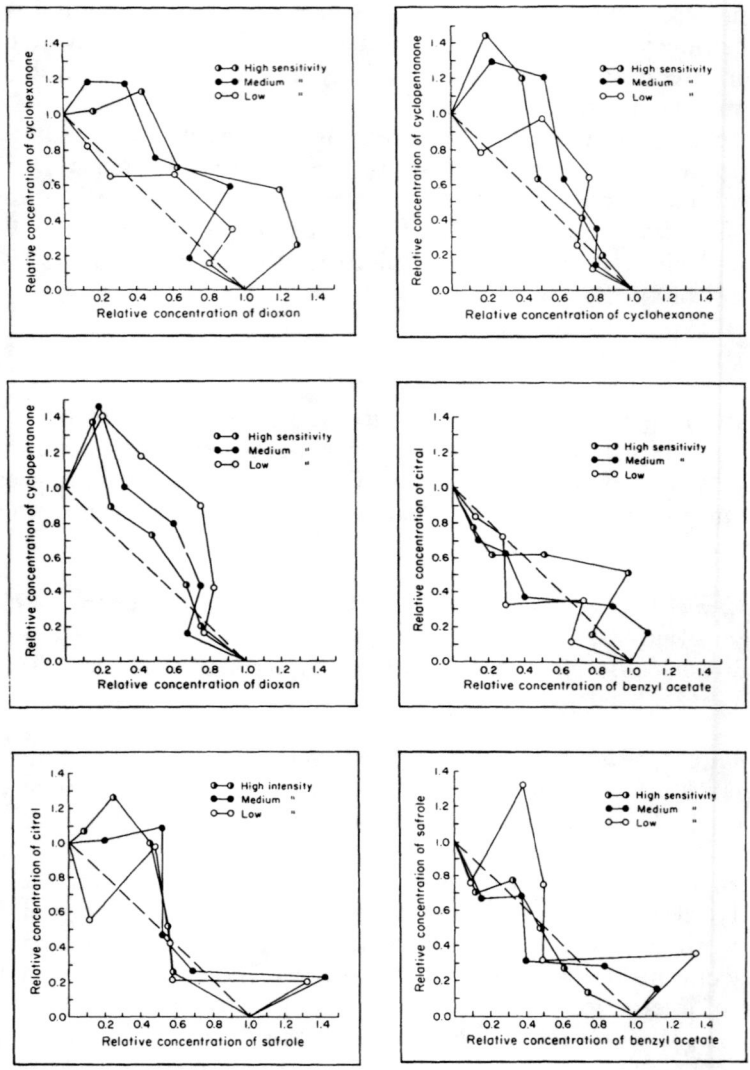

Fig. 12.2. Experimental psychophysical data from Koster (1969). The three levels 'low', 'medium' and 'high', correspond to 30%, 50% and 70% of positive responses respectively. The dotted lines are considered to be no-interaction references. The areas under these lines are synergistic, and those above these lines are inhibitory. Important changes in reciprocal relative concentrations of the components of mixtures along the iso-perceived intensity curves are observed. In fact, the interactions defined on this basis can be only *apparent* (see text). An absence of similarity between the shapes of curves from one level of perception to another is also observed.

equation linking the sensorial variables ψ_{AB}, ψ_A and ψ_B was proposed for a mixture of two sulfides.

The second really new event was put forward by Berglund, Berglund Lindvall & Svensson (1973), who suggested that the perception of odors should be treated with the same rule as that used to add vectors, like a parallelogram of forces. Under these conditions, according to the authors, a unique number (cos α), characterizes a given pair of odorants, whatever their respective concentrations and their level of intensities. The concept of an angle characterizing a given pair of substances was previously applied to describe qualitative odorous similarity by Eckman *et al.*, (1964), but it was formally discussed and applied to intensities in 1973 by Berglund *et al.*

The vector model was immediately found to be a good predictive tool (Cain & Drexler, 1974; Cain, 1975). The goodness of fit was observed, not only by using global correlations, but also when taking into account the diversity of experimental results such as those reported in Figure 12.1 (second comment). The phenomenon is explained in Figure 12.3, in which two examples of vectorial construction with the same value of α (110°) are drawn. Depending on the respective values of the component vectors ψ_A and ψ_B, the sum vector ψ_{AB} may be smaller than one of the two components (left side) or greater than both (right side). In other words,

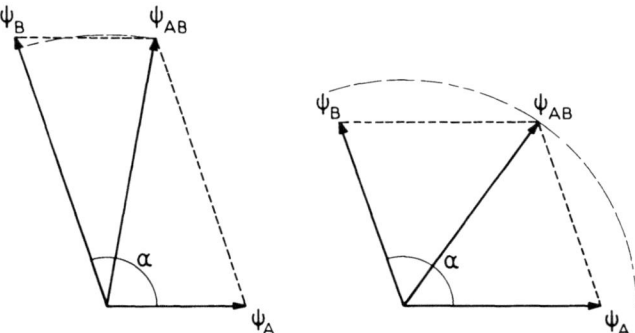

Fig. 12.3. According to Berglund *et al.* (1973) the olfactory perception of mixtures (ψ_{AB}) can be predicted by using a vector model. Here ψ_A and ψ_B vectors represent the perception of pure components *A* and *B*. In the two diagrams, with the same angle of 110°, it can be seen that the vector model may explain, according to the respective values of the vectors, both hypo-addition (on the right) and substraction (on the left), in agreement with experimental data (see second comment of Fig. 12.1).

the two apparently different situations of hypo-addition and subtraction may be expressed, at least in some cases, with the same α value of the vector model.

IV.   Third period: 1979–82

When two weak odors are added, the resulting odor will most often be weak, and when two strong odors are added, the odor mixture will generally be strong. Therefore, in many cases, all models will provide strong correlations between predicted and perceived odor strength. In order to overcome this difficulty when comparing models, two concepts were defined by Patte and Laffort (1979): σ (sigma), or *synergy*, which evaluates the deviation from additivity, and τ (tau), phonetically a French word meaning *proportion*, since it can be defined as a proportion of perceived intensity.

σ and τ are given by equations 12.1 and 12.2

$$\sigma = \frac{\psi_{AB}}{\psi_A + \psi_B} \tag{12.1}$$

$$\tau = \frac{\psi_B}{\psi_A + \psi_B} \tag{12.2}$$

At the same time, these authors suggested two possible alternative models, named V and U, and defined by equations 12.5 and 12.7. These three general models cover all the possibilities which involve only the perceived intensities of components, and particularly the most usual simple ones. For example, the strongest component model is the V model with a value of $\cos \alpha = -0.5$; the additivity model is the vector model when $\cos \alpha = 1$ or the U model with $\cos \alpha = 0$; finally, the euclidian additivity model is the vector model with $\cos \alpha = 0$. By using equations 12.1 and 12.2, the three general models can be expressed as functions of σ versus τ, as follows:

Vector model

$$\psi_{AB} = \sqrt{\psi_A^2 + \psi_B^2 + 2\psi_A \psi_B \cos \alpha} \tag{12.3}$$

$$\sigma = \sqrt{1 + 2\tau(\tau - 1)(1 - \cos \alpha)} \tag{12.4}$$

V model

$$\psi_{AB} = \psi_A + \psi_B + 2 \min (\psi_A, \psi_B) \cos \alpha \qquad (12.5)$$

$$\sigma = 1 + 2 \min [\tau, (1 - \tau)] \cos \alpha \qquad (12.6)$$

U model

$$\psi_{AB} = \psi_A + \psi_B + 2\sqrt{\psi_A \psi_B} \cos \alpha \qquad (12.7)$$

$$\sigma = 1 + 2 \sqrt{\tau (1 - \tau)} \cos \alpha \qquad (12.8)$$

The letter ψ (psi) is used by psychophysicists who developed the first models to describe perception. All these equations are of course also valid to describe physiological responses. In the latter case ψ will also stand for millivolts or number of impulses by second.

Figure 12.4 represents the three families of general models for several hypotheses of cos α. In addition, for the vector model, curves for cos α values greater than 1 (and therefore called κ) have been drawn to represent synergistic situations. The names V and U of the two new models are derived from the shape of the curves represented here; the vector model elicits curves that are intermediate between the other two curves. By comparing these theoretical curves and the available experimental data with those of Cain and Drexler (1974), and Ca n (1975), Patte and Laffort (1979) concluded that the V model was not suitable at all, and that the U model was better or equivalent to the vector model, depending on the cases, but never worse.

Further studies (Laffort & Dravnieks, 1982; Olsson, 1986) showed that the differences between U and vector models in fitting experimental data were usually negligible, with both ψ_{AB} and σ expressions. However, two difficulties encountered with the vector model must be underlined. The first concerns the confidence intervals of cos α values directly established by using equation 12.3: they are considerably greater than those obtained in establishing cos α values for the U model with equation 12.7, as pointed out by Patte and Laffort (1979) and Laffort and Dravnieks (1982). To overcome this difficulty, authors using the vector model generally limit the experimental data allowing calculation of cos α to equally strong components, the chosen criterion being : $0.8 < \psi_A/\psi_B < 1.25$ (Olsson, 1986). With this condition, equation 12.3 may be modified as follows:

$$\psi_{AB} = (\psi_A + \psi_B) \cos \alpha/2 \qquad (12.9)$$

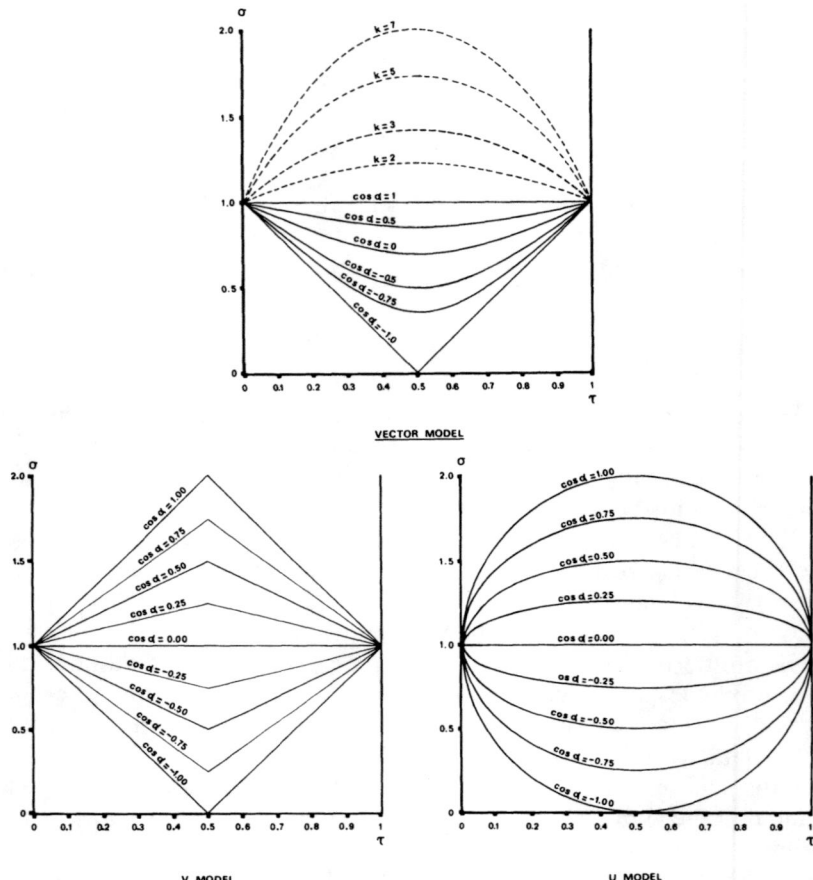

Fig. 12.4. Representation of families of curves corresponding to three models of interaction in a plot of σ (sigma) versus τ(tau), for several hypotheses of cos α or *K* values, according to equations 12.1 to 12.8 (from Patte & Laffort, 1979). Comparison of these theoretical curves with experimental data expressed in the same plot makes it possible to compare models in a more suitable way than direct comparison between predicted and experimental perceived intensities.

The second difficulty encountered with the vector model concerns the mixtures with more than two components. In 1974, Berglund suggested an extended equation of the vector model for general mixtures, and successfully applied it to experiments with ternary mixtures:

$$\psi(p) = \sqrt{\sum_{i=1}^{p} \psi i^2 + \sum_{i=1}^{p-1} \sum_{j=i+1}^{p} 2\psi i \psi j \cos \alpha_{ij}} \qquad (12.10)$$

As pointed out by Moskowitz (1979), and Laffort and Dravnieks (1982), equation 12.10 provides an irrational predicted value of ψ_{AB} (i.e. square root of a negative value), whenever the following condition is not observed:

$$\cos \alpha \geq \frac{1}{1 - p}$$

(12.11)

where p is the number of components.

For example, for mixtures with an infinite number of components, $\cos \alpha$ must not be smaller than 0, or, in other words, α must not be greater than 90°. This contradicts all currently known experimental data: α is generally equal to or greater than 110°. Experimentally, numerous irrational predicted perceived intensities were obtained by applying the extended vector model to data for ternary, quaternary and quinternary mixtures obtained by Moskowitz and Barbe (1977), as pointed out by Laffort and Dravnieks (1982).

Of course, other more suitable extensions of the vector model may perhaps be defined, other than that given by equation 12.10. However, in the meantime, the extension of the U model proposed by Laffort and Dravnieks (1982) does not present this disadvantage and fits the experimental data rather well. This extension is given by the following equations:

$$\psi_{(p)} = \frac{1}{p} \left(\sum_1^p \psi_{(p-1)} + \sum_1^p \psi_1 \right) + \frac{2 \cos \alpha \, (p)}{p} \sum_1^p \sqrt{\psi_{(p-1)^\bullet} \psi_1}$$

(12.12)

$$\cos \alpha(p) = \frac{\sum\limits_{BIN=1}^{p(p-1)/2} \cos \alpha_{(BIN)^\bullet} \psi_{(BIN)}}{\sum\limits_{BIN=1}^{p(p-1)/2} \psi_{(BIN)}}$$

(12.13)

Explanations and details concerning the relatively complicated equations 12.10, 12.12 and 12.13 can be found in Laffort and Dravnieks (1982). They are not further described here.

V. Fourth period: 1982–87

One main objection may be made about both U and vector models: They do not reflect a degree of mutual interaction, since by adding a substance to itself, $\cos \alpha$ values different from 1 with the vector model and different from zero with the U model are obtained. This is due to the general power law, commonly applied in sensory physiology to link the physical stimulus (the concentration C) and the perceived intensity ψ. When it is limited to low and medium levels, that is, below the plateau level, this power function reflects well all the stimulus-response curves obtained in both psychophysical and physiological experiments. It is expressed as follows:

$$\psi = \left(\frac{C}{C_o}\right)^n \tag{12.14}$$

where C_o and n are constants.

In olfaction, the exponent is generally smaller than 1, and therefore the rule of additivity of perception cannot be applied when a substance is added to itself. From equations 12.1, 12.2 and 12.4, the rule can be written:

$$\psi_{AA} = \left(\frac{C_A + C'_A}{C^o_A}\right)^{n_A} \tag{12.15}$$

$$\sigma = [\tau^{1/n_A} + (1 - \tau)^{1/n_A}]^{n_A} \tag{12.16}$$

Figure 12.5 shows families of σ versus τ curves, for several hypotheses of the power law exponent n. They are not unlike that obtained with the vector model in Figure 12.3. In other words, when true mixtures do not generate a horizontal line in a diagram of σ versus τ, this is not necessarily due to an interaction, but at least partially to the power function.

In order to overcome this difficulty, Laffort and Dravnieks (1982) suggested that the power law exponents and perceived intensity of the components be combined in a model named UPL (for U model and Power Law). By comparing U and UPL models, these authors were able to define a unique index of true interaction γ (gamma), which has a value of 1 when a substance is added to itself. In true mixtures a γ value equal to 1 was interpreted to show an absence of interaction, and a value different from 1, a true interaction.

Another model of olfactory mixtures proposed by Schutte (1985) also includes the power law exponents of the components. Without entering into details, the following objections may be made about this model. First, it generates two indices of interaction for any pair of odorants, instead of one. Secondly, the model is based on the assumption that the power law exponents of all odorants have the same value and this

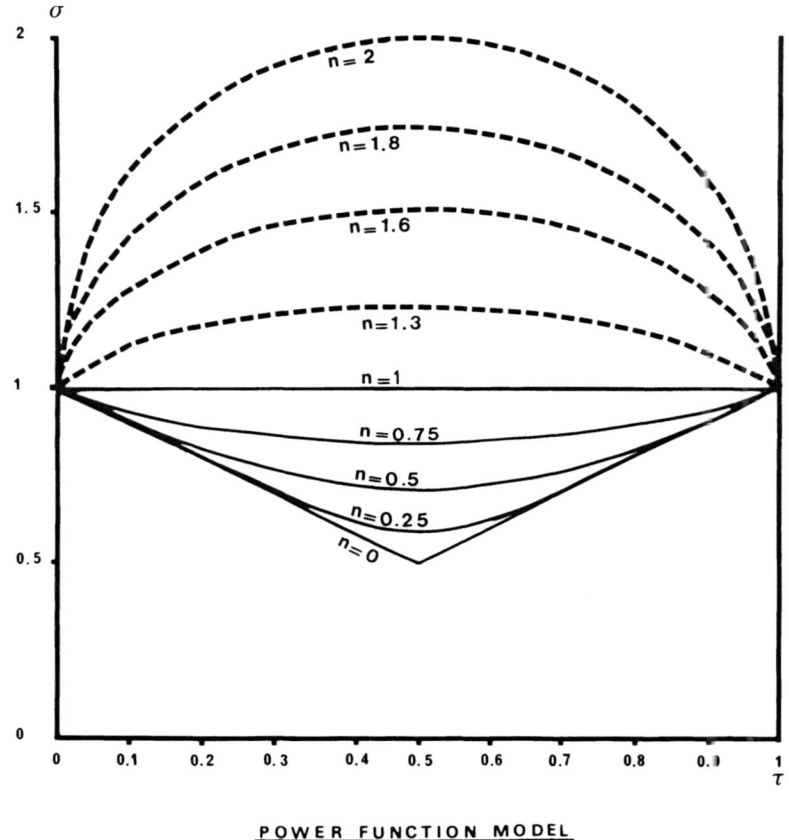

POWER FUNCTION MODEL

Fig. 12.5. Similar representation of theoretical curves as Figure 12.4, when one substance is added to itself, with the hypothesis of the power function alone, for several values of the n exponent, according to equations 12.14 to 12.16. When experimental points of true mixtures are not placed along the horizontal line (i.e. additivity), this may be due, at least in part, to the power function.

assumption is not true (a review of a large amount of experimental data was reported in Patte *et al.* 1975).

The definition of the γ index of Laffort and Dravnieks (1982) was slightly modified by Laffort, Etcheto, Patte and Marfaing (in press), and, for this reason, is now named Γ (capital gamma). In fact, because γ is a particular case of Γ for equally strong components, it seemed more suitable to keep a definition valid in all circumstances.

The Γ index is defined by the following equation:

$$\Gamma = f \left. \begin{array}{l} \nearrow \psi_{AB}, \psi_A, \psi_B \\ \searrow n_A, n_B, \psi_A, \psi_B \end{array} \right. = \frac{1 + \cos \alpha_u}{1 + \cos \alpha_{UPL2}}$$

$$\begin{array}{l} \Gamma = 1 \rightarrow \text{no interaction} \\ \Gamma > 1 \rightarrow \text{synergy} \\ \Gamma < 1 \rightarrow \text{inhibition} \end{array} \qquad (12.17)$$

where:

1. $\cos \alpha_U$ is derived from equation 12.7:

$$\cos \alpha_u = \frac{\psi_{AB} - \psi_A - \psi_B}{2\sqrt{\psi_A \psi_B}} \qquad (12.18)$$

2. $\cos \alpha_{UPL2}$ requires the prior definitions of P, $\cos \alpha_A$ and $\cos \alpha_B$ (UPL2 stands for second version of UPL model).

$$P = \frac{\psi_B^{1/n_B}}{\psi_A^{1/n_A} + \psi_B^{1/n_B}} \qquad (12.19)$$

$$\cos \alpha_A = \frac{1 - P^{n_A} - (1 - P)^{n_A}}{2 \, P^{n_A/2} * (1 - P)^{n_A/2}} \qquad (12.20)$$

$$\cos \alpha_A = \frac{1 - P^{n_A} - (1 - P)^{n_A}}{2 \, P^{n_A/2} * (1 - P)^{n_A/2}} \qquad (12.21)$$

$$\cos \alpha_B = \frac{1 - P^{n_B} - (1 - P)^{n_B}}{2 \, P^{n_B/2} * (1 - P)^{n_B/2}} \qquad (12.22)$$

Leaving aside the mathematical formalism of equations 12.18 to 12.22, it is possible to get to the core of the principle of the Γ index by using only its simplified definition in equation 12.17, and by illustrating it with concrete examples:

1. In Figure 12.1, which plots the experimental data of Cain and Drexler (1974) cited above, three theoretical curves representing three models have been drawn: the best, the U model; the worst, the additivity model; and the UPL2 model, intermediate between the other two. It can clearly be seen that the Γ index represents, as it were, a distance between the U and UPL2 model, and that cos α_U represents the distance between the two extreme curves (third comment in Figure 12.1).

2. Values of Γ for several pairs of odorants can be established from global electrophysiological responses. Their comparison at several levels of integration in a given species of animal may provide information about the site of the interaction. At the moment, only values obtained by electroantennograms in the honeybee have been published by Etcheto *et al.*, (1982) and Laffort *et al.*, (1987). These authors found Γ values of 1 for 12 out of the 18 pairs of odorants studied. Because simultaneous depolarization and hyperpolarization potentials were observed only for the six remaining pairs, Laffort *et al.*, (in press) suggested that a mutual electrical neutralization of positive and negative potentials may occur in the global measurements, but not in the information transmitted to the brain. These authors concluded that there is probably no true olfactory interaction at the periphery in the honeybee. This hypothesis could be verified by single unit recording.

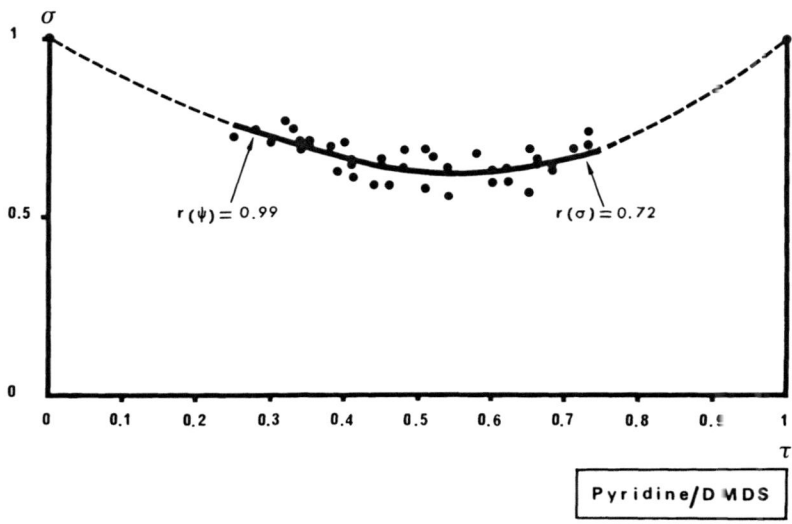

Fig. 12.6. Representation in a plot of σ versus τ of 41 experimental psychophysical data reported by Olsson (1986). The solid curve was drawn using the index models (equations 12.17 to 12.22); the dotted curves have been extrapolated. The theoretical curve, as well as the distribution of the points, are dissymmetric, unlike the U or vector model curves. The data fit is therefore significantly better when using the Γ index than the U or vector models.

3. The Γ index is also useful, as its value for a pair of odorants in a given experiment is more constant that the corresponding cos α value (from the U or vector model). This fact is illustrated in Figure 12.6, which gives the experimental psychophysical responses obtained by Olsson (1986) for the pyridine-dimethyl disulfide (DMDS) mixture, in a plot of σ versus τ. The 41 experimental points are clearly distributed dissymmetrcially. A symmetrical curve would be obtained by using any one of the models of Figure 12.4. The solid dissymmetrical line drawn in Figure 12.6 was obtained by using the hypothesis that the Γ index is constant.

In practice, Γ may be calculated by applying equations 12.17 to 12.22 to the 41 points. A value of 0.94 ± 0.06 is obtained. Then, for each point, the particular value of cos α_{UPL2} and the value of 0.94 for Γ produces a predicted value of cos α_U (Equation 12.17), from which a predicted value of ψ_{AB} (equation 12.7) and σ_{AB} (equation 12.1) may be derived. The solid line is formed with the 41 predicted values of ψ_{AB}. The correlation coefficients between the predicted and experimental values of ψ_{AB} are equivalent when using the U model, vector model or Γ index, that is 0.98 or 0.99. On the contrary, the correlations between the predicted and experimental σ_{AB}, give a value of $r = 0.72$ when using the Γ index, and only $r = 0.61$ with the U or vector models. It may be noted that if $\Gamma = 1$ instead of 0.94, the results are almost identical in the present case (r for $\sigma_{AB} = 0.71$).

4. As pointed out by several authors (Cain 1975; Moskowitz & Barbe, 1977; Laffort & Dravnieks, 1982; Schutte 1985; Gregson 1986; Laffort *et al.*, (in press), apparent synergies and inhibitions can result from values of the power law exponent different from 1, when the perceived intensities of mixtures are expressed as functions of the concentrations of components. This phenomenon, according to Gregson (1986) has been known in other sensory phenomena for more than a century and is called the 'Fechner paradox'. Laffort *et al.*, (in press) have drawn several families of curves, by using computer simulations, with the hypotheses of a true partial inhibition ($\Gamma = 0.75$) and of an absence of a true interaction ($\Gamma = 1$). Four families of curves with the latter hypothesis are illustrated in Figure 12.7. A wide variety of shapes are observed and some are not unlike that of Figure 12.2 (from Koster, 1969). The comparison of these two figures clearly shows why before 1971 it was so difficult to characterize the interaction phenomenon when this was expressed in terms of the concentrations of the components.

5. When the power law exponent is close to 1, as often happens in taste, the equations used to calculate Γ are considerably simplified. Laffort *et*

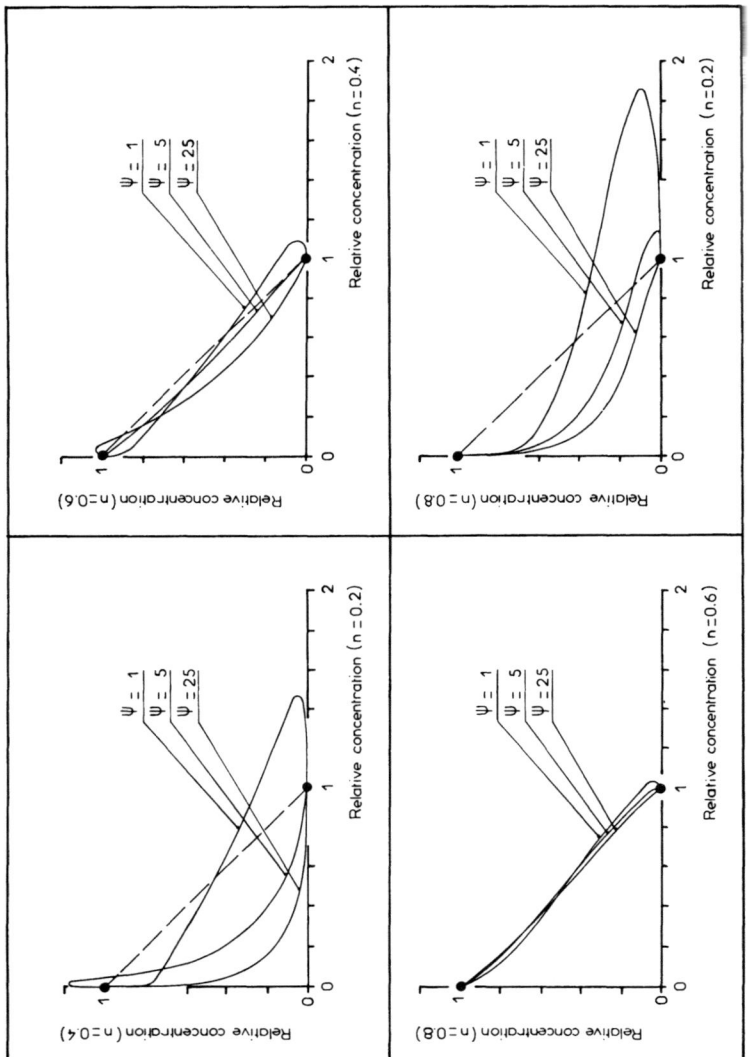

Fig. 12.7. Theoretical olfactory isointensity curves for binary mixtures, obtained by computer simulation, with the hypothesis that $\varGamma = 1$. The three levels of perceived intensity correspond to one, five and 25 times the perceived intensity at threshold. The dotted line is obtained by adding a substance to itself. The shape of the curves depends on the values and on the difference of values of the two power law exponents, n. The curves are similar to some experimental curves in Figure 12.2 (From Laffort *et al.*, in press).

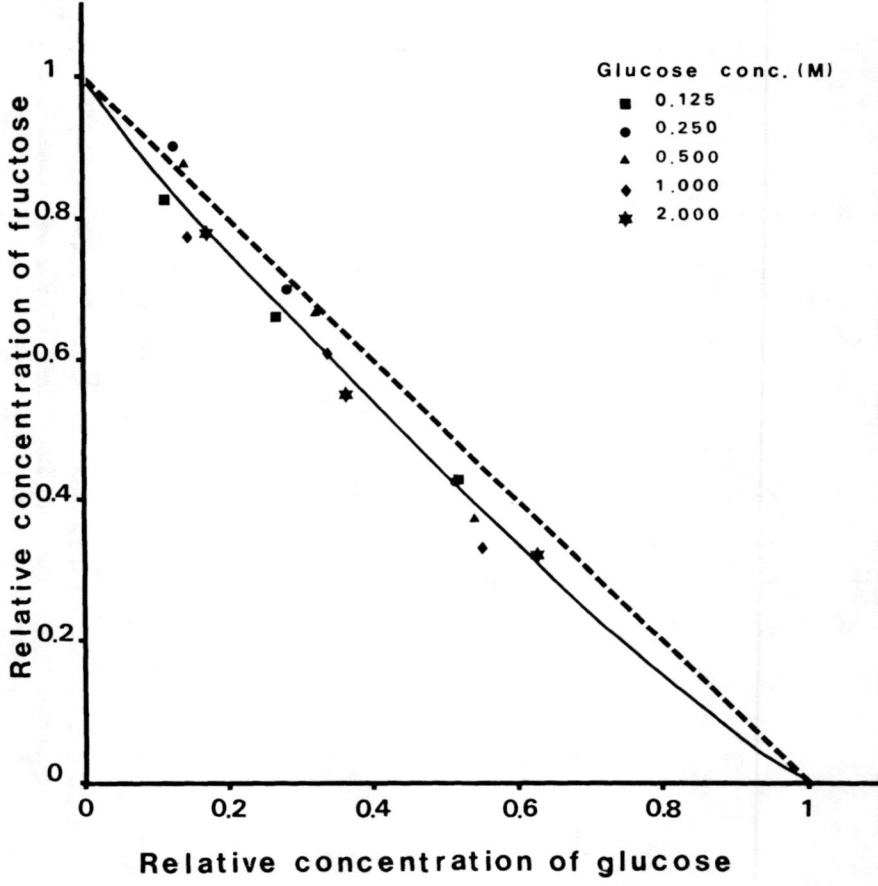

Fig. 12.8. Isointensity plots for sweet taste psychophysical responses to glucose–fructose mixtures at five levels of perception (De Graaf & Frijters, 1986). All points are placed on the same curve, in contrast to the olfactory points of Figure 12.2. This fact reflects that the power law exponents of fructose and glucose are similar and probably close to 1. On the other hand, the very slight true synergy found by calculation ($\Gamma = 1.06$) is visualized here by the proximity between the curve and the diagonal straight line.

al., (in press) showed that, in this case, $\cos \alpha_{UPL2} = 0$. On the other hand, for iso-intensitive curves, Γ can be expressed by using the following equation:

$$\Gamma = 1 + \frac{1 - C_A/C_A^0 - C_B/C_B^0}{2 \sqrt{CA/C_A^0 * CB/C_B^0}} \tag{12.23}$$

Equation 12.23 has been applied to psychophysical responses to the fructose–glucose mixture (De Graaf & Frijters, 1986). A value of $\Gamma = 1.06$ is obtained, i.e. a slight synergy. This has never yet been obtained in olfaction. The problem is open to consider if the Γ value is significantly different from 1 or not: the answer is *yes* if the chosen criterion is the standard error of mean (0.01), and it is *no* if the chosen criterion is the standard deviation (0.05). The statistical rules for this type of result are not completely clear. Figure 12.8 shows the 15 experimental points of De Graaf and Frijters (1986) and the two curves corresponding to both hypotheses: a dotted line for $\Gamma = 1$ and a solid line for $\Gamma = 1.06$. The considerable difference between the curves obtained in this figure and those obtained in olfactory investigations should be pointed out (Figure 12.2). The fact that for the fructose–glucose mixture the experimental points of five different levels of perceived intensity are all placed on the same curve seems to confirm, as suggested by Figure 12.7, that the assumption of an exponent equal to 1 in this case is very probably justified. It may also be noted, as demonstrated by Laffort *et al.*, (in press), that in this case of iso-intensity curves for taste, the model suggested by Beidler (1971) is a particular case of equation 12.23, with $\Gamma = 1$. This example shows that the Γ index makes useful predictions when applied to taste.

VI. Conclusion

The aim of a useful model for olfactory mixtures, as described in the introduction, has apparently been reached: a unique number can describe, in a given type of experiment, the olfactory interaction for each pair of odorants. This model may also be applied to taste. The different steps taken to reach this result have been made progressively over the last sixteen years. Progress in understanding sensory mixtures will probably now be made by accumulating experimental Γ index values.

References

Backman, E. L. (1918). The olfactology of the methyl-benzol series. *Onderzoekingen gedaan in het Physiologisch laboratorium der Rijksuniversiteit te Utrecht* 5, 349–364.

Baker, R. A. (1964). Response parameters including synergism-antagonism in aqueous odor measurement. *Annals of the New York Academy of Sciences* 116, 495–503.

Beidler, L. M. (1971). Taste-receptor stimulation with salts and acids. In L. M. Beidler, ed., *Handbook of sensory physiology*, vol. 4, part 2, 'Taste', pp. 200–220. Springer-Verlag, New York.

Berglund, B. (1974). Quantitative and qualitative analysis of industrial odors with human observers. *Annals of the New York Academy of Sciences* 237, 35–51.

Berglund, B., Berglund, U., & Lindvall, T. (1971). On the principle of odor interaction. *Acta Psychologica* 35, 255–268.

Berglund, B., Berglund, U., & Lindvall, T. (1976). Psychological processing of odor mixtures. *Psychological Reviews* 83, 432–441.

Berglund, B., Berglund, U., Lindvall, T., & Svensson, L. T. (1973) A quantitative principle of perceived intensity summation in odor mixtures. *Journal of Experimental Psychology* 100, 29–38.

Borelli, F., & Angleraud, O. (1965). Studi practici sull'odorizzazione intensita di odore di differenti gas. *Gas* (Roma), suppl. 7–8, 27pp.

Cain, W. S. (1975). Odor intensity: mixtures and masking. *Chemical Senses and Flavour* 1, 339–352.

Cain, W. S., & Drexler, M. (1974). Scope and evaluation of odor counteraction and masking. *Annals of the New York Academy of Sciences* 237, 427–439.

De Graaf, C., & Frijters, J. E. R. (1986). A psychophysical investigation of Beidler's mixture equation. *Chemical Senses* 11, 295–314.

Ekman, G., Engen, T., Künnapas, T., & Lindman, R. (1964). A quantitative principle of qualitative similarity. *Journal of Experimental Psychology* 68, 530–536.

Etcheto, M., Pichon, Y., & Laffort, P. (1982). Sensibilité antennaire de l'abeille a des mélanges binaires de substances odorantes. *Journal de Physiologie* (Paris) 78, 266–269.

Gregson, R. A. M. (1980). A model of paradoxical odour mixture perception. *Chemical Senses and Flavour* 5, 257–269.

Gregson, R. A. M. (1986). Quantitative and qualitative intensity components of odour mixtures. *Chemical Senses and Flavour* 11, 455–470.

Guadagni, D. G., Buttery, R. G., Okano, S., & Burr, H. K. (1963). Additive effect of sub-threshold concentrations of some organic compounds associated with food aromas. *Nature* 200, 1288–1289.

Jones, F. N., & Woskow, M. H. (1964). On the intensity of odor mixtures. *Annals of the New York Academy of Sciences* 116, 484–494.

Kendall, D. A., & Neilson, A. J. (1966). Sensory and chromatographic analysis of mixtures formulated from pure odorants. *Journal of Food Science* 31, 268–274.

Köster, E. P. (1968). Relative intensity of odour mixtures at supra-threshold level. *Olfactologia (Supplement Cahiers Oto-Rhino-Laryngologie III)* 1, 29–41.

Köster, E. P. (1969). Intensity in mixtures of odorous substances. In C. Pfaffmann, ed., *Olfaction and taste III*, pp. 142–149. The Rockefeller University Press, New York.

Köster, E. P., & Macleod, P. (1975). Psychophysical and electrophysiological experiments with binary mixtures of acetophenone and eugenol. In D. G. Moulton, A. Turk & J. W. Johnston Jr, eds, *Methods in olfactory research*, pp. 431–444. The Rockefeller University Press, New York.

Laffort, P. (1968). Interactions quantitatives dans un mélange d'odeurs: niveau liminaire. *Olfactologia (Supplement Cahiers Oto-Rhino-Laryngologie III)* 1, 95–104.

Laffort, P., & Dravnieks, A. (1982). Several models of suprathreshold quantitative olfactory interaction in humans applied to binary, ternary and quaternary mixtures. *Chemical Senses* 7, 153–174.

Laffort, P., Etcheto, M., Patte, F., & Marfaing, P. (in press). Implications of power law exponent in synergy and inhibition of olfactory mixtures. *Chemical Senses*.

Laing, D.G., & Willcox, M. E. (1983). Perception of components in binary odor mixtures. *Chemical Senses* 7, 249–264.

Laing, D. G., Panhuber, H., Willcox, M. E., & Pittman, E. A. (1984). Quality and intensity of binary odor mixtures. *Physiology & Behavior* 33, 309–319.

Macleod, P. (1968). Interactions quantitatives dans un mélange d'odeurs. Etude électrophysiologique. *Olfactologia, (Supplement Cahiers Oto-Rhino-Laryngologic III)* 1, 23–27.

Moskowitz, H. R. (1979). Utility of the vector model for higher-order mixtures: a correction. *Sensory Processes* 3, 366–369.

Moskowitz, H. R. & Barbe, C. D. (1977). Profiling of odor components and their mixtures. *Sensory Processes* 1, 212–226.

Muller, W. (1979). Effectiveness of mixtures of odorants in the frog's olfactory receptors. *Chemoreception Abstracts* 7, 953.

Olsson, M. (1986). *Perceptual models of odor interaction for binary mixtures.* Report no. 8/1986. National Institute of Environmental Medicine, Stockholm.

Passy, J. (1895). Revue générale sur les sensations olfactives. *Année Psychologique* 2, 363–410.

Patte, F., & Laffort, P. (1979). An alternative model of olfactory quantitative interaction in binary mixtures. *Chemical Senses and Flavour* 4, 267–274.

Patte, F., Etcheto, M., & Laffort, P. (1975). Selected and standardised values of suprathreshold odor intensities for 110 substances. *Chemical Senses and Flavour* 1, 283–305.

Rosen, A. A., Peter, J. B., & Middleton, F. M. (1962). Odor thresholds of mixed organic chemicals. *Journal of Water Pollution Control Federation* 34, 7–14.

Schutte, L. (1985). A new model for describing interactions in odour mixtures. In J. Adda, ed., *Progress in Flavour Research* pp. 3–13. Elsevier, Amsterdam.

Zwaardemaker, H. (1907). Uber die proportion der geruchskompensation. *Archiv für Anatomie und Physiologie* (Leipzig) 31, suppl. 59–70.

Zwaardemaker, H. (1925). *L'Odorat.* G. Doin, Paris.

13

Is taste mixture suppression a peripheral or central event?

Jan H. A. Kroeze

Psychological Laboratory,
Utrecht University, The Netherlands

I. Introduction

Tastants contained in a mixture often evoke responses different from those elicited when the tastants are presented unmixed. This is apparent from mixture suppression and mixture enhancement, both being phenomena of a quantitative nature. Besides intensity differences, subjects may perceive quality differences between mixed and unmixed stimuli (Moskowitz, 1972; Kuznicki & Ashbaugh, 1979, 1982).

Some basic concepts

In this chapter, mixture suppression and adaptation will be mentioned frequently. Therefore, the meaning of these terms will be discussed first.

1. Mixture suppression

Suppose subjects estimate the sweetness of a 0.32 M sucrose solution a number of times. From these estimates a sweetness value can be calculated. In another experimental treatment, the taste solution contains sodium chloride as a second substance. From the estimates of this mixture, a mean sweetness value is also calculated. In most cases this mixture sweetness value is significantly lower than the first value. Apparently, the salt in the sugar solution depresses sweetness. This phenomenon, suppression of the intensity of a certain taste quality in the presence of one or more other taste substances, is well documented and called *taste mixture suppression* (Fabian & Blum, 1943; Beebe-Center, Rogers, Atkinson & O'Connell, 1959; Sjöström & Cairncross, 1953;

Copyright © 1989 by Academic Press Australia.
All rights of reproduction in any form reserved.

Kamen, Pilgrim, Gutman & Kroll, 1961; Moskowitz, 1972; Pangborn, 1960, 1961, 1962; Pangborn & Trabue, 1967; Bartoshuk, 1975). Alternatively, subjects may estimate the overall intensity of a mixture. In this way mixture suppression may also be demonstrated. However, a problem with such a definition is how to combine the sensations of the unmixed substances. Should the mixture estimates be compared with the simple sum of the separate estimates? Or would it be more appropriate to combine the separate sensations in some non-linear way? In other words, what does the term *independent* mean? It would not be considered correct, at least with non-linear psychophysical functions, to define independence as simple summation. The definition of the reference value (with which the mixture value must be compared) already implies a certain combinatory theory.

We will restrict ourselves mainly to mixture experiments in which subjects judge the intensity of a previously specified quality. Furthermore, the taste stimuli used were single, dominant taste qualities with either very weak or negligible side tastes. Sucrose sweetness and sodium chloride saltiness are such pure tastes. Quinine sulfate and quinine-hydrochloride are well-known, purely bitter-tasting substances. Citric acid is often used as a purely sour stimulus.

Mixture suppression does not always occur. In about 20% of subjects, no mixture suppression occurred in our experiments. In some subjects, mixture suppression occurred in one session but not in another. Some of the subjects displayed mixture enhancement rather than mixture suppression. However, it could be shown that enhancement in these cases was related to side tastes. For example, subjects added the bitter side taste of a salty mixture component to the bitter component (Kroeze, 1982*d*). This is not to say that all cases of mixture enhancement result from side tastes. For example, enhancement in mixtures with mono-sodium glutamate is not related to side taste and probably occurs at the receptor level (Yamaguchi, 1967; Yamaguchi, Yoshikawa, Ikeda & Ninomiya, 1968). Furthermore, mixture enhancement may occur with near-threshold stimuli (Kroeze, 1977), but its extent and nature need further clarification.

2. Adaptation

Another way of reducing the perceived intensity of a taste stimulus is to adapt the taste system. Psychophysicists define taste adaptation as the gradual decline in the detectability or reported sensation intensity of a taste stimulus when it is applied continuously to the taste sensitive tissue. When the test stimulus to be detected or estimated after continuous exposure to an inducing stimulus is identical to the inducing (exposure) stimulus, the resulting adaptation is called *self-adaptation*. When the adapting stimulus and test stimulus are different, the resulting adaptation is called *cross-adaptation*. The latter case can be subdivided into cross-adaptation *within* the same quality (e.g. do different salty tasting salts adapt one another?) and *between* different qualities (e.g. does exposure to sucrose induce adaptation to sodium chloride?)

Throughout the taste literature, four methods of assessing taste adaptation can be distinguished.

The threshold method. In this method absolute sensitivity is measured immediately after exposure to the adapting stimulus.

The subjective intensity method. The subjective intensity of a stimulus is measured immediately after exposure to the adapting stimulus.

The point-of-disappearance method. Subjects indicate when, during continuous stimulation, the taste sensation disappears. This procedure provides the researcher with adaptation times only. Adaptation time is the elapsed time between the beginning of stimulation and the indicated disappearance of the sensation (Abrahams, Krakauer & Dallenbach, 1937; Krakauer & Dallenbach, 1937).

The course-of-adaptation method. During one and the same stimulus exposure, subjects repeatedly estimate the sensation strength (Meiselman, 1968; Gent & McBurney, 1978; Ganzevles & Kroeze, 1987), or, with an adjustable device, track the sensation continuously (Yoshida, 1986; Overbosch, 1986; Overbosch, Van den Ende & Keur, 1986).

3. Locus of adaptation
Combined psychophysical and electrophysiological measurements in humans (Diamant, Oakley, Ström & Zotterman, 1965; Borg, Diamant, Oakley, Ström & Zotterman, 1967a; Borg, Diamant, Ström & Zotterman, 1967b) indicate that *chorda tympani* responses to taste stimuli correspond to psychophysical judgments of the same stimuli. Response decrease during prolonged NaCl-stimulation appeared to be similar with whole-nerve recordings and magnitude estimation. Referring to these results, Zotterman (1971) stated, 'Thus we may conclude that the human psychophysical observation of rapid and complete salt adaptation can be accounted for by diminished activity in the *chorda tympani* nerve. There is no need to postulate the existence of a central adaptation mechanism'. It should be noted that Zotterman does not say that the adaptation mechanism is located in the first-order neuron. It may be even more

peripheral (i.e. in the synapse or at the receptor cell level). In an experiment with frogs, Sato (1971) found that the course of adaptation in the first-order neuron (*glossopharyngal* nerve) was much steeper than in the receptor cell. He therefore suggested (Sato, 1971, 1972) that the greater degree of adaptation seen in the glossopharyngal nerve is due to a mechanism in the first-order neuron or at the synaptic level. However, Sato's result may be explained in a different way: if receptor potential and first-order neuron activity are related by an accelerated function, then precisely such a result may be expected.

II.   Is mixture suppression central or peripheral?

The question of where mixture suppression is located in the taste system cannot be answered simply by 'peripheral' or 'central'. The peripheral-central stereotype may unintentionally lead to conceptualization of the taste system as a receptor structure and an awareness structure, connected by a passive afferent nerve bundle. So rather than asking what is going on centrally or peripherally, one should ask: *what* happens *where* in the continuous pathway between impact site and sensation (or response) site? Taking time into consideration, the question may be rephrased as 'what happens *when*? The commonsense understanding of temporal and spatial characteristics is that they correspond (i.e. early events are more peripheral than late events). However, this view may be misleading. It is quite conceivable that a peripheral event occurs later than a central event. This insight is expressed in the concept of reafferentiation (Holst & Mittelsteadt, 1950). In particular, during development, active perception results in perceptual abilities different from the abilities of individuals who had only the opportunity of passive perception of stimuli (Held & Hein, 1963). Feedback channels in perceptual systems, such as those involved in accommodation and convergence in vision, may not just be a convenient extra, but may be a characteristic without which perception would not be possible. Thus 'peripheral' may not necessarily mean 'early'. A similar concept is 'top-down processing' (Lindsay & Norman, 1977). Stimuli not only hit the organism; depending upon internal state (Cabanac & Duclaux, 1970; Cabanac, 1979), subjects also actively seek stimuli and, in so doing, display selective attention. The hypothesis that there are peripheral (late!) sensory changes which are centrally induced cannot be ruled out.

Neither can taste perception be understood solely on the basis of stimulus-receptor interaction, or on the basis of the action potential. Psychophysical and other behavioral data are needed to improve understanding of central taste mechanisms. A productive strategy is to combine, in a controlled way, separate stimulus attributes (or simple stimuli) into complex stimuli. This strategy of 'controlled complexity' may be achieved using several different tactics.

A.   Adaptation to a taste mixture before tasting one of its components

The following experiment (Kroeze, 1978) was carried out to determine the relative location of mixture suppression in the taste system. This experiment made it possible to specify the order in which adaptation and mixture suppression took place. It consisted of five conditions, specified in Figure 13.1A. In each condition the subjects estimated the saltiness of the stimuli placed on their tongues, by moving a needle along an ungraduated linear scale, 300 mm long. The needle was reset to zero by the experimenter after each trial. The stimuli were a NaCl–sucrose mixture and its two components. These stimuli were applied to a restricted area of the left half of the anterior tongue. In order to ensure that the same area was repeatedly stimulated, the tongue was inserted in a tongue box that allowed exposure of an area 1.5 to 1.8 cm^2. The stimuli were presented by a gravitational flow system, built from glass reservoirs and silicone tubing. The outflow rate was 0.8 mL/s and the stimulus

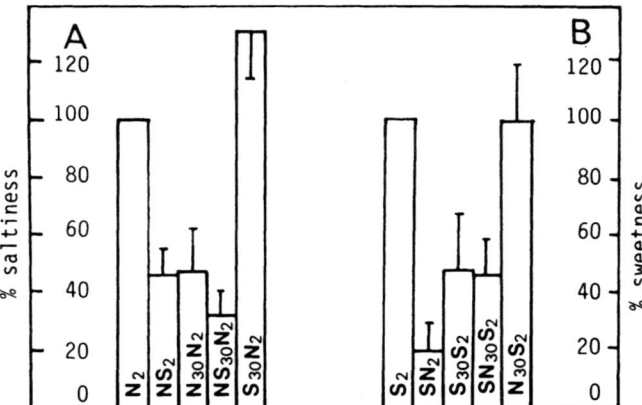

Fig. 13.1. Two experiments (A and B). In A, subjects estimated saltiness; in B they estimated sweetness. N_2: unmixed 0.32 M NaCl stimulus presented during 2 s. NS_2: mixture of NaCl and sucrose applied for 2 s. $N_{30}N_2$: A 2 s NaCl stimulus estimated after an adapting NaCl stimulus of 30 s. $NS_{30}N_2$: estimation of a 2 s NaCl stimulus after adaptation to the mixture for 30 s. $S_{30}N_2$: saltiness of a 2 s NaCl stimulus after 30 s of adaptation to 0.32 M sucrose. B is the exact reverse of A.

duration 2 s, except in the conditions with a continuous adapting stimulus. The adapting stimuli lasted for 30 s. The interstimulus intervals were 60 s with the exception of the zero intervals between a test stimulus and its preceding adapting stimulus. An interval of 60 s with a rinse is sufficient to recover from adaptation (Hahn, 1934; Kroeze, 1982c). Before each trial the tongue was rinsed with distilled water.

In the crucial condition of the experiment, subjects estimated the saltiness of a NaCl test stimulus after adaptation to a NaCl-sucrose mixture ($NS_{30}N_2$ in Figure 13.1). Saltiness of the test stimulus was significantly lowered, which is not surprising since the mixture contained salt, to which the taste sense adapted. However, the meaningful outcome was that the saltiness of the test stimulus was lowered to the same degree as in the condition in which *unmixed* NaCl preceded as the adapting stimulus ($N_{30}N_2$ in Figure 13.1). Apparently self-adaptation to NaCl was not affected by the presence of sucrose in the mixture. This appeared true despite strong mixture suppression of saltiness by sucrose. Such a result is compatible only with a mixture suppression mechanism that is located centrally to the adaptation mechanism. Regression analysis showed a nearly perfect correlation ($r = -0.97$) between saltiness estimates after self-adaptation and saltiness after adaptation to the NaCl-sucrose mixture. This was taken as additional evidence that the sucrose part of the mixture did not play a role as far as adaptation was concerned. Furthermore, it appeared impossible to predict adaptation on the basis of mixture suppression ($r = -0.04$). Failure of the suppressor to 'cripple' the adapting power of the other mixture component cannot be ascribed to cross-adaptation, as there was no cross-adaptation (Figure 13.1).

Similar results were found in a second experiment (Figure 13.1B) in which sucrose, instead of NaCl, was the test stimulus, and in which sweetness was estimated (Kroeze, 1979).

The conclusions may be summarized as follows:

1. Under the experimental conditions of these experiments, adaptation and mixture suppression have different locations in the taste system, with mixture suppression more centrally located.
2. Adaptation and mixture suppression are independent (*i.e.* it is impossible to predict from a person's adaptation result his or her mixture suppression result).

The results do not support location of both the adaptation and suppression mechanisms in the first-order neuron, unless the unlikely assumption is made that both mechanisms operate independently in the same uninterrupted neural fibers. The statistical independency of the behavioral effects of the two mechanisms disallow such an assumption. Thus, if the adapting mechanism resides entirely or partially in the *chorda tympani* nerve, then mixture suppression must reside in a higher structure. If, on the other hand, mixture suppression is located in the *chorda tympani*, then adaptation must be located in a more peripheral structure.

Lawless (1982) repeated these experiments with a 4 mL/s flow on the whole tongue and found identical results in the case of NaCl-saltiness after adaptation to the NaCl-sucrose mixture. In his experiment, as in ours, NaCl in the mixture adapted the taste system to the same degree as unmixed NaCl. Therefore, he inferred, as we did, that saltiness suppression by sucrose is central to the locus of adaptation. However, sweetness was adapted a little less by the NaCl-sucrose mixture than by unmixed sucrose. The flow rate appeared crucial. When Lawless lowered the flow rate from 4 mL/s to our rate of 0.8 mL/s, the results of our experiments were exactly reproduced. Thus, besides more centrally located mixture suppression, increased flow rate revealed mixture suppression peripheral to adaptation. The fact that *flow* is important may indicate a possible role of perireceptor events. These are events not belonging to the receptor process as such, but contributing to the effectiveness with which the stimulus approaches the receptor

B. Adaptation to one of the mixture components

Instead of adapting subjects to mixtures and then testing them on one of the components, the reverse procedure can be used. Adapting the subject first to one of the components prevents mixture suppression, because the adapted mixture component is removed peripherally. Lawless (1979), Bartoshuk (1979) and Bartoshuk and Seibyl (1982) used this procedure. First Lawless adapted his subjects to sucrose and then presented a sucrose-quinine sulfate mixture. Under this condition the bitterness of quinine sulfate was not suppressed by sucrose. Bartoshuk (1979) called this undoing of mixture suppression *suppression release*. The suppressed component is, so to speak, released from suppression by removing the suppressor during the preceding adaptation period. In a second experiment, Lawless (1979) treated his subjects' tongues with *Gymnema Sylvestre*, a taste modifier that removes sugar sweetness at the receptor level. After this treatment, a sucrose-quinine sulfate mixture tasted just as bitter as unmixed quinine sulfate. Lawless concluded that the results of these experiments support neural explanations of mixture suppression.

III.   Central mixture suppression

The experiments to be discussed next assume as a working hypothesis that mixture suppression between qualitatively different stimuli of supra-threshold intensity is located in the central nervous system. If this is true, then a procedure which affects taste centrally rather than peripherally may be a better way to assess the location. The suppression-release experiments mentioned in the previous section made use of adaptation. However, as discussed above, adapting to a stimulus means that its effect is removed peripherally, (i.e. the message is not even present in the afferent pathway). Thus a suppression-release procedure in which adaptation is used gives only an indication of the location. Even the assertion that suppression release confirms the neural nature of mixture suppression (Lawless, 1979) may be questionable; the truth of that statement depends on how peripheral adaptation is.

A.   *Habituation and suppression*

Habituation is generally defined as the decrement in response to an initially novel stimulus, when it is given repeatedly (Cotman & McGaugh, 1980). Processes responsible for habituation are central rather than peripheral. Habituation must be distinguished from adaptation and effector fatigue, both of which are peripheral processes (Thompson & Spencer, 1966). Habituation has been measured physiologically as a decreasing galvanic skin response with repeated stimulations, or as a change in the size of evoked potentials over stimulus repetitions. It has also been measured behaviorally. Habituation has been known for centuries but was not named habituation: Pavlov (1927) was the first to give a more detailed account of behavioral habituation. He mentioned the example of a dog which, when first encountering a certain event (such as a person entering the room), pricks up its ears and moves its head in the direction of the stimulus. But when the same person enters the room in the same way a second time, there are fewer behavioral signs of the event registering. After a limited number of repetitions, the behavioral response has completely disappeared: the dog has habituated behaviorally to the stimulus.

Fisher and Fisher (1969) were the first to measure habituation of the non-specific electrodermal response (EDR) with repeated taste stimuli. They observed rapid habituation to sucrose and NaCl, but no habituation to quinine sulfate. The authors ascribed the habituation failure in the case of quinine sulfate to the aversive nature of this bitter stimulus.

In our experiment, instead of preceding a NaCl-sucrose mixture by an adapting sucrose stimulus, we preceded it by a repetitive sucrose

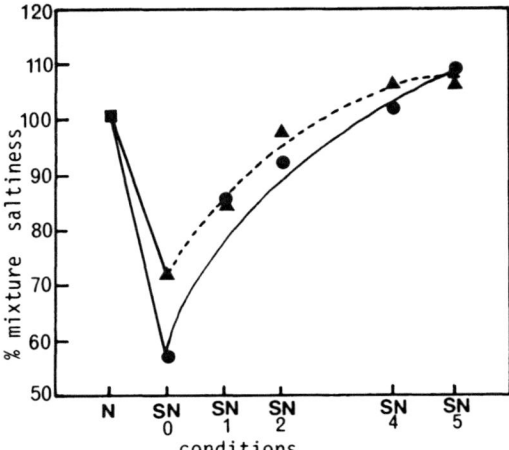

Fig. 13.2. Saltiness of an NaCl–sucrose mixture after different numbers (ind cated by the subscript) of preceding sucrose stimuli. The figure shows the suppression-release functions of two independent experiments.

stimulus (Kroeze, 1982a). It was offered for 2 s only, then stopped and followed by a rinse. Exactly 60 s later the same stimulus was offered again in the same way. Finally, after a certain number of repetitions, the mixture was offered. Subjects were instructed to estimate after each stimulus presentation the perceived saltiness. Suppression release was observed and the amount of suppression release appeared proportional to the number of preceding repetitions (Fig. 13.2). In other words, as the number of preceding sucrose stimuli increased, the amount of saltiness suppression decreased. By the fifth repetition, mixture suppression had completely disappeared.

The intervals between repetitions completely prevented the build-up of adaptation, as was shown in a separate experiment with different intervals and rinses (Kroeze, 1982c). It was thus impossible to ascribe suppression release to adaption. The reverse result was obtained as well: after repeated exposure to a salty stimulus, the sweetness of a salty-sweet mixture was no longer suppressed (Kroeze, 1983). These results were taken as evidence for a central location of mixture suppression.

B. Procedural factors and the repetition effect

Measurement of mixture suppression requires at least two different stimuli: an unmixed and a mixed stimulus. The amount of suppression is then defined as the difference between the estimates of those two stimuli. The results showed that the amount of suppression actually calculated was sensitive to the relative frequencies of mixed and unmixed stimuli in the series. Without an unmixed stimulus it is, by definition, impossible to calculate mixture suppression. The same is true when there are no mixed stimuli. The ideally balanced experiment contains an equal number of mixed and unmixed stimuli. This requirement seems reasonable, since the frequency weights of both types of stimuli exactly cancel each other. With unequal frequencies the degree of imbalance fits Helson's adaptation level principle (Kroeze, 1982b).

Another procedural effect essential to this type of habituation is the way in which subjects are instructed to respond. Habituation implies a loss of attention to a certain stimulus attribute. Thus, as soon as the subject's attention is directed towards the stimulus the subject will either not habituate or habituate much less (Van Olst, 1971; Hulstijn, unpublished doctoral dissertation, University of Nijmegen, Nijmegen (Holland), 1978; Bernstein, 1979). For example, when subjects are instructed to estimate the sweetness of a repeatedly offered sucrose stimulus the estimates will not show habituation. Subjects are not likely to habituate to the stimulus series in scaling tasks for three reasons:

1. The main stimulus attribute has relevance for the required response.
2. Stimulus intensity varies.
3. This variation is unpredictable.

In the habituation experiment described above, the stimulus attribute to which habituation was assumed to occur had no relevance for the required response. This is an essential characteristic of this type of experiment, without which it will not work. The situation could be compared with someone reading a book in a silent room with a quietly ticking clock. The person does not hear the clock unless told to count the ticks or to judge their intensity. If a stimulus is sufficiently regular and monotonous, a subject may have difficulty in paying attention to it even when instructed to do so. It may require real effort not to habituate to such a stimulus.

C.   Habituation and temporal contrast

There appear to be two ways in which a repeated stimulus can cause suppression release. These are illustrated in the diagram in Figure 13.3.

The habituation explanation (A) supposes that the repeated sucrose stimulus (S) leads to habituation of the suppressing effect of S on sodium chloride (N). The contrast explanation (B) assumes that the sweet experience built up over repetitions constitutes a background against

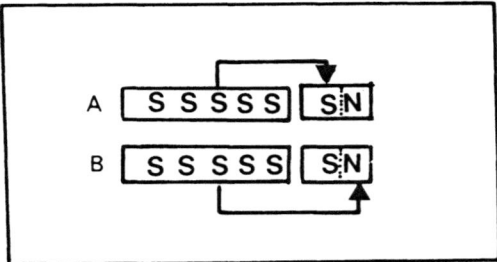

Fig. 13.3. Two possible explanations of suppression release. In A, it is assumed that the repeated presentation of sucrose (S) inhibits its suppressing effect on NaCl-saltiness. As a consequence, NaCl-saltiness is no longer suppressed. In explanation B, NaCl-saltiness is overestimated against the sweet background built up over repeated sucrose stimuli.

which N-saltiness is overestimated. In both cases, estimated saltiness will increase after repeated sucrose stimulation. The contrast explanation, however, does not require a mixture as a test stimulus. Thus we may substitute an unmixed NaCl stimulus of about the same saltiness as the mixture (i.e. lower physical concentration). Contrast, if responsible, should lead to overestimation of saltiness and thus to a suppression-release-like function. Such experiments would challenge the habituation explanation. The results of such an experiment (Kroeze, 1983) are shown in Figure 13.4.

Panel A of Figure 13.4 gives the saltiness suppression-release function with increasing number of preceding sucrose stimuli, and panel B gives the saltiness contrast after increasing numbers of sucrose stimuli. Panel C shows the pure suppression release function, obtained by subtracting per subject (N = 20) the contrast effect from the mixture effect. The results show that, even after removal of temporal contrast, a significant suppression-release effect remains. Therefore, successive contrast cannot explain suppression release. Although an experiment like this does not *prove* the correctness of the habituation hypothesis, it becomes more likely after the temporal contrast explanation has been excluded.

D. Split-tongue experiments

1. Suprathreshold mixtures
The repetition experiments suggest a central locus of mixture suppression. But there is yet another way to learn about the possible location of a mixture-suppression mechanism. The left and right halves of the tongue do not have direct neural connections with each other (Vij & Kanagasuntheram, 1972). Each half of the tongue projects ipsilaterally to

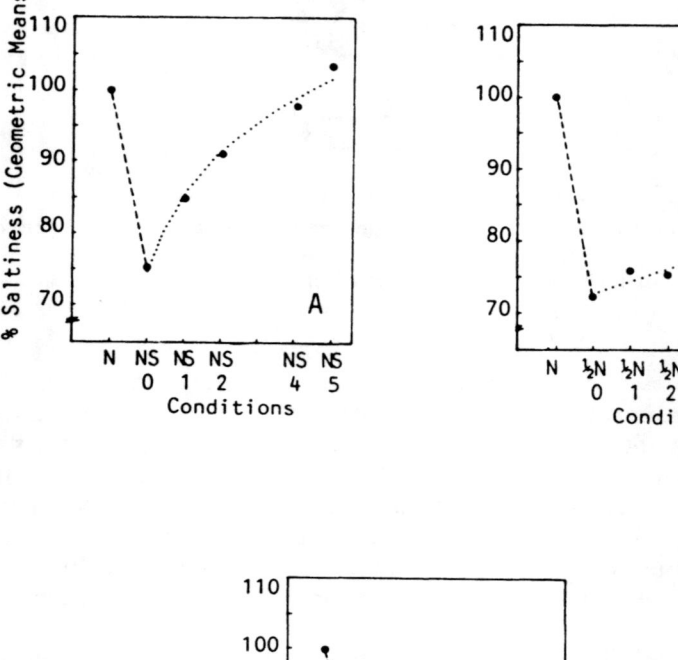

Fig. 13.4. Suppression release of saltiness (panel C) is obtained by subtracting per subject (N=20) temporal contrast (panel B) from the mixture saltiness values in panel A. The figure shows that temporal contrast cannot explain suppression release. Previous matching revealed that the group mean saltiness of the mixture (NS) was about as intense as half the NaCl concentration used in that mixture.

the central parts of the brain (Norgren, Grill & Pfaffmann, 1970; Norgren & Leonard, 1973; Rollin, 1973). One can take advantage of this structural feature and present each side of the tongue with a different mixture component. The magnitude estimates can then be compared with a condition in which the components are mixed on the tongue.

Figure 13.5 shows such an experiment (Kroeze & Bartoshuk, 1985). Stimuli were 0.32 M NaCl (N), 0.32 M sucrose (S), 0.001 M QHCl (Q), a quinineHCl-sucrose mixture (Q-S) and a mixture of quinineHCl and NaCl (Q-N). From the conditions created in this experiment, only the three relevant to the present discussion are reproduced in Figure 13.5. In the first condition, unmixed Q was presented to one half of of the tongue and the solvent (deionized water) to the contralateral side. In the second condition, the suppressor was mixed with Q and presented to the same side, while deionized water was applied to the contralateral side. In the third condition the suppressor, either sucrose (panel A) or NaCl (panel B) was presented to the contralateral side.

After each stimulus presentation, the subjects estimated the bitterness. In the case of Q-S, the amount of mixture suppression in the split-tongue condition (peripheral suppression excluded) was equal to that in the mixed condition. Moreover, the estimates from both conditions were

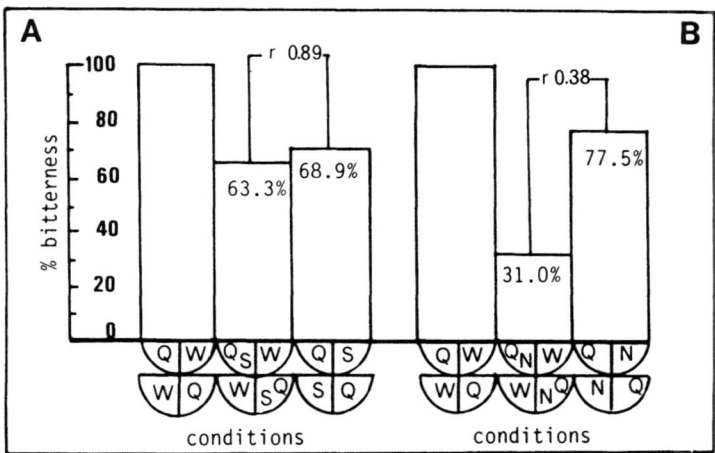

Fig. 13.5. Bitterness of quinine-sucrose and quinine–NaCl combinations in different split-tongue conditions. Diagrams of the tongue with the stimulus combinations are shown at the base of the histogram. This figure shows that bitterness of quinineHCl–sucrose mixtures is central, whereas in the case of quinineHCl–NaCl, bitterness suppression is partially central and partially peripheral.

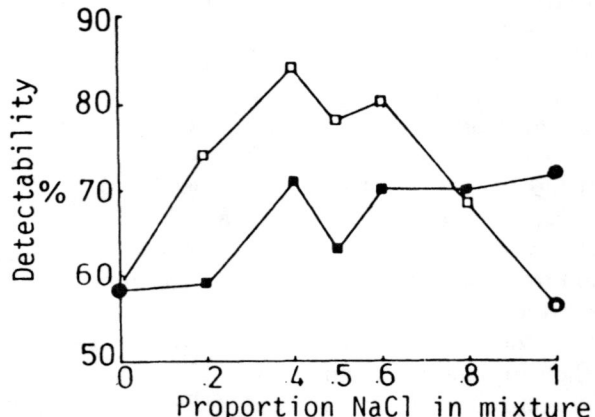

Fig. 13.6. The detectability (indicated on the vertical axis as a percentage) of a near-threshold NaCl-sucrose combination is higher with mixed components (unfilled points) than with components split across the tongue's midline (filled points). In particular, with about equal proportions of both components the effect is significant. The shape of the 'mixed' detection values suggests near-threshold enhancement of detectability.

correlated ($r = 0.89$, $n = 10$, $p < .0001$), indicating a common source. This result suggests that bitterness suppression in a supra-threshold QHCl-sucrose mixture is central. With the Q-N mixture, however, significant bitterness suppression was obtained in both the split-tongue and the mixture condition. In this case, both conditions differed significantly and were only slightly related ($r = .38$, $n = 10$, $p > .10$). It was concluded that the bitterness decrease in the Q-N mixture was the combined effect of about one-third central and two-thirds peripheral mixture suppression.

Gillan (1982), using pieces of soaked filter paper, found that ipsilaterally separated NaCl and sucrose produced the same amount of mixture suppression as compounds separated across the tongue's midline. We conclude from such results that suppression among ipsilaterally spatially separated stimuli may also be of central origin.

2. Near-threshold mixtures
In the following experiment, the split-tongue technique has been used with NaCl and sucrose stimuli of near-threshold intensity. (Kroeze, 1977). For each subject of a group, the individual detection thresholds of sucrose and NaCl were determined by the method of constant stimuli (Engen, 1971). Then, for each subject, these two threshold concentrations were used to construct a series of seven stimuli. Each series ranged from

unmixed sucrose to unmixed NaCl in the following way: (%vol S/%vol.N) 100/0, 80/20, 60/40, 50/50, 40/60, 20/80, 0/100.

Within the series the stimuli were randomized and offered to the appropriate subject in a temporal two-alternative forced choice detection task, either mixed on the tongue or separated by the medial plane. The result is depicted in Figure 13.6. These mixture components, generally showing suppression at high concentrations, displayed enhancement at threshold concentrations. The split-tongue arrangement reveals that this enhancement has no central origin; its mechanism must reside somewhere in the periphery.

3. Theoretical implications

Theories specifying the nature of mixture suppression must take into account that at least two qualitatively different causative mechanisms exist. Most theories that explain suppression at receptor level are competition theories. In a competition theory it is assumed that the molecules in the mixture compete for the same receptor sites. Perhaps the best-known peripheral mixture suppression theory is Beidler's occupation theory (Beidler, 1954, 1971, 1978). It predicts activity in the taste system from the relative occupation by different molecules in the equilibrium phase of the response, (i.e. when the number of molecules binding to a site is balanced by the number that dissociates from the site). Another well-known receptor theory is Heck and Erickson's rate theory (Heck & Erickson, 1973). This uses the *rate* of occupation as a predictor of the response. If the rate increases, the response increases correspondingly. The rate formulation allows combination of different stimuli (De Graaf & Frijters, 1986) and can also predict response magnitude on the basis of competition.

Receptor theories cannot make valid predictions of psychophysical measurements when a substantial part of the total mixture suppression is of non-peripheral origin. Such theories can only be applied when the neural contribution to total suppression is first calculated and subtracted.

IV. A preliminary model of suppression release

The experimental results obtained so far lead us to the model of binary mixture suppression outlined in Figure 13.7. We assume distinct central systems for each of the four so-called qualities. When these systems are excited simultaneously, and with sufficient strength, they suppress each other.

The model must be constructed in such a way that the central taste centers do not inhibit each other directly. Direct mutual inhibition would

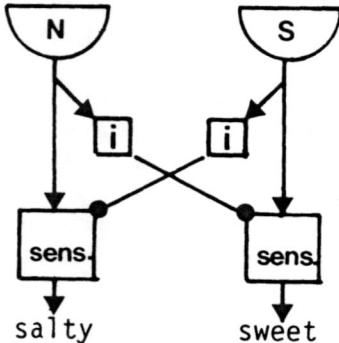

salty   sweet

Fig. 13.7. A hypothetical circuit that might be involved in suppression release. Two channels are depicted: salty (N) and sweet (S). When both channels are simultaneously activated, they inhibit each other through collateral excitation of inhibitory structures (*i*), each of which projects to the other channel. Suppression release by adaptation can take place independently from suppression release by habituation. In the latter case, the repetitious stimulus is assumed to decrease the inhibitory lateral effect.

make it impossible to differentiate between suppression release by adaptation and suppression release by habituation. The model specifies the minimal requirements to explain suppression release by adaptation and by habituation. When two qualitatively different stimuli are presented, the central taste structure of each is excited. But a collateral line excites an inhibitory structure, which in turn has inhibitory projections to the other taste centers. In the case of adaptation, suppression release is caused by decreased peripheral activity. However, a repeated stimulus gradually deactivates the central inhibitory influence directed at the neighbouring taste systems. Psychophysical experiments can never demonstrate the physical existence of neural structures, but they can specify the requirements to be fulfilled, in order to understand behavior.

References

Abrahams, A., Krakauer, D., & Dallenbach, K. M. (1973). Gustatory adaptation to salt. *American Journal of Psychology* **49**, 462–469.

Bartoshuk, L. M. (1975). Taste mixtures: is mixture suppression related to compression? *Physiology & Behavior* **14**, 643–649.

Bartoshuk, L. M. (1979). *Taste interactions in mixtures of NaCl with QHCl and sucrose with QHCl.* Poster presented at the 4th annual meeting of the Society for Neuroscience, Atlanta, Georgia.

Bartoshuk, L. M., & Seibyl, J. P. (1982). *Suppression of bitterness of QHCl in mixtures: possible mechanisms.* Poster presented at the 4th annual meeting of the Association for Chemoreception Sciences, Sarasota, Florida.

Beebe-Center, J. G., Roger, M. S., Atkinson, W. H., & O'Connell, D. N. (1959). Sweetness and saltiness of sucrose and NaCl as a function of solutes. *Journal of Experimental Psychology* **57**, 231–234.

Beidler, L. M. (1954). A theory of taste stimulation. *Journal of General Physiology* **38**, 133–139.

Beidler, L. M. (1971). Taste receptor stimulation with salts and acids. In L. M Beidler., ed. *Handbook of sensory physiology*, vol. 4, part 2, 'Taste'. pp. 200–220. Springer Verlag, New York.

Beidler, L. M. (1978). Biophysics and chemistry of taste. In E. C. Carterette & M. P. Friedman, eds, *Handbook of perception*, vol. 6-A, pp. 21–49. Academic Press, New York.

Bernstein, A. S. (1979). The orienting response as novelty and significance detector. *Psychophysiology* **16**, 263–273.

Borg, G., Diamant, H., Oakley, B., Ström, L., & Zotterman, Y. (1967a). A comparative study of neural and psychophysical responses to gustatory stimuli. In T. Hayashi, ed. *Olfaction and taste II*, pp. 253–264. Pergamon Press, London & New York.

Borg, G., Diamant, H., Ström, L., & Zotterman, Y. (1967b). The relation between neural and perceptual intensity: a comparative study on the neural and psychophysical response to taste stimuli. *Journal of Physiology* **192**, 13–20.

Cabanac, M. (1979). Gustatory pleasure and body needs. In J. H. A. Kroeze, ed.. *Preference behaviour and chemoreception*, pp. 275–288. Information Retrieval Limited, London.

Cabanac, M., & Duclaux, R. (1970). Specificity of internal signals in producing satiety for taste stimuli. *Nature* **227**, 966–967.

Cotman, C. W., & McGaugh, J. L. (1980). *Behavioral neuroscience, an introduction*, pp. 255–295. Academic Press, New York.

De Graaf, C., & Frijters, J. E. R. (1986). A psychophysical investigation of Beidler's mixture equation. *Chemical Senses* **11**, 295–314.

Diamant, H., Oakley, B., Ström, L., & Zotterman, Y. (1965). A comparison of neural and psychophysical responses to taste stimuli in man. *Acta Physiologica Scandinavica* **64**, 67–74.

Engen, T. (1971). Psychophysics, I. Discrimination and detection. In J. W. Kling & L. A. Riggs, eds, *Experimental psychology*, pp. 11–46. Rinehart & Winston, New York.

Fabian, F. W., & Blum, H. B. (1943). Relative taste potency for some basic food constituents and their competitive and compensatory action. *Food Research* **8**, 179–193.

Fisher, G., & Fisher, B. (1969). Different rates of GSR-habituation to pleasant and unpleasant sapid stimuli. *Journal of Experimental Psychology* **82**, 339–342.

Ganzevles, P. G. J., & Kroeze, J. H. A. (1987). Cross adaptation in taste measured with a filter paper method. *Chemical Senses* **12**, 341–353.

Gent, J., & McBurney, D. H. (1978). Time course of gustatory adaptation. *Perception & Psychophysics* **23**, 171–175.

Gerebtzoff, M. A. (1939). Les voies centrales de la sensibilité et du gout et leurs terminaisons thalamiques. *Cellule* **48**, 91–146.

Gillan, D. J. (1982). Mixture suppression: the effect of spatial separation between sucrose and NaCl. *Perception & Psychophysics* **32**, 504–510.

Hahn, H. (1934). Die Adaptation des Geschmackssinnes. *Zeitschrift für Sinnesphysiologie* **65**, 105–145.

Heck, G. L., & Erickson, R. P. (1973). A rate theory of gustatory stimulation *Behavioral Biology* **8**, 687–712.

Held, R., & Hein, A. (1963). Movement-produced stimulation in the development of visually guided behavior. *Journal of Comparative and Physiological Psychology* **56**, 872–876.

Holst, E. von, & Mittelstaedt, H. (1950). Das Reafferenzprinzip (Wechselwirkungen zwischen Zentral Nervensystem und Peripherie). *Naturwissenschaften* 37, 464–476.

Kamen, J. M., Pilgrim, F. J., Gutman, N. J., & Kroll, B. J. (1961). Interactions of suprathreshold stimuli. *Journal of Experimental Psychology* 62, 348–356.

Krakauer, D., & Dallenbach, K. M. (1937). Gustatory adaptation to sweet, sour and bitter. *American Journal of Psychology* 49, 469–475.

Kroeze, J. H. A. (1977). Taste thresholds for bilaterally and unilaterally presented mixtures of sugar and salt. In J. LeMagnen & P. MacLeod, eds, *Olfaction and taste VI*, p. 486. IRL Press, London.

Kroeze, J. H. A. (1978). The taste of sodium chloride: masking and adaptation. *Chemical Senses and Flavour* 3, 443–449

Kroeze, J. H. A. (1979). Masking and adaptation of sugar sweetness intensity. *Physiology & Behavior* 22, 347–351.

Kroeze, J. H. A. (1982a). After repetitive sucrose stimulation saltiness suppression in NaCl–sucrose mixtures is diminished: implications for a central mixture suppression mechanism. *Chemical Senses* 7, 81–92.

Kroeze, J. H. A. (1982b). The influence of relative frequencies of pure and mixed stimuli on mixture suppression in taste. *Perception & Psychophysics* 31, 276–278.

Kroeze, J. H. A. (1982c). Reduced sweetness and saltiness judgments of NaCl–sucrose mixtures depend on a central inhibitory mechanism. In J. E. Steiner & J. R. Ganchrow, eds, *Determination of behaviour by chemical stimuli*, pp. 161–174. IRL Press, London.

Kroeze, J. H. A. (1982d). The relationship between the side tastes of masking stimuli and masking in binary mixtures. *Chemical Senses* 7, 23–37.

Kroeze, J. H. A. (1983). Successive contrast cannot explain suppression release after repetitious exposure to one of the components of a taste mixture. *Chemical Senses* 8, 211–223.

Kroeze, J. H. A., & Bartoshuk, L. M. (1985). Bitterness suppression as measured by split-tongue taste stimulation in humans. *Physiology & Behavior* 35, 779–783.

Kuznicki, J. T., & Ashbaugh, N. (1979). Taste quality differences within the sweet and salty taste categories. *Sensory Processes* 3, 157–182.

Kuznicki, J. T., & Ashbaugh, N. (1982). Space and time separation of taste mixture components. *Chemical Senses* 7, 39–62.

Lawless, H. T. (1979). Evidence for neural inhibition in bitter-sweet taste mixtures. *Journal of Comparative & Physiological Psychology* 93, 538–547.

Lawless, H. T. (1982). Adapting efficiency of salt-sucrose mixtures. *Perception & Psychophysics* 2, 419–422.

Lindsay, P. H., & Norman, D. H. (1977). *Human information processing*, 2nd edn, p. 251. Academic Press, New York.

Meiselman, H. L. (1968). Magnitude estimation of the course of gustatory adaptation. *Perception & Psychophysics* 4, 193–196.

Moskowitz, H. R. (1972). Perceptual changes in taste mixtures. *Perception & Psychophysics* 11, 257–262.

Norgren R., & Leonard, C. M. (1973). Ascending central gustatory pathways. *Journal of Comparative Neurology* 150, 217–238.

Norgren, R., Grill, H. J., & Pfaffmann, C. (1970). Projections of taste to the dorsal pons and limbic system with correlated studies of behavior. In Y. Katsuki *et al.*, eds, *Food intake and chemical senses*, pp. 233–243. University Park Press, Baltimore.

Overbosch, P. A. (1986). Theoretical model for perceived intensity in human taste and smell as a function of time. *Chemical Senses* 11, 315–329.

Overbosch, P., Van den Ende, J. C., & Keur B. M. (1986). An improved method for measuring perceived intensity/time relationships in human taste and smell. *Chemical Senses* 11, 331–338.

Pangborn, R. M. (1960). Taste interrelationships. *Food Research* 25, 245–256.

Pangborn, R. M. (1961). Taste interrelationships II: suprathreshold solutions of sucrose and citric acid. *Journal of Food Science* 26, 648–655.

Pangborn, R. M. (1962). Taste interrelationships III: suprathreshold solutions of sucrose and sodium chloride. *Journal of Food Science* 27, 495–500,

Pangborn, R. M. & Trabue, I. M. (1967). Detection and apparent taste intensity of salt-acid mixtures in two media. *Perception & Psychophysics* 2, 503–509.

Pavlov, I. P. (1927). *Conditioned reflexes.* Clarendon Press, Oxford.

Rollin, H. (1973). Geschmacksausfälle nach Operation von Kleinhirnbrücken-winkeltumoren. *Hals, Nase und Ohrenheilkunde (HNO)* 21, 237–240.

Sato, T. (1971). Site of gustatory neural adaptation. *Brain Research* 31, 385–388.

Sato, T. (1972). Adaptation of primary gustatory nerve responses in the frog. *Comparative Biochemistry & Physiology* 43, 1–12.

Sjöström, L. B., & Cairncross, S. E. (1953). Role of sweetness in food flavor. *Advances in biochemistry series*, pp. 108–113. American Chemical Society, Washington, DC.

Thompson, R. F., & Spencer, W. A. (1966). Habituation, a model phenomenon for the study of neural substrates of behavior. *Psychological Review* 73, 16–43.

Van Olst, E. H. (1971). *The orienting reflex*, pp. 122–131. Mouton Publishing Company, The Hague.

Vij, S., & Kanagasuntheram, R. (1972). Development of the nerve supply to the human tongue. *Acta Anatomica* 81, 466–477.

Yamaguchi, S. (1967). The synergistic taste effect of monosodium glutamate and disodium 5'-inosinate. *Journal of Food Science* 32, 473–478.

Yamaguchi, S., Yoshikawa, T., Ikeda, S., & Ninomiya, T. (1968). The synergistic taste effect of monosodium glutamate and disodium 5'guanylate. *Journal of the Agricultural Chemistry Society of Japan* 43, 378.

Yoshida, M. (1986). A microcomputer (PC9801/MS Mouse) system to record and analyze time-intensity curves of sweetness. *Chemical Senses* 11, 105–118.

Zotterman, Y. (1971). Recording of the electrical response from human taste nerves. In L. M. Beidler, ed., *Handbook of sensory physiology*, vol. 4, part 2, 'Taste,' pp. 102–115. Springer-Verlag, New York.

14

Modeling taste mixture interactions in equiratio mixtures

Jan E. R. Frijters and Cees De Graaf

Department of Food Science and Department of Market Research,
The Netherlands' Agricultural University, Wageningen, The Netherlands

I. Introduction

In their review, Pfaffmann, Bartoshuk and McBurney (1971) distinguished between two kinds of psychophysical taste mixture studies: investigations concerned with mixtures of compounds eliciting different taste qualities, and those concerned with mixtures composed of substances having similar taste qualities. Although the results of mixture studies reported before 1970 did not permit unanimous conclusions, Pfaffmann *el al.* (1971) made two important generalizations The first was that, in mixtures of compounds that elicit similar taste qualities, synergy or addition usually occurs. The total taste intensity of such a mixture is greater than or equal to the sum of the intensities of the components. The second generalization was that, in mixtures of components having dissimilar taste qualities, suppression seems to be the rule. In this case, the mixture's total taste intensity is less than the sum of the perceived taste intensities of the components. Later, Moskowitz (1972) suggested that the magnitude of the suppression effect is inversely proportional to the similarity between the taste qualities (cf. Bartoshuk & Cleveland, 1977). Obviously, of all potential parameters, the taste qualities of the compounds were considered to be the most critical parameters in determining the nature of taste mixture interactions.

In a number of papers, Moskowitz (1973, 1974*a*, 1974*b*) presented two psychophysical models for predicting the taste intensity of mixtures; the addition model and the substitution model. In the majority of the earlier studies, only a limited number of mixtures had been investigated, and these had often been composed of more or less arbitrarily chosen

Copyright © 1989 by Academic Press Australia.
All rights of reproduction in any form reserved.

concentrations of the components. In contrast to this somewhat unsystematic approach, Moskowitz initiated the use of psychophysical functions to study the relationship between component intensities and the intensity of the mixture. He developed quantitative mixture models on the basis of the psychophysical functions of the individual components. Although psychophysical modeling might even at present be considered premature in view of the knowledge available, models are meritorious because they provide the opportunity to derive testable predictions.

The aim of the present paper is to discuss a few methodological developments since the introduction of the addition and substitution models. Discussion of these models will be omitted here, because their status has recently been reviewed elsewhere (Frijters & Oude Ophuis, 1983; Frijters, 1987). Here, the concept of 'equiratio mixture type' and the equiratio taste mixture model will be explained. Finally, it will be demonstrated how ideas regarding equiratio mixtures can be usefully combined with a functional measurement approach to the psychophysical study of perceptual interactions in mixtures of tastants.

II.   General considerations

When designing a mixture experiment, the investigator is confronted with the question as to how the mixtures should be prepared. The choice of the unit to express the concentrations of the tastants may at first seem unimportant, but, as Pfaffmann *et al.* (1954) and Myers (1982) have pointed out, it is in fact crucial. Some of the various units currently in use are non-linearly related. A correct unit to express physical stimulus intensity would be some measure of 'degree of activity', but such a unit has not yet been proposed. At present, molarity is preferred to other concentration units used in (flavor) chemistry and food science. Although, molarity is not identical to activity, solutions of equal molarity at least contain an equal number of molecules and these are the active agents in stimulating receptor sites.

Another consideration in planning a mixture investigation is the type of mixture that should be constructed: which mixtures are the most interesting from a theoretical point of view? In a review of the taste mixture literature, Bartoshuk and Gent (1985) noted that mixtures are not usually prepared by actual physical mixing of solutions, but by adding solutes to solvents. For example, a mixture of 1 M sucrose and 1 M fructose is prepared by adding 342.3 g sucrose plus 180.2 g fructose to water. In the laboratory, the substances are first dissolved in a certain amount of water. More water is subsequently added, so that the total volume of the solution becomes 1 L. Thus, the total concentration of the

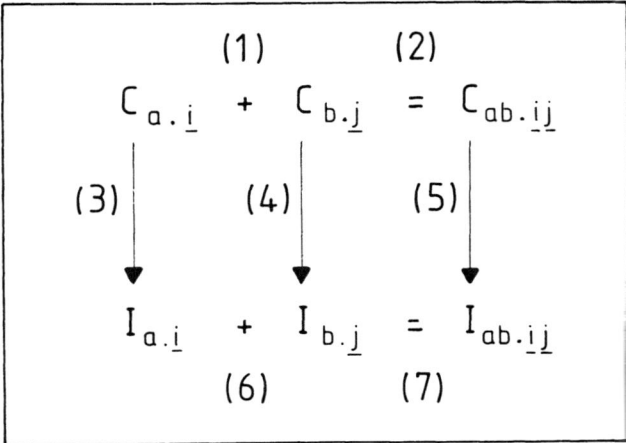

Fig. 14.1. A simple additive mixture 'model' for a binary mixture of tastants.

final solution is equal to the sum of the concentrations of the substances, which is also equal to the sum of the concentrations of the individual components outside the mixture.

This 'standard' method of mixture construction has had an important influence on the study of perceptual taste mixture interactions, which is not often realised. As a result, when mapping the relations between the physical concentrations onto the perceptual level, research questions meant to be 'mixture' questions were actually dealt with as if these were 'addition' questions. The tacitly assumed isomorphic relation between the physical and corresponding perceptual entities is illustrated by the diagram in Figure 14.1. This conceptual scheme contains seven relations; two additive ($+$; 1 & 6), two identity ($=$; 2 & 7) and three psychophysical ones (i.e. the functional relationship between the concentrations $C_{a.i}$, $C_{b.j}$, & $C_{ab.ij}$, and their corresponding sensations $I_{a.i}$, $I_{b.j}$, & $I_{ab.ij}$. Note that these sensations have different status. $I_{a.i}$ & $I_{b.j}$, are the perceived taste intensities of the components each tasted alone, whereas $I_{ab.ij}$, is the total intensity of the mixture. The diagram does not contain the component sensations perceived in the mixture.

Two of the seven relations are of no concern. These are the physically additive relationship between $C_{a.i}$ and $C_{b.j}$ (1), and the identity relationship between ($C_{a.i} + C_{b.j}$) and $C_{ab.ij}$ (2). Whether these relationships are empirically correct can easily be verified by physical measurement. The other five relationships are the ones of psychophysical and perceptual interest. The relations (3), (4) and (5) between $C_{a.i}$, $C_{b.j}$, $C_{ab.ij}$ and $I_{a.i}$, $I_{b.j}$, $I_{ab.ij}$, respectively, are of a representational nature. These should be established on the basis of empirical psychophysical functions. It should be borne in mind that such functions cannot be used to determine whether the quality and intensity of the sensation of a component assessed outside the mixture is identical to that elicited by the component in the mixture. The relationship between the taste intensity of a component outside the mixture and the taste intensity of the same component in the mixture is not dealt with at all in the above scheme. Hence it does not describe how distinguishable taste sensations in the mixture's percept are integrated.

From a measurement point of view, there are two prerequisites to be met by the above mixture 'model'. Firstly, identical units should be used to express the intensities of the sensations elicited by the components and by the mixture; secondly, ratio scales should be used to measure the sensations. If these criteria are not met, the investigation of any relation between $I_{a.i}$ and $I_{b.j}$ is meaningless. The assumption that this relation is additive seems to derive from the existence of such an operation in the physical realm, and not because of insight into the perceptual capacity of the gustatory system. This relation [(6)] and the identity relation [(7)] between ($I_{a.i} + I_{b.j}$) and $I_{ab.ij}$ are reciprocal from a theorectical point of view. Conclusions that two compounds interact synergistically, additively or suppressively are meaningful only because of the (arbitrarily) postulated additive operation of the gustatory system, when processing mixtures of tastants. A particular weakness of the paradigm depicted in Figure 14.1 is that it easily leads to circularity. Non-linearity of the psychophysical functions (3) and (4) is a sufficient condition to violate the reciprocity of the relations (6) and (7). Conversely, non-additivity can be explained as a result of non-linear psychophysical functions, as pointed out by Bartoshuk (1975). Non-linearity of psychophysical taste functions may be of interest for other reasons, but in itself it does not explain anything about integration of single sensations into a mixture percept.

Frijters and Oude Ophuis (1983) analyzed the addition and substitution models. Both were developed for the prediction of the taste intensity of mixtures containing an amount of substance that is equal to the summed quantities of solute of the components tasted outside the mixture. These authors argued that the models are inadequate, either because of lack of generality or because of internal inconsistency. Apart from the deficiencies from a modeling point of view, both of these models fail to accommodate some of the subsequent experimental findings.

In the first place, there is the effect of cross-adaptation on perceived intensity. Substances which cross-adapt must interact peripherally (e.g. McBurney, 1972; McBurney, Smith & Shick, 1972). If molecules of different substances compete for adsorption at the same receptor sites, they inhibit each other's activity. The consequence is that the taste sensations of the components in the mixture are less intense than the sensations elicited by the single compounds outside the mixture. This kind of inhibition might be called peripheral suppression. It can best be demonstrated by 'mixing' a solution of a certain compound with another solution of the same substance, as explained by Bartoshuk (975). If a particular concentration $C_{a,i}$ elicits a taste intensity of $I_{a,i}$ then the concentration $C_{a,2i}$ might elicit a taste intensity smaller than $2 \times I_{a,i}$, as a result of self-inhibition. This fact cannot be accommodated by the addition model because it is based on the assumption of peripheral independency of the tastants in a mixture.

A second phenomenon that cannot be reconciled by the addition or the substitution model is that peripherally independent substances having different taste qualities suppress the perceived intensity of each other (Kroeze, 1978). As in the case of peripheral suppression, the consequence of central mixture suppression is also that the perceived intensities of the individual taste sensations in a heterogeneous mixture percept are less intense than the corresponding taste sensations of the components tasted outside the mixture.

III. The concept of 'equiratio mixture type'

Frijters and Oude Ophuis (1983) concluded that no valid model was available for the prediction of the taste intensity of a mixture of sensorily dependent substances having similar tastes. For this reason, they designed a new model based on a different approach to mixture construction and on the concept of the 'equiratio mixture type'. The characteristics of such a mixture type can best be explained on the basis of the mixture scheme used by Frijters and Oude Ophuis, which is depicted in Figure 14.2. In constrast to a full factorial mixture design, which is obtained when all concentrations of substance A are mixed with all concentrations of substance B, mixing occurs according to a certain 'triangular design'. Figure 14.2 shows that the mixture triangle has an X- and Y-axis. The X-axis represents the concentration of glucose, the Y-axis the concentration of fructose. The hypotenuses represent equimolar lines. All solutions, represented as points on a particular hypotenuse, contain the same total amount of substance expressed in M (mol/litre). However, the ratio of the concentrations of the two substances depends

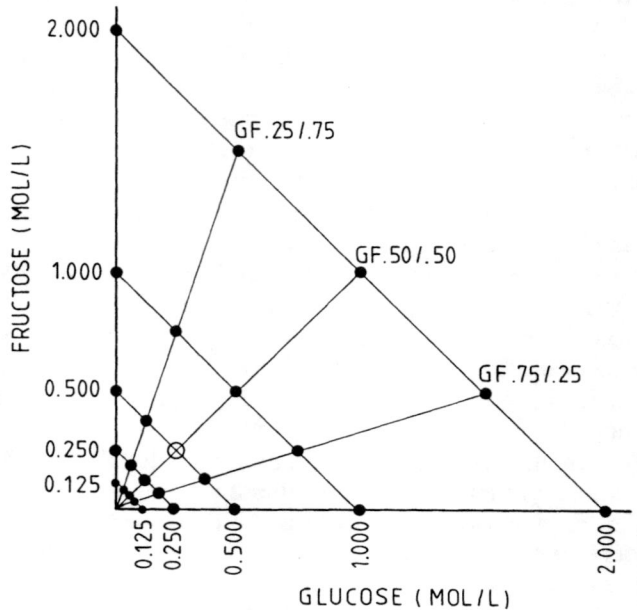

Fig. 14.2. Mixture scheme for preparing equiratio mixtures. Each point on the triangle represents a stimulus. The circle denotes the stimulus used as standard stimulus ($C_{ab.s.0.50/0.50}$) by Frijters and Oude Ophuis (1983). The lines running from the origin represent three different equiratio mixture types of glucose and fructose.

on the location of the point on the hypotenuse. Each line extending from the origin represents *a series* of mixtures of different total concentration, but the ratio between the concentrations of the components is the same for all of these mixtures. Such a series is called an 'equiratio mixture type'. In principle, an unlimited number of different equiratio mixture types can be constructed for each pair of substances. Which type will be obtained depends on the mixture ratio. This ratio is manipulated by changing the angle of the equiratio line in the mixture scheme (i.e. by changing the ratio of the mixture components).

The main advantage of this form of mixture construction is the simplicity of establishing a psychophysical mixture function. In a complete factorial mixture design, the perceived intensity of a mixture is conceived as being a composite function of the concentrations of the components (i.e. $I_{ab.ij} = f(C_{a.i}, C_{b.j})$. Fitting such a function by regression analysis requires the use of two independent variables, $C_{a.i}$ and $C_{b.j}$, and the dependent variable, $I_{ab.ij}$. If a power function is assumed to be an appropriate type of function to describe a psychophysical relationship between concentration and perceived taste intensity, modeling the relation between perceived mixture intensity and concentrations of the compounds in the mixture becomes quite complex. Since each of the two power functions contains two parameters (i.e. the exponent and the constant), such a model would at least contain four constants. In case of an equiratio mixture type, a psychophysical power function can be established simply on the basis of experimental data, in the same way as is done for a single unmixed compound. This can be done because the total concentration of an equiratio mixture ($C_{ab.ij.pq} = C_{a.i} + C_{b.j}$) is taken as the (only) independent variable. The general equation describing a psychophysical power function for an equiratio mixture type is given by:

$$I_{ab.ij.pq} = k_{ab.pq} \, C_{ab.ij.pq}^u \tag{14.1}$$

As they explain in their paper, Frijters and Oude Ophuis (1983) did not conceive the sensory response as a linear representation of the perceived intensity. Hence perceived intensity denoted as $I_{ab.ij.pq}$ in equation 14.1 was replaced by sensory response denoted as $R_{ab.ij.pq}$ to obtain the following expression:

$$R_{ab.ij.pq} = k_{ab.pq} \, C_{ab.ij.pq}^u \tag{14.2}$$

In these equations the subscripts a and b refer to the substances A and B in a binary mixture, the subscripts i and j refer to the concentrations of the substances A and B, respectively, and the subscripts p and q refer to the mixing ratio ($p + q = 1$).

IV.   The equiratio taste mixture model

As previously explained, an equiratio mixture type consists of a series of mixtures, but it is dealt with as if it were a series of solutions of an un-mixed compound. A psychophysical function for any mixture type can be established in the same way as is done for a single compound. The question of interest posed by Frijters and Oude Ophuis was: is it feasible to predict such a function for a binary equiratio mixture type of two com-pounds on the basis of the psychophysical functions of the unmixed compounds? The equiratio mixture model was designed for this purpose. It should be remembered that this model is limited in its generality; it was designed for equiratio mixture types of peripherally competitive sub-stances having similar tastes, so that the individual mixtures elicit homogeneous taste percepts. Probably, such mixtures are the most simple ones, because there is no central mixture suppression between taste qualities and no complicated psychological integration to obtain the total taste intensity.

Assuming that a power function is an appropriate general form for describing a psychophysical relation between concentration and sensory response, the exponent $\hat{Q}$ of an equiratio mixture function is predicted by

$$\hat{u} = pm + qn \qquad (14.3)$$

in which m and n are the exponents of experimentally determined psychophysical power functions of the substances A and B, respectively, and p and q are the proportions of the concentrations of the substances A and B in a particular equiratio mixture type. The constant $\hat{K}_{ab.pq}$ in equation 14.2 can be obtained by

$$\hat{K}_{ab.pq} = \frac{pk_a/C_{a.s} + qk_b/C_{b.s.}}{p/C_{a.s.} + q/C_{b.s.}} \qquad (14.4)$$

where k_a and k_b are the constants of the psychophysical functions of the substances A and B, respectively, and $C_{a.s}$ and $C_{b.s}$ are the concentrations of the substances A and B, respectively, which give rise to a sensory response of the same magnitude as the response given to the standard stimulus in the magnitude estimation experiment (i.e. $R_{a.s} = R_{b.s.}$). These equi-intense concentrations can be determined in various ways: firstly, by equating the psychophysical functions, provided they have a common standard stimulus; secondly, by application of the method of constant stimuli. In the latter case, one of the two concentrations $C_{a.s}$, and $C_{b.s}$, has to be used as a reference stimulus.

This model has been tested in three experiments; two with sweet-tasting substances (Frijters & Oude Ophuis, 1983; Frijters, De Graaf & Koolen, 1984) and one with sour-tasting compounds (Frijters & Stevens, 1986). The validity of the model is tested by simply comparing the predicted and experimentally determined psychophysical functions for equiratio mixtures. In Table 14.1, predicted and experimentally determined functions are given for three equiratio mixture types of glucose and fructose (Frijters & Oude Ophuis, 1983). Clearly, they are very similar; the greatest prediction error being 4% with respect to the exponents, and 3% with respect to the constants. In a similarly designed study on sorbitol–sucrose mixtures (Frijters et al., 1984), the predictions were slightly less precise. Probably, this was due to the non-linearity of the log-log plotted psychophysical function of sucrose (see Fig. 14.3).

Table 14.1. Experimentally determined and predicted psychophysical power functions for three equiratio mixture types of glucose and fructose.

Substance or mixture type	Experimentally determined power function	Predicted power function
Glucose	$R = 14.48\ C^{1.55}$	
GluFru 0.75/0.25	$R = 19.70\ C^{1.40}$	$R = 19.03\ C^{1.46}$
GluFru 0.50/0.50	$R = 22.85\ C^{1.37}$	$R = 22.27\ C^{1.36}$
GluFru 0.25/0.75	$R = 24.88\ C^{1.27}$	$R = 24.70\ C^{1.27}$
Fructose	$R = 26.58\ C^{1.18}$	

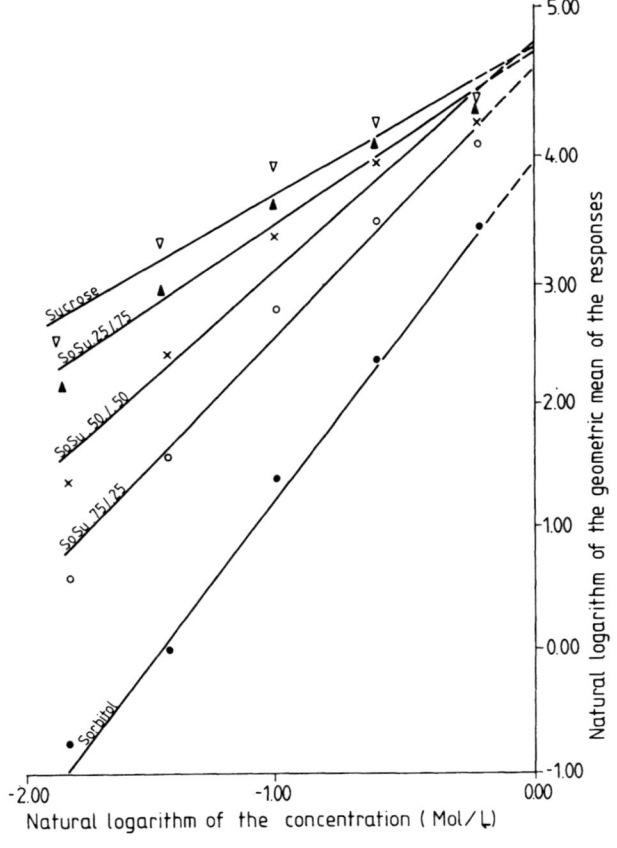

Fig. 14.3. Psychophysical power functions of sucrose, sorbitol and three equiratio mixture types, plotted on log-log coordinates. (From Frijters *et al.*, 1984, with permission.)

Despite this complication, it was concluded that the model has good predictive validity. In a follow-up study, Frijters and De Graaf (1987) investigated whether the equiratio taste mixture model could be extended to predict the sensory response to mixtures consisting of more than two components. Since the model was originally developed for binary mixture types, equation 14.2 had to be expanded. As an example, this is demonstrated here for a quaternary mixture type.

Let the psychophysical functions of the substances A, B, C and D be given by

$$R_{a.i} = k_a c^v_{a.i}$$

$$R_{b.j} = k_b c^w_{b.j}$$

$$R_{c.k} = k_c c^x_{c.k}$$

$$R_{d.l} = k_d c^y_{d.l}$$

Analogous to equation 14.2 for the binary case, the general formula for a quaternary equiratio mixture type is given by:

$$R_{abcd.ijkl.pqrs} = k_{abcd.pqrs} c^z_{abcd.ijkl.pqrs} \tag{14.5}$$

where $R_{abcd.ijkl.pqrs}$ represents the sensory response to a mixture of concentration i of A plus j of B plus k of C plus l of D; the total concentration is $C_{abcd.ijkl.pqrs}$; and the subscripts p,q,r and s refer to the proportions of the concentrations of the individual compounds (i.e. $p + q + r + s = 1$).

According to the generalized equiratio taste mixture model (equation 14.5), the exponent z can be estimated by

$$\hat{z} = pv + qw + rx + sy \tag{14.6}$$

and the constant $k_{abcd.ijkl.pqrs}$ by

$$k_{abcd.pqrs} = \frac{pk_a/C_{a.s.} + qk_b/C_{b.s} + rk_c/C_{c.s.} + sk_d/C_{d.s}}{p/C_{a.s} + q/C_{b.s} + r/C_{c.s} + s/C_{d.s}} \tag{14.7}$$

In their study, De Graaf and Frijters (1987) investigated the predictability of the psychophysical functions of quaternary and octonary equiratio mixture types. Table 14.2 contains the results for three equiratio mixture types composed of arbitrarily chosen sugars and sugar-alcohols. From this it can be concluded that the predicted power

Table 14.2. Experimentally determined psychophysical power functions o˜ eight sweet-tasting compounds, two quaternary and one octernary equiratio mixture type. The power functions predicted on the basis of the generalized form of the equiratio ˜aste mixture model are also given.

Substance or mixture type	Experimentally determined power function	Predicted power function
Fructose	$R = 32.70 \, C^{1.10}$	
Sucrose	$R = 46.15 \, C^{0.90}$	
Xylitol	$R = 22.03 \, C^{1.32}$	
Maltose	$R = 28.55 \, C^{1.51}$	
Sorbitol	$R = 16.60 \, C^{1.39}$	
Glucose	$R = 16.28 \, C^{1.54}$	
Xylose	$R = 15.39 \, C^{1.53}$	
Galactose	$R = 17.77 \, C^{1.51}$	
FSXIM 0.25/0.25/0.25/0.25	$R = 36.22 \, C^{1.14}$	$R = 35.70 \, C^{1.21}$
SoGXeGa 0.25/0.25/0.25/0.25	$R = 15.80 \, C^{1.53}$	$R = 16.54 \, C^{1.49}$
FSXIMSoGXeGa 0.125/0.125/ 0.125/0.125/0.125/0.125 0.125/0.125	$R = 29.16 \, C^{1.35}$	$R = 29.23 \, C^{1.35}$ (1) $R = 29.48 \, C^{1.35}$ (2)

The mean of the 95% confidence intervals of the 11 exponents is $\times \pm 0.24$.

functions are virtually identical to those that were expe˜imentally determined. It can also be seen that it makes no difference whether this function is predicted on the basis of the eight single functions [col 3., (1)], or on the basis of the two quaternary mixtures using the equi˜atio taste mixture model for binary mixture [col 3., (2)]. In the latte˜ instance the quaternary equiratio mixture types were considered to be single substances.

From these experiments it can be concluded that the response to a mixture is some kind of 'average' of the values of the sensory responses to the unmixed compounds of a molar concentration equal to t˜at of the mixture. Because the substances used for mixture composition are sensorily dependent substances, competition for absorption at the same receptor sites seems to be adequate to explain the 'averaging' prir.ciple. However, such a peripheral mutual inhibition does not necessarily preclude the possibility of 'averaging' at a neural or central level. The generality of this 'averaging' mechanism was shown by De Graaf and Frijters (1987), who reanalyzed the data of previously publist.ed sweetener mixture studies. In the majority of these studies the intensity of a mixture lay in between the intensities of the unmixed compounds, when the comparisons were made on the basis of isol-molar concentrations.

If the above-mentioned averaging is the result of periphera1 competition between the molecules of the different substances in the mixture, then it is hypothesized that the equiratio taste mixture model will not correctly predict the sensory response to all possible mixtures. For example, mixtures of substances that operate independently at the periphery (i.e. they do not show cross-adaptation), such as sodium chloride and sucrose, might not be expected to follow the averaging rule. Furthermore, the sensory response to the total intensity of mixtures of substances of different taste qualities may also be expected to be lower than would be predicted by the Equiratio Taste Mixture Model. This is due to the phenomenon of central mixture suppression, which has been shown to occur in these types of mixture (Kroeze, 1978, 1979; Kroeze & Bartoshuk, 1985).

V.   Psychophysical predictions by Beidler's mixture equation

A psychophysical power function is usually written as a relation between two absolute quantities, that is, the concentration of a stimulus and the corresponding sensory response. Stevens (1975) conceived the sensory response as being identical to perceived intensity. However, as pointed out by several authors (e.g. Baird & Noma, 1978), it can also be written as a relation between ratios. Such an expression for a compound A is given by

$$R_{a.i}/R_{a.s} = (C_{a.i}/C_{a.s})^n \tag{14.8}$$

in which $C_{a.s}$ is the standard stimulus concentration and $R_{a.s}$ is the response to this stimulus. The latter value usually corresponds, more or less, to the value assigned to the standard stimulus in the instructions given to the subjects (if the magnitude estimation experiment is carried out with a fixed standard and modulus). $C_{a.s}$ and $R_{a.s}$ can be interpreted as concentration 'unit' and response 'unit', respectively (cf. Frijters & Oude Ophuis, 1983). For another compound, B, the corresponding equation would be

$$R_{b.j}/R_{b.s} = (C_{b.j}/C_{b.s})^m \tag{14.9}$$

The responses to the stimuli of the substances A and B are expressed in the same units when the values of $R_{a.s}$ and $R_{b.s}$ are equal. These values are associated with the standard concentrations $C_{a.s}$ and $C_{b.s}$, respectively. The tastes of these concentrations should therefore be equi-intense. If a particular concentration of A is chosen by the experimenter as the standard stimulus $C_{a.s}$, the concentration of compound B that is equally strong in taste (to be denominated as $C_{b.s}$) can be experimentally determined by the method of constant stimuli or another matching procedure.

If $R_{a.s} = R_{b.s}$ and the values of $C_{a.s}$ and $C_{b.s}$ are known, the psychophysical power functions for the substances A and B can be directly established on the basis of experimental data that have been obtained by magnitude estimation. This would simply mean that equations 14.8 and 14.9 are used.

A psychophysical power function for a particular equiratio mixture type of the substances A and B can likewise be written as

$$R_{ab.ij.pq}/R_{ab.s.pq} = (C_{ab.ij.pq}/C_{ab.s.pq})^u \qquad (14.10)$$

where $C_{ab.s.pq}$ is the concentration of an equiratio mixture with a proportion p of substance A and a proportion q of substance B, which gives rise to the response $R_{ab.s.pq}$.

The functions specified by equations 14.8 to 14.10 are directly comparable only if they incorporate the same units. This means that the values of $R_{a.s}$, $R_{b.s}$ and $R_{ab.s.pq}$ must be equal. These values have corresponding concentrations of $C_{a.s}$, $C_{b.s}$ and $C_{ab.s.pq}$, respectively. If these equi-intense concentrations have been determined experimentally, in addition to the psychophysical functions of the substances A and B specified by equations 14.8 and 14.9, the power function for a particular equiratio mixture type of the two substances, as given by equation 14.11, can be predicted. The exponent u in this equation can be predicted using equation 14.3.

It would be theoretically desirable, and also more elegant, if the concentration $C_{ab.s.pq}$ need not be determined experimentally, but could be predicted on the basis of the other two equi-intense concentrations $C_{a.s}$ and $C_{b.s}$. De Graaf and Frijters (1986) derived an expression from Beidler's mixture equation (Beidler, 1962, 1971) for this purpose. This expression predicts the composition and concentration of all possible mixtures of A and B which will elicit a sensory response that is equal to the responses elicited by the equi-intense concentrations, $C_{a.s}$ and $C_{b.s}$. It is given by

$$X + Y = C_{a.s} \times C_{b.s}/(qC_{a.s} + pC_{b.s}), \qquad (14.11)$$

in which X is the molar concentration of A and Y the molar concentration of B in the mixture $C_{ab.s.pq}$ (see Fig. 1 in De Graaf & Frijters, 1986).

Predictions made by this equation were tested by De Graaf and Frijters (1986) at five different concentration levels, in an experiment

Table 14.3. Concentrations of fructose and three equiratio mixture types of glucose and fructose, equal in perceived intensity to five concentration levels of glucose.

Glucose concentration (M)	Comparison substance or mixture	Equi-intense concentration exp. determined	Equi-intense concentration predicted
0.125	GluFru 0.75/0.25	0.0873	0.0896
0.125	GluFru 0.50/0.50	0.0666	0.0699
0.125	GluFru 0.25/0.75	0.0558	0.0573
0.125	Fructose	0.0485	——
0.250	GluFru 0.75/0.25	0.1817	0.1840
0.250	GluFru 0.50/0.50	0.1439	0.1456
0.250	GluFru 0.25/0.75	0.1206	0.1204
0.250	Fructose	0.1027	——
0.500	GluFru 0.75/0.25	0.3639	0.3917
0.500	GluFru 0.50/0.50	0.3118	0.3219
0.500	GluFru 0.25/0.75	0.2682	0.2733
0.500	Fructose	0.2374	——
1.000	GluFru 0.75/0.25	0.7729	0.8461
1.000	GluFru 0.50/0.50	0.6550	0.7334
1.000	GluFru 0.25/0.75	0.5928	0.6471
1.000	Fructose	0.5790	——
2.000	GluFru 0.75/0.25	1.6310	1.7992
2.000	GluFru 0.50/0.50	1.4552	1.6351
2.000	GluFru 0.25/0.75	1.4007	1.4984
2.000	Fructose	1.3828	

using glucose, fructose and mixtures of these substances. The concentrations of glucose (substance A) used as $C_{a,s}$ were 0.125, 0.25, 0.50, 1.00 and 2.00 M. Table 14.3 shows that the corresponding equi-intense concentrations ($C_{b,s}$) of fructose (substance B), as experimentally determined, are 0.0485, 0.1027, 0.2374, 0.5790 and 1.3828 M. This table also shows the predicted as well as the experimentally determined equi-sweet concentrations of three equiratio mixture types of glucose and fructose. From these results it can be concluded that, at low sweetness levels (0.125 & 0.25 M glucose), equation 14.11 predicts with great precision. The difference between predicted and experimentally obtained concentrations is on average −2%. However, at high taste intensity levels (1.00 and 2.00 M glucose), the experimentally determined concentrations are significantly lower than predicted, with a mean deviation of about −9%. At the level of 0.50 M glucose, the mean deviation is −4%.

As mentioned earlier, equation 14.11 is a special case of Beidler's mixture equation; it is therefore also based on the assumption of mutual peripheral competition between molecules of different substances for absorption at one type of receptor site. It seems that equation 14.11 is valid at low concentration levels only. One might argue that fructose and glucose have similar taste qualities at low intensities, but that taste quality differences emerge at high concentrations (Kuznicki & Ashbaugh, 1979), so that in mixtures of high concentrations of these substances central mixture suppression occurs. If this is correct, equation 14.11 must predict incorrectly, because it does not take central taste mixture suppression into account. However, this reasoning cannot be correct. The three concentrations of the glucose–fructose mixtures that give rise to a perceived intensity equal to the taste intensity of 1.00 and 2.00 M glucose (Table 14.3) are lower than predicted. If central mixture suppression had occurred, these experimentally obtained concentrations would have been higher, not lower. The gustatory system is obviously more efficient than expected on the basis of Beidler's mixture equation. De Graaf and Frijters (1986) hypothesized, in line with the multiple sweetener receptor theory (Faurion, Saito & MacLeod, 1980; Schiffman, Cahn & Lindley, 1981; Lawless & Stevens, 1983), that glucose and fructose share one common kind of receptor site, but either one or both substances have additional and different binding mechanisms.

Further investigation of equation 14.11 with mixtures of other substances is encouraged.

VI. Functional measurement with equiratio mixtures

Interpretation of the equiratio model will depend on the meaning that is attached to magnitude estimation as a scaling procedure, since the model incorporates power functions obtained through this technique. As mentioned earlier, according to Stevens (1975), the numerical response obtained using magnitude estimation instructions is a direct and unbiased estimate of the perceived intensity of the stimulus. This view represents the behavioristic, S-R conceptualization of psychophysics (Shepard, 1981; McKenna, 1985). In contrast, the S-O-R paradigm of psychophysical judgment proposes a psychophysical stage-relating stimulus to sensation (psychophysical input function), and a judgmental stage relating sensation to overt response (judgmental output function) (Torgerson, 1961; Attneave, 1962; Treisman, 1964). Investigators who have adopted the S-O-R view have shown that the judgment function generated when using magnitude estimation is a non-linear and positively accelerating function of the internal sensation (e.g. Curtis,

Attneave, & Harrington, 1968; Rule, Curtis, & Markley, 1970; Rule & Curtis, 1977; Weiss, 1972; Veit, 1978). The 'bias' (i.e. the degree of non-linearity of this function) is dependent on the experimental conditions.

The existence of a non-linear response output function is of no consequence to the validity of the equiratio taste mixture model, because it is an S-R model that predicts the sensory response to a mixture on the basis of sensory responses to unmixed compounds. It does not predict its preceived taste intensity. If an equiratio mixture function is predicted on the basis of psychophysical power functions (meaning here S-R functions) of single substances, *determined in the same experiment under identical conditions*, then the predicted power function is 'biased' to the same degree as the power functions of the single compounds. This follows from the fact that the reported taste intensities of the solutions of both substances and the mixtures are 'biased' by the same response output function. Although the psychophysical input function is substance and mixture specific, the response output function is not. The results of the equiratio mixture studies carried out to date show that the shape of a particular equiratio mixture function does vary from experiment to experiment. However, from the results it can also be concluded that the *relationship* between the functions of the single substances and that of an equiratio mixture type is independent of the stimulus context.

Measurement of the psychophysical function between stimulus and sensation, independently of the judgment function between sensation and response, clearly requires some methodology other than ratio scaling techniques. Klitzner (1975), McBride (1986) and Frank and Archambo (1986) have used functional measurement in the study of mixtures of tastants. De Graaf, Frijters and Van Trijp (1987) employed a method by which psychophysical relationships (S-O) for equiratio mixture types can be determined, and by which the linearity of the response output function (O-R) can be independently verified. This method is also based on the principles of functional measurement (Anderson, 1981).

In this conceptual framework, use is made of factorial designs. In the case of mixtures of taste substances, such a design can be constructed at two different levels; at the physical level and at the judgmental level. There is a fundamental difference between these two. A *factorial mixture design* was used by McBride (unpublished doctoral dissertation, Macquarie University, Sydney, 1982; 1986), where each of a number of concentrations of sucrose, fructose and glucose was added to each of a number of concentrations of the other sugars, yielding sucrose–fructose, sucrose–glucose, and fructose–glucose mixtures. The perceived intensities of *single* stimuli, each composed of two substances, were rated on a 13-point category scale. It was found that the factorial plots of the responses did not exhibit parallel lines. All plots showed a convergent set of lines denoting a significant interaction between the two sugars. Interpretation of such a pattern in terms of mixture interactions is less straightforward than it may seem. We hold the view that, due to the nature of this type of experimental design, it is logically impossible to separate non-parallelism that results from a non-linear judgment function, from non-parallelism that results from non-additive taste intensity integration.

In order to achieve the disentanglement of non-linear sensory integration and non-linearity resulting from judgment processes, equiratio mixture functions for glucose–fructose mixture types were indirectly obtained using a two stimulus task, as frequent y used in combination with functional measurement by Birnbaum and colleagues (e.g. Birnbaum & Veit, 1974; Veit, 1978; Mellers, Davis & Birnbaum, 1984). This means that a *factorial judgment design* was used, where subjects compared the taste intensities of each concentration level of one substance with each of the concentration levels of a particula equiratio mixture type. Their task was to rate the difference between the intensities of the two stimuli. A factorial plot of the difference-responses shows the nature of the (difference) response output function. In the nine designs used in the study by De Graaf *et al.* (1987), the plots appeared to consist

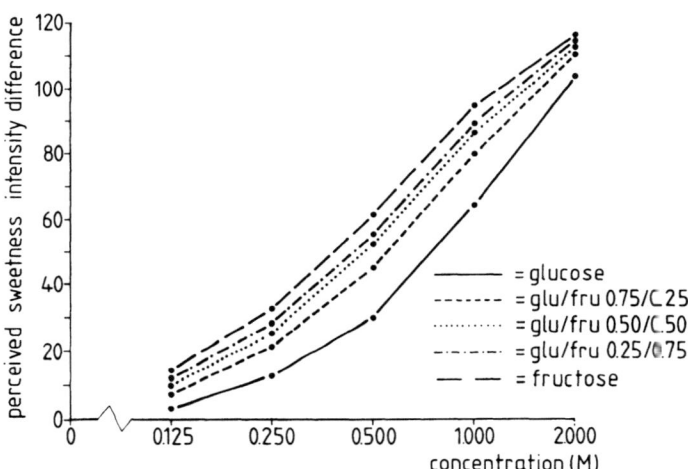

Fig. 14.4. Functions of fructose, glucose and three equiratio mixture types composed of these substances. The functions were obtained with functional measurement in combination with a two-stimulus procedure. (From De Graaf *et al.*, 1987, with permission.)

of sets of parallel lines. It was therefore concluded that in all cases the response output function was linear, so that the difference responses given by the subjects were linear representations of the difference in perceived taste intensity of the pairs of stimuli. The observed parallelism implied that the marginal means of the row and column stimuli of the factorial matrix are valid estimates of the sweetness intensities of the stimuli. By assuming that the sweetness intensity of water is equal to zero, De Graaf *et al.* (1987) established psychophysical functions on a ratio scale for glucose, fructose and the three equiratio mixture types of these substances, that had been earlier investigated by Frijters and Oude Ophuis (1983). As can be seen in Figure 14.4, the psychophysical functions (S-O) of the equiratio mixtures lie between the functions of unmixed glucose and fructose. This experiment shows the usefulness of the approach based on Birnbaum's combination of a two stimulus procedure and functional measurement in conjunction with the concept of Equiratio Taste Mixture Type. Future experimentation is required to test its applicability to the study of mixtures of peripherally independent substances of dissimilar taste qualities.

References

Attneave, F. (1962). Perception and related areas. In E. Koch, ed., *Psychology: a study of a science, Study II,* vol. 4, pp. 619–659. McGraw-Hill, New York.

Anderson, N. H. (1981). *Foundations of information integration theory.* Academic Press, New York.

Baird, J. C., & Noma, E. (1978). *Fundamentals of scaling and psychophysics.* Wiley, New York.

Bartoshuk, L. M. (1975). Taste mixtures: is mixture suppression related to compression? *Physiology & Behavior* 14, 643–649

Bartoshuk, L. M., & Cleveland, C. T. (1977). Mixtures of substances with similar tastes. *Sensory Processes.* 1, 177–186.

Bartoshuk, L. M., & Gent, J. F. (1985). Taste mixtures. An analysis of synthesis. In W. Pfaff., ed., *Taste, olfaction and the central nervous system,* pp. 210–232. The Rockefeller Press, New York.

Beidler, L. M., (1962). Taste receptor stimulation. *Progress in Biophysics and Biophysical Chemistry* 12, 107–151.

Beidler, L. M. (1971). Taste receptor stimulation with salts and acids. In L. M. Beidler, ed., *Handbook of sensory physiology,* vol. 4, part 2, 'Taste' pp. 200–220. Springer-Verlag, New York.

Birnbaum, M. H., & Veit, C. T. (1974). Scale convergence as a criterion for rescaling: information integration with difference, ratio and averaging tasks. *Perception & Psychophysics* 15, 7–15.

Curtis, D. W., Attneave, F., & Harrington, T. L. (1968). A test of a two-stage model of magnitude estimation. *Perception & Psychophysics* 3, 25–31.

De Graaf, C., & Frijters, J. E. R. (1986). A psychophysical investigation of Beidler's mixture equation. *Chemical Senses* 11, 295–314.

De Graaf, C., & Frijters, J. E. R. (1987). Sweetness intensity of a binary sugar mixture lies between sweetness intensities of its components, when each is tasted alone and at the same molarity as the mixture. *Chemical Senses* 12, 113–129.

De Graaf, C., Frijters, J. E. R., & Van Trijp, H. C. M. (1987). Taste interaction between glucose and fructose assessed by functional measurement. *Perception & Psychophysics* 41, 383–392.

Faurion, A., Saito, S., & MacLeod, P. (1980). Sweet taste involves several distinct receptor mechanisms. *Chemical Senses* 5, 107–121.

Frank, R. A., & Archambo, G. (1986). Intensity and hedonic judgments of taste mixtures: an information integration analysis. *Chemical Senses* 11, 427–438.

Frijters, J. E. R. (1987). Psychophysical models for mixtures of tastants and mixtures of odorants. *Annals of the New York Academy of Sciences* 510, 67–78.

Frijters, J. E. R., & De Graaf, C. (1987). The equiratio taste mixture model successfully predicts the sensory response to the sweetness intensity of complex mixtures of sugars and sugar alcohols. *Perception* 5, 615–628.

Frijters, J. E. R., & Oude Ophuis, P. A. M (1983). The construction and prediction of psychophysical power functions for the sweetness of equiratio sugar mixtures. *Perception* 12, 753–767.

Frijters, J. E. R. & Stevens, D. A., (1986). *Psychophysical taste functions for equiratio acid mixtures.* Internal report, Clark University, Worcester, Massachusetts, USA.

Frijters, J. E. R., De Graaf, C., & Koolen, H. C. M. (1984). The validity of the equiratio taste mixture model investigated with sorbitol–sucrose mixtures. *Chemical Senses* 9, 241–248.

Klitzner, M. D. (1975). Hedonic integration: test of a linear model. *Perception & Psychophysics* 18, 49–54.

Kroeze, J. H. A. (1978). The taste of sodium chloride: masking and adaptation. *Chemical Senses & Flavour* 3, 443–449.

Kroeze, J. H. A. (1979). Masking and adaptation of sugar sweetness intensity. *Physiology & Behavior* 22, 347–351.

Kroeze, J. H. A., & Bartoshuk, L. M. (1985). Bitterness suppression as revealed by split-tongue taste stimulation in humans. *Physiology & Behavior* 35, 779–783.

Kuznicki, J. T., & Ashbaugh, N. (1979). Taste quality differences within the sweet and salty taste categories. *Sensory Processes* 3, 157–182.

Lawless, H. T., & Stevens, D. A. (1983). Cross adaptation of sucrose and intensive sweeteners. *Chemical Senses* 7, 309–315.

McBride, R. L. (1986). The sweetness of binary mixtures of sucrose, fructose and glucose. *Journal of Experimental Psychology: Human Perception & Performance* 12, 584–594.

McBurney, D. H. (1972). Gustatory cross adaptation between sweet-tasting compounds. *Perception & Psychophysics* 11, 225–227.

McBurney, D. H., Smith, D. V., & Shick, T. R. (1972). Gustatory cross adaptation: sourness and bitterness. *Perception & Psychophysics* 11, 228–232.

McKenna, F. P. (1985). Another look at the 'new psychophysics'. *British Journal of Psychology* 76, 97–109.

Mellers, B. A., Davis, D. M., & Birnbaum, M. H. (1984). Weight of evidence supports one operation for 'ratios' and 'differences' of heaviness. *Journal of Experimental Psychology: Human Perception & Performance* 10, 216–230.

Moskowitz, H. R. (1972). Perceptual changes in taste mixtures. *Perception & Psychophysics* 11, 257–262.

Moskowitz, H. R. (1973). Models of sweetness additivity. *Journal of Experimental Psychology* 99, 88–98.

Moskowitz, H. R. (1974a). Models of additivity for sugar sweetness. In H. R. Moskowitz, B. Scharf & J. C. Stevens, eds, *Sensation and measurement: papers in honor of S.S. Stevens*, pp. 378–388. Reidel, Dordrecht, The Netherlands.

Moskowitz, H. R. (1974b). The sourness of acid mixtures. *Journal of Experimental Psychology* **102**, 640–647.

Myers, A. K. (1982). Psychophysical scaling and scales of physical stimulus measurement. *Psychological Bulletin* **92**, 203–214.

Pfaffmann, C., Young, P. T., Dethier, V. G., Richter, C. P., & Stellar, E. (1954). The preparation of solutions for research in chemoreception and food acceptance. *Journal of Comparative & Physiological Psychology* **47**, 93–96.

Pfaffmann, C., Bartoshuk, L. M., & McBurney, D. H. (1971). Taste psychophysics. In L. M. Beidler, ed., *Handbook of sensory physiology*, vol. 4, part 2, 'Taste', pp. 75–101. Springer-Verlag, New York.

Rule, S. J., & Curtis, D. W. (1977). Subject differences in input and output transformations from magnitude estimations of differences. *Acta Psychologica* **41**, 61–65.

Rule, S. J., Curtis, D. W., & Markley, R. P. (1970). Input and output transformations from magnitude estimation. *Journal of Experimental Psychology* **86**, 343–349.

Shepard, R. N. (1981). Psychological relations and psychophysical scales: on the status of 'direct' psychophysical measurement. *Journal of Mathematical Psychology* **24**, 21–57

Schiffman S. S., Cahn, H., & Lindley, M. G. (1981). Multiple receptor sites mediate sweetness: evidence from cross-adaptation. *Pharmacology, Biochemistry & Behavior* **15**, 377–388.

Stevens S. S. (1975). *Psychophysics*. Wiley, New York.

Torgerson, W. S. (1961). Distances and ratios in psychophysical scaling. *Acta Psychologica* **19**, 201–205.

Treisman, M. (1964). Sensory scaling and the psychological law. *Quarterly Journal of Experimental Psychology* **16**, 11–22.

Veit, C. T. (1978). Ratio and subtractive process of psychophysical judgment. *Journal of Experimental Psychology: General* **107**, 81–107.

Weiss, D. J. (1972). Averaging, an empirical validity criterion for magnitude estimation. *Perception & Psychophysics* **12**, 385–388.

15

Three models for taste mixtures

Robert L. McBride

CSIRO Division of Food Processing,
Sydney, Australia

I. Introduction

Taste psychophysics has traditionally been concerned with the psycho-physical functions of *single* stimuli—for example, how does sweetness vary with sucrose concentration? This is understandable, indeed logical as a first step, but such studies have little ecological validity: people do not drink solutions of sucrose in water, nor solutions containing any other single taste (cf. Lawless, 1986). For taste psychophysics to accrue genuine usefulness outside the psychological laboratory, it must come to grips with the perception of *taste mixtures.*

So far, research on mixture perception has progressed little beyond the identification of major phenomena, such as taste suppression (Bartoshuk, 1975). In the 1970s, according to Kroeze (unpublished doctoral dissertation, University of Utrecht, 1982, p. 43), 'little more was known about suppression in taste mixtures than that it occurred'. The empirical database on taste mixtures is meager compared with that for single stimuli, and much of what has been done is unsophisticated (Pangborn, 1987); the selection of stimuli has tended to be idiosyncratic and confined to low concentrations. Moreover, uncertainty over the validity of direct scaling has clouded the interpretation of data (McBride, 1986a, 1986b, 1987a).

Clearly, taste research has lacked a systematic paradigm for mixture analysis. As a consequence, the work to date has been largely empirical and insufficiently theoretical — in the words of Lawless (1986), too largely 'top down' (studies in food science with focus on the product) rather than 'bottom up' (studies in psychophysics with focus on the perceptual process). In particular, taste researchers have not availed

Copyright © 1989 by Academic Press Australia.
All rights of reproduction in any form reserved.

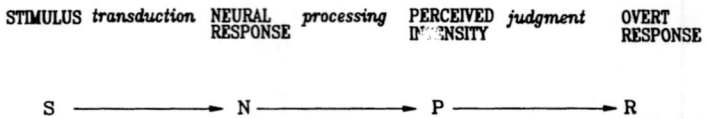

Fig. 15.1. A basic process model for the taste perception of a single stimulus. The stimulus, S, is transduced on the tongue into a neural response, N; the neural response is processed into the taste sensation, or percept, P; this percept is then judged, producing the behavioral response, R.

themselves of conceptual *process models* (as distinct from mathematical/statistical models), such as have advanced knowledge in other areas of psychophysics (MacRae, 1987). The main aim of this article is to present process models for taste mixture perception, structured within the framework of integration psychophysics (Anderson, 1981; McBride, 1986*b*; McBride & Anderson, unpublished data; McBride & Johnson, 1987).

A.   The basic model

The basic process model for the perception of a single stimulus is given in Figure 15.1. In this schema the stimulus, S, is transduced on the tongue into a neural response, N; the neural response is processed into taste sensation, or percept, P; this percept is then judged, producing the behavioral response, R.

A major problem for psychophysics has been uncertainty over the nature of the judgment function, that is, the link between P and R. Psychophysicists are fundamentally interested in the relationship between S and P. However, P is a private event, not directly accessible; only R, the overt response, is measurable. Until we can be sure that R is a linear (unbiased) measure of P, progress is severely handicapped. This issue is dealt with in depth elsewhere (Anderson, 1981, ch. 5; De Graaf, Frijters & Van Trijp, 1987; McBride, 1986*a*, 1987*a*) and will not be recounted here. Suffice it to say there is now considerable evidence to support rating as a linear judgment technique.

One strand of evidence embodies the convergence between neurophysiological and psychophysical measures of taste intensity (McBride, 1987*a*, 1987*b*). When varying concentrations of a taste stimulus are presented to an animal (or human), and the resulting neural response is recorded in the taste nerve, the response versus concentration curve is given by the Beidler equation (Beidler, 1954):

$$\frac{N}{N_{max}} = \frac{CK}{1 + CK} \tag{15.1}$$

where N is neural response, N_{max} is the maximum (saturated) neural response, C is (molar) stimulus concentration, and K the association (binding) constant of the stimulus. However, equation 15.1 not only accounts for the neural taste response; significantly, it also holds for the *psychophysical* taste response in humans (Beidler, 1987; McBride, 1987b). In terms of Figure 15.1, recording the R response gives the same result as recording the N response; in terms of equation 15.1, R/R_{max} may be substituted for N/N_{max}. The simplest interpretation of this accord is that both the N-P and P-R links in Figure 15.1 are linear.

Equation 15.1, with R/R_{max} substituted on the left-hand side, is an advance over previous psychophysical formulations. First, it is not merely descriptive, since the parameters K and R_{max} have underlying mechanistic significance: K relates to the binding efficiency of the stimulus—the larger the K, the more efficient the binding and the lower the taste threshold; R_{max}, on the other hand, relates to the maximum possible (saturated) intensity, which may well depend on the number of receptor sites available: the greater the number of sites, the larger the R_{max}. A second feature of the Beidler equation is the capacity to express taste intensity in terms of a taste index (or coefficient), R/R_{max}. Zero is the lower limit, only theoretically attainable, and 1 (or 100 if convention would prefer) is the upper limit of complete saturation, again only theoretically attainable. Thus, we have something approaching a 'Kelvin' scale of taste, with true ratio properties (McBride, 1987b).

B.   Two types of taste mixtures

Equipped with a basic model, the next step is to extend Figure 15.1 to accommodate simple (i.e. binary) taste stimuli. First, it is useful to draw a distinction between *homogeneous* and *heterogeneous* taste mixtures (cf. Frijters, 1987).

Homogeneous mixtures are those that produce a unitary taste percept. Sweetener mixtures are a good example: a mixture of sucrose and fructose evokes sweetness only, as does a mixture of sucrose and aspartame. Thus, the term refers to a homogeneity in the sensation, not to any physical (molecular) commonality. This categorization is something of an oversimplification because no two taste stimuli evoke *exactly* the same sensation; nevertheless, it has general plausibility and facilitates model construction.

Heterogeneous mixtures, in contrast, are those that clearly evoke more than one type of taste sensation. For instance, a mixture of sucrose and citric acid produces three percepts and their corresponding responses: total intensity (irrespective of taste quality), sweetness, and acidity (sourness).

II.   Homogeneous mixtures

Figure 15.2A depicts a *single-site model* for homogeneous mixtures. In this case there is only one type of receptor site and the two stimuli, S_1 and S_2 (e.g. sucrose and fructose), compete for absorption at these same sites. Thus, 'interaction' occurs at the surface of the tongue; once the receptor site has been stimulated, the processing proceeds as for the single stimulus in Figure 15.1. Other ways of conceptualizing this operation include a 'competition for sites' or 'substitution' process: the two stimuli differ in their affinity for the receptor site (K value), but otherwise the taste system does not differentiate between them (e.g. see Moskowitz, 1973). Assuming that R_{max} is constant for components and their mixtures

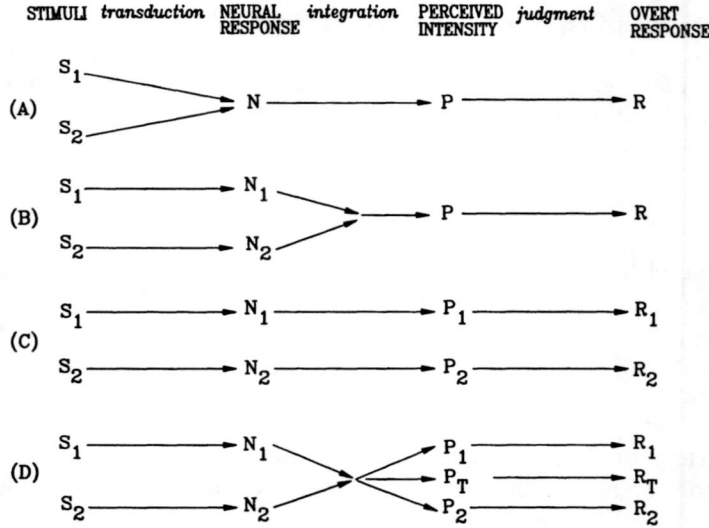

Fig. 15.2.   Process models for the perception of (binary) taste mixtures. (A) Single-site model for homogeneous mixtures. (B) Separate-sites model for homogeneous mixtures. (C) Independent-processing model for heterogeneous mixtures. (D) Interaction model for heterogeneous mixtures. For explanation see text.

(McBride, 1987*b*), a quantitative formulation of the single-site model may be given by the Beidler mixture equation (Beidler, 1962, 1987):

$$\frac{R_{AB}}{R_{max}} = \frac{C_A K_A + C_B K_B}{1 + C_A K_A + C_B K_B} \tag{15.2}$$

where R_{AB} is the taste response to a mixture of stimuli A and B, R_{max} is the maximum (saturated) taste response, K_A and K_B are, respectively, the association constants of stimuli A and B, and C_A and C_B are molar concentrations.

The alternative Figure 15.2B depicts a *separate-sites model* for homogeneous taste mixtures. Here, the taste system *does* distinguish between the two stimuli: they are transduced independently at separate receptor sites, then subsequently integrated in the taste system. The mathematical formulation for the model may be given as (Jakinovich, 1982; McBride, 1986*b*, 1987*a*, 1988):

$$R_{AB} = R_A + R_B - \left(\frac{R_A \cdot R_B}{R_{max}}\right) \tag{15.3}$$

where R_{AB} is the response to the mixture of A and B, R_{max} is maximum response, and R_A and R_B are, respectively, the responses to components A and B. Values for R_A and R_B are obtained from equation 15.1.

A. Which model?

Which model more closely underpins the reality of mixture perception? A recent overview of sweetness reception (Bartoshuk, 1987) suggests a preponderance of support for a separate-sites mechanism. The following experiment, reported in detail elsewhere (McBride, 1988), further confirms this position.

The experimental rationale was simple. Equations 15.2 and 15.3 make different predictions about the intensity of a homogeneous mixture relative to the intensity of its components, and this permits an empirical check. Figure 15.3 gives the individual psychophysical functions for the sugars sucrose and fructose, with percentage weight per volume (% w/v) as the stimulus unit. These functions follow the Beidler equation (K values are 5.3 for sucrose, 3.3 for fructose) and have been confirmed by both Just Noticeable Difference (JND) cumulation and category rating (McBride, 1987*b*). The other two curves in Figure 15.3 are theoretical predictions of the sweetness intensity for an equal-parts (by weight)

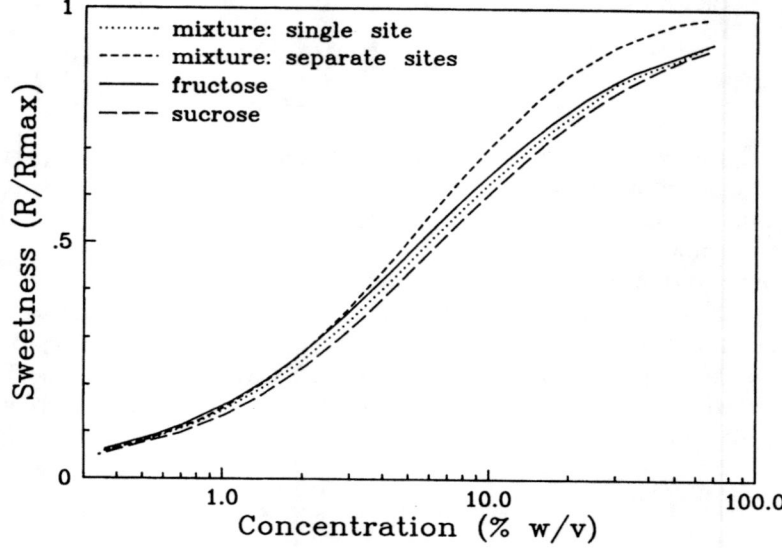

Fig. 15.3. Psychophysical functions for sucrose, fructose, and for an equal-parts (by weight) mixture of the two as specified by the single-site and separate-site reception models (% w/v = percentage weight per volume).

mixture of the two sugars, as specified by equation 15.2 (single-site model) and equation 15.3 (separate-sites model). The abscissa units represent total weight: at the 10% mark, for example, the concentration of sucrose alone is 10%, fructose alone is 10%, and the mixture consists of 5% sucrose + 5% fructose.

There is little disparity between the two models at low stimulus levels (< 2%), with both predicting the intensity of the mixture to lie between equivalent concentrations of its components. As the concentration increases, however, the curves diverge. According to the single-site model, the sweetness of the mixture always lies between its components, regardless of total concentration, whereas the separate-sites model specifies an increase in sweetness over the fructose alone, with the crossover predicted to occur at 2.2% fructose. These predictions were checked by several series of paired-comparison tastings (i.e. 'which is sweeter?'), the most salient of which was comparison of the sucrose-fructose mixture versus fructose alone at several concentrations (1.0%, 3.2%, 10.0%, 17.8%, 31.8%; 40 judgments at each).

The results clearly supported the separate-sites model and disconfirmed the single-site model. At the 1.0% level, the split of the 40 responses showed fructose to be sweeter than the mixture (32/8, $p < .05$); at 3.2%, the split was equivocal (21/19); and, at 10%, the split indicated the mixture to be sweeter than the fructose (9/31, $p < .05$). So, within the space of a 1 log-unit jump in total concentration (1% to 10%), there was a clear reversal in relative sweetness, consistent with the separate-sites model in Figure 15.3. At the two highest concentration levels (17.8% and 31.6%) the mixture was still sweeter than the fructose alone (26/14 and 24/16, respectively); however, the split was not as decisive as would be predicted from the separation of the curves in Figure 15.3, suggesting that equation 15.3 overpredicts mixture sweetness at high concentrations (McBride, 1988).

A similar outcome was achieved in an investigation of fructose-glucose mixtures, further supporting the separate-sites model, although again equation 15.3 tended to overpredict the sweetness of the mixture at high concentration. Slight reformulation is necessary.

B. 'Synergism' explained

'Synergism' has often been noted with mixtures of sweeteners (cf. Lawless, 1986). The exact definition of this phenomenon varies (an issue in itself), but the general consensus is that the sweetness of a mixture is somehow greater than might be rationally expected given the sweetnesses of the individual components. The above experiment provides a good example: fructose is sweeter than sucrose at all concentrations, yet, when half of a 10% fructose solution is replaced by the less sweet sucrose, the sweetness of the mixture *increases* rather than decreases.

The separate-sites model readily accounts for this effect. It occurs because the psychophysical functions of the single sugars are negatively accelerated: the sweetness at 5% is greater than half the sweetness at 10% (see Fig. 15.3). So, when the sweetness of 5% sucrose is combined with the sweetness of 5% fructose according to equation 15.3, they together exceed the sweetness of 10% fructose alone. Note that this effect is *not* synergistic in that it exceeds additivity; on the contrary, as is evident from equation 15.3, the sweetness of the mixture is actually less than the sum of its components.

III. Heterogeneous mixtures

Figure 15.2C displays an *independent-processing model* for heterogeneous mixtures. The two stimuli are independently transduced at separate receptor sites; the respective neural responses are then independently processed into separate percepts; the percepts are then independently processed into separate responses. There is no interaction or integration anywhere in the system, so the response for either component is as if it were presented alone. It follows, furthermore, that because there is no integration, the total intensity of the mixture must correspond to the intensity of the stronger, or dominant component.

Figure 15.2D depicts an *interaction model* for heterogeneous taste mixtures. The two stimuli are transduced independently at separate sites, after which there is provision for interaction of the two channels before they are perceived. The two percepts, P_1 and P_2, are then transformed to their respective outputs, R_1 and R_2. In this model there is also provision for perceived total intensity, P_T, to be an integrated percept of N_1 and N_2 (cf. Fig. 15.2B).

The following experiment, conducted within the protocol of integration psychophysics, sought to investigate independent-processing (Fig. 15.2C) and interaction (Fig. 15.2D) as models for the perception of heterogeneous mixtures.

A.   Experiment

Fourteen employees (eight men, six women) of the CSIRO Food Research Laboratory participated voluntarily. All had had some experience in sensory testing. Stimuli were mixtures of sucrose and citric acid (both reagent grade) dissolved in distilled water. Five levels of sucrose were combined factorially with five levels of citric acid, giving 25 stimuli in all. The concentrations of sucrose were 0, 0.080, 0.172, 0.371, and 0.800 Molar (M); of citric acid, 0, 0.002, 0.006, 0.017, and 0.050 M (both series were geometrically spaced). Thus, the weakest stimulus consisted of distilled water only, and the strongest of 0.800 M sucrose *and* 0.050 M citric acid in 1 L of distilled water. Solutions were made up at least 24 hours before testing and stored at 5°C. At evaluation, each sample consisted of 10 mL of solution served at ambient temperature (20°C) in a small glass tumbler.

Subjects made their responses in a booklet of response sheets, one sheet per stimulus. Each sheet contained a set of three, 150 mm graphic rating scales, one each for the assessment of (a) total intensity, (b) sweetness and (c) acidity. For total intensity, the descriptors 'No taste at all', 'Moderate', and 'Extremely strong' were positioned 0 mm, 75 mm and 150 mm, respectively, from the left-hand end of the scale. Corresponding descriptors for sweetness were 'No sweetness at all', 'Moderately sweet', and 'Extremely sweet'; and, for acidity, 'No acid at all', 'Moderately acid', and 'Extremely acid'.

Each subject assessed all 25 stimuli at each of three separate sessions (replicates) held a few days apart, providing a total of 42 observations per stimulus. Subjects were required to 'sip and spit' each stimulus according to the (random) order specified in the response booklet, and to rate the total intensity (sweetness, acidity) of each on the appropriate graphic rating scale. Rinsing with distilled water was mandatory between stimuli.

B. Perception of total intensity

Mean ratings for total intensity are given as a factorial plot in Figure 15.4. Sucrose concentration is given on the abscissa (log scale) and each curve corresponds to a different level of citric acid. Analysis of variance revealed large main effects for sucrose and for citric acid, $F_{(4, 698)} \geq 182.91$ $p < .001$, and, more pertinently, a large sucrose $\times$ citric acid interaction, $F_{(16, 698)} = 26.51$ $p < .001$. Any two data points separated vertically by more than 1 LSD (least significant difference) are significantly different, $p < .05$.

The sucrose $\times$ citric acid interaction confirms that total intensity is not the simple algebraic sum of sweetness and sourness; visually, this is evident from the non-parallel, convergent pattern of the curves in Figure 15.4 (Anderson, 1981; McBride, 1986b). In fact, the simplest interpretation of these data is that they conform to a model of independent processing (Fig. 15.2C), that is, to a model of no integration at all.

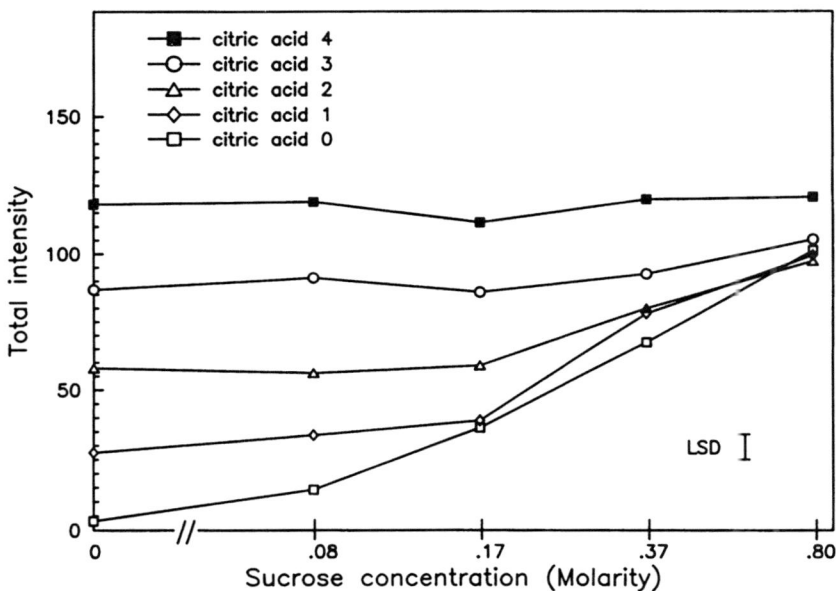

Fig. 15.4.   Total intensity. Mean ratings for 25 sugar–acid combinations are given on the ordinate, sucrose concentration on the (log) abscissa, and each curve represents a different level of citric acid. Data points separated (vertically) by more than 1 LSD are significantly different ($p < .05$).

For example, note the mean rating for the total intensity of the highest level of citric acid alone (top data point on the ordinate). When increasing concentrations of sucrose are added to this level of citric acid (i.e. we move along the top curve in the plot), there is no increase in total intensity at all; the addition of sucrose, even at high (0.80 M) concentration, does nothing to augment the total intensity of the mixture. This pattern also holds for the second highest level of citric acid, with the possible exception of the mixture on the extreme right of the curve. The middle concentration of citric acid, likewise, dictates the total intensity of the mixture, *until* the sucrose concentration reaches 0.37 M; at this point and at the next (0.80 M), the total intensity of the mixture is almost the same as the total intensity of the sucrose when tasted alone (bottom curve) — in these two cases it is the sucrose that dictates the intensity of the mixture. The curve second from the bottom displays the same 'crossover' pattern: total intensity is dictated by citric acid at the three lowest levels of sucrose, but by sucrose at the two highest levels of sucrose.

The simplest way to conceptualize this outcome is in terms of a *dominant component model*: the (subjectively) dominant component determines the total intensity of the mixture. That is, if R_1 and R_2 are the respective intensities of the mixture components when tasted alone, the intensity of the mixture is equal to R_1 when $R_1 > R_2$, and R_2 when $R_1 < R_2$. Very similar patterns have been reported in another recent integration study of sucrose-citric acid and sucrose-sodium chloride mixtures (Frank & Archambo, 1986), and also for sucrose-citric acid in lemon drink (McBride & Johnson, 1987).

More generally, the concept of the dominant component has now been established with neurophysiological data (Smith, this volume). It was invoked by Cometto-Muniz (1981) in a description of odor/taste mixtures, and may also be applicable to certain odor mixtures (Laing & Willcox, 1983, Fig. 3) and mixtures of olfactory and trigeminal stimuli (Cain & Murphy, 1980). It follows, of course, that acceptance of Figure 15.2C as a model for total intensity rules out total intensity as an integrated percept (i.e. there can be no P_T and R_T in Fig. 15.2D).

C.   Sweetness

Sweetness responses are given in Figure 15.5A. There was, predictably, a highly significant main effect of sucrose, $F(4, 698) = 1301.55$ $p < .001$, but also a main effect of citric acid, $F(4, 698) = 11.51$ $p < .001$. Increasing concentrations of sucrose increased sweetness; increasing

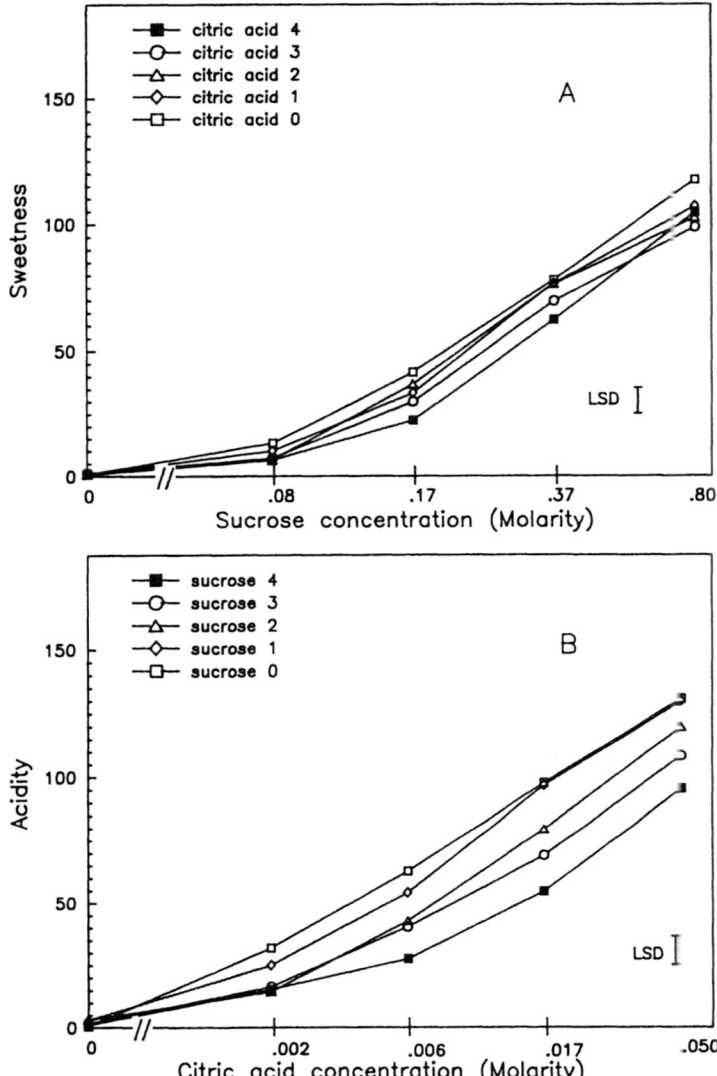

Fig. 15.5. Sweetness and acidity. Panel A contains mean sweetness ratings for the 25 sugar–acid combinations; sucrose concentration is given on the (log) abscissa, and each curve represents a different level of citric acid. Panel B contains mean ratings for acidity; citric acid concentration is given on the (log) abscissa and each curve represents a different level of sucrose. Data points separated (vertically) by more than 1 LSD are significantly different ($p < .05$).

concentrations of citric acid decreased sweetness. The sucrose × citric acid interaction was also significant, $F(16, 698) = 2.21$ $p < .01$; however, it disappeared when the data were reanalyzed without the two lowest levels of sucrose, $F(8, 627) = 1.40$.

Citric acid had only a slight suppressive effect on the sweetness of sucrose; substantively, the suppression is noticeable only for the highest concentration of citric acid (the curves for all but the highest levels of citric acid are generally within an LSD of the curve for sucrose alone). This result is identical to that obtained with sucrose-citric acid mixtures in lemon-juice drink (McBride & Johnson, 1987), and consistent with other previous studies (Frank & Archambo, 1986; Gordon, 1965; Pangborn, 1960, 1961).

D.   Acidity

Acidity responses are given in Figure 15.5B. The main effect of citric acid was highly significant, $F(4, 698) = 997.98$ $p < .001$, as was the main effect of sucrose $F(4, 698) = 53.11$ $p < .001$: increasing concentrations of citric acid increased acidity; increasing concentrations of sucrose decreased acidity. The sucrose × citric acid interaction was significant, $F(16, 698) = 5.66$ $p < .01$, but dropped out when the data were reanalyzed without the two lowest levels of citric acid, $F(8, 627) = 0.91$.

Sucrose (sweetness) suppressed the acidity of citric acid; indeed, as is apparent from Figure 15.5, the suppression of acidity by sucrose is considerably greater than the suppression of sweetness by citric acid. Suppression was mutual, but not balanced. Again, this result is identical to that obtained in lemon-juice drink (McBride & Johnson, 1987), and consistent with other previous studies (Frank & Archambo, 1986; Gordon, 1965; Pangborn, 1960, 1961). The priority or 'primacy' of sweetness over acidity may be related to a recent finding from taste coding (Scott & Giza, 1987) that sweetness is neurally the most distinct of taste qualities.

E.   Suppression as subtraction

As noted above, when data for the two bottom abscissa values are omitted from the sweetness and acidity judgments (i.e. data matrices in Fig. 15.5 are reduced to 3 × 5 factorials), the sucrose × citric acid interactions drop out. This implies the interactions are due solely to a 'squeezing' of the curves at the lower end. This squeezing is inevitable. Taste sensation cannot be negative, so when a weak stimulus is totally suppressed, the ratings are perforce squeezed together at the bottom of the response scale. For example, when 0.002 M citric acid (weak) is mixed with either 0.37 M or 0.80 M sucrose (strong), the acidity rating is no different from the acidity rating for distilled water (see Fig. 15.5B).

However, at the top three abscissa concentrations, where suppression is only partial, the curves are parallel as determined by analysis of variance: the magnitude of the suppression depends *only* on the intensity of the suppressor and is independent of the intensity of the component suppressed. Figure 15.5B shows, for instance, that the highest level of sucrose (sucrose 4) exerts the same degree of suppression whether it is mixed with 0.006 M, 0.017 M or 0.050 M citric acid.

This same pattern is evident in the data of Lawless (1977), on sucrose–quinine mixtures, and in mixtures of sucrose-citric in lemon drink (McBride & Johnson, 1987). It raises the intriguing possibility that taste suppression might, at least in some instances, follow a simple, subtractive principle. In terms of Figure 15.2D, the integration function would be algebraic subtraction.

IV. Three models for taste mixtures

The foregoing analysis suggests the perception of binary taste mixtures might be accounted for by the following three models:

1. *a separate-sites model* (Fig. 15.2B) for the perceived total intensity of homogeneous mixtures;
2. *a dominant component model* (Fig. 15.2C) for the perceived total intensity of heterogeneous mixtures;
3. *an algebraic integration model* (Fig. 15.2D with P_T and R_- omitted) for the perceived intensity of the components in a heterogeneous mixture.

A. *Locus of integration*

This issue is far from resolved, although there are indications that the locus of integration may depend upon mixture type. For homogeneous mixtures, Jakinovich (1982) invoked equation 15.3 on the basis of the neural (*chorda tympani*) response of the gerbil to mixtures of sucrose and saccharin. This implies the integration occurs peripherally, the separate neural responses converging in a common effector system.

But for heterogeneous mixtures there are indications, consistent with other work (Kroeze unpublished doctoral dissertation, University of Utrecht, 1982; Kroeze & Bartoshuk, 1985; Lawless, 1979), that integration takes place at a higher, central level. First, the apparent lack of integration in judgment of total intensity (Fig. 15.4) would seem to

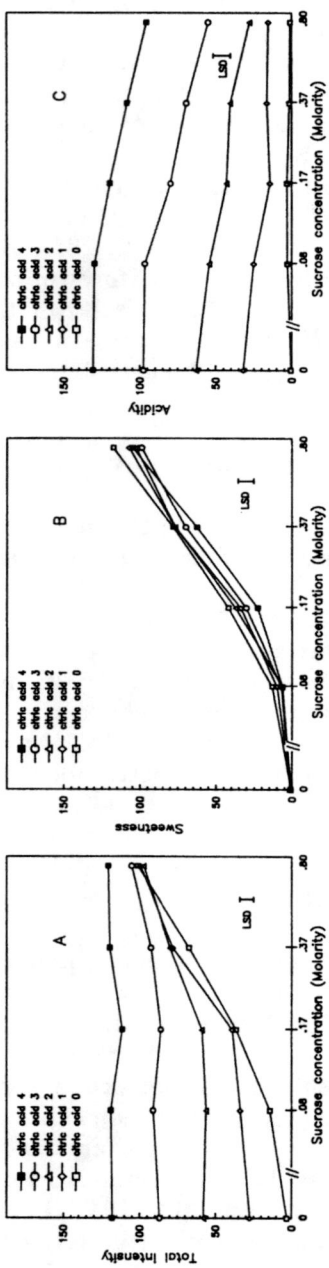

preclude peripheral interaction. Second, the parallelism in Figure 15.5 resembles the patterns of cognitive algebra (Anderson, 1981): by association, therefore, there is implication of a high-order, central effect. Third, it seems unlikely a peripheral mechanism could account for the following, paradoxical effect.

B. *The intensity/quality contradiction*

Figure 15.6 is a replot of the sucrose–citric acid experiment reported above. Panels A and B are as per Figures 15.4 and 15.5A, respectively, and the data in panel C are replotted from Figure 15.5B, but with sucrose concentration on the abscissa. This puts all panels in Figure 15.6 on an equivalent basis for comparison.

As noted earlier, at the top level of citric acid the acid alone determines total intensity, and this remains constant regardless of sucrose addition (top curve in Fig 15.6A is horizontal). However, according to the top curve in Figure 15.6C, not all of this total intensity is perceived as acidity: above 0.08 M sucrose, the acid intensity is suppressed and begins to drop away. It is as if the total intensity of the acid component is transmitted to consciousness unaffected by the sweetness, yet, at the same time, not all of this intensity is perceptible as acidity. This contradiction might possibly be reconciled if suppression were a high-level, 'attentional' phenomenon — that is if one component could not be perceived at the exclusion of the other (cf. Kroeze, this volume).

C. *Practical implications*

Food manufacturers have long been aware of the practical implications of mixture mechanisms. For instance, they have taken advantage of the supplemental action that occurs with sweeteners and with other types of homogeneous mixtures (e.g. the umami stimuli MSG and GMP; Rifkin & Bartoshuk, 1980). If equation 15.3 holds, even as a good approximation, the food technologist will be able to predict the intensity of a homogeneous mixture from the physical concentrations of its components.

◁ **Fig. 15.6.**   Total intensity/sweetness/acidity: comparison of ratings for the three percepts. In each panel (log) sucrose concentration is given on the abscissa and the curves represent different levels of citric acid.

For heterogeneous mixtures, too, practical application of mixture mechanisms is of long standing: the suppression of sourness and bitterness by sweetness is a well-established principle in food preparation (e.g. the addition of sugar to grapefruit or coffee). Furthermore, the intensity/quality contradiction may be an important, albeit unrecognized, principle in recipe formulation.

Overall taste intensity — 'impact' in the mouth — is important in the acceptability of food and drink (McBride & Booth, 1986). In a lemon drink, overall intensity is just as important as the actual sugar/acid balance (McBride & Johnson, 1987). Figure 15.6 implies that a desirably high level of total intensity may be obtained in a product without the dominant component necessarily imparting all of its character to the product: the acid component of a soft drink may provide high flavor intensity, but sweetness will moderate the perceived acidity. Similarly, the addition of salt to a bland product (e.g. a cereal) will dramatically increase the overall taste intensity, but, if sugar is added as well, as it is in the case of porridge and breakfast cereals, the sweetness will suppress much of the saltiness. (At certain levels the salt may even enhance the sweetness; Frank & Archambo, 1986.) The result is a product of high taste intensity that tastes only moderately sweet and moderately salty.

D.   *Process models in mixture research*

Of models in sensory analysis, MacRae (1987, p. 7) wrote:

> Models are often regarded as the concern of the theoretician and not of the practical person, but that is far from being true. Models are involved at various stages of any investigation, though they are not always recognized. At the very least, they should be considered explicitly whenever dealing with a new situation.

The aim of this article has been to present process models for the perception of taste mixtures. Those presented appear promising, although their actual veracity is to some extent beside the point. The models are simply hypotheses, open for scrutiny and challenge; they will stimulate further research and, as knowledge accumulates, they will be revised. Process models enable mixture research to progress beyond undirected empiricism.

References

Anderson, N. H. (1981). *Foundations of information integration theory.* Academic Press, New York.

Bartoshuk, L. M. (1975). Taste mixtures: is mixture suppression related to compression? *Physiology & Behavior* **14**, 643–649.

Bartoshuk, L. M. (1987). Is sweetness unitary? An evaluation of the evidence for multiple sweets. In J. Dobbing, ed., *Sweetness*, pp. 33–47. Springer-Verlag, Berlin.

Beidler, L. M. (1954). A theory of taste stimulation. *Journal of General Physiology* 38, 133–139.

Beidler, L. M. (1962). Taste receptor stimulation. *Progress in Biophysics* 12, 109–151.

Beidler, L. M. (1987). Vertebrate taste receptors. In R. F. Chapman, E. A. Bernays & J. G. Stoffolano, Jr, eds, *Perspectives in chemoreception and behavior*, pp. 47–58. Springer-Verlag, New York.

Cain, W. S., & Murphy, C. L. (1980). Interaction between chemoreceptive modalities of odour and irritation. *Nature* 284, 255–257.

Cometto-Muniz, J. E. (1981). Odor, taste, and flavor perception of some flavoring agents. *Chemical Senses* 6, 215–223.

De Graaf, C., Frijters, J. E. R., & Van Trijp, H. C. M. (1987). Taste interaction between glucose and fructose assessed by functional measurement. *Perception & Psychophysics* 41, 383–392.

Frank, R. A., & Archambo, G. (1986). Intensity and hedonic judgments of taste mixtures: an information integration analysis. *Chemical Senses* 11, 427–438.

Frijters, J. E. R. (1987). Psychophysical models for mixtures of tastants and mixtures of odorants. *Annals of the New York Academy of Sciences* 510, 67–78.

Gordon, J. (1965). Evaluation of sugar-acid-sweetness relationships in orange juice by a response surface approach. *Journal of Food Science* 30, 903–907.

Jakinovich Jr, W. (1982). Stimulation of the gerbil's gustatory receptors by saccharin. *Journal of Neuroscience* 2, 49–56.

Kroeze, J. H. A., & Bartoshuk, L. M. (1985). Bitterness suppression as revealed by split-tongue taste stimulation in humans. *Physiology & Behavior* 35, 779–783.

Laing, D. G., & Willcox, M. E. (1983). Perception of components in binary odour mixtures. *Chemical Senses* 7, 249–264.

Lawless, H. T. (1977). The pleasantness of mixtures in taste and olfaction. *Sensory Processes* 1, 227–237.

Lawless, H. T. (1979). Evidence for neural inhibition in bittersweet taste mixtures. *Journal of Comparative and Physiological Psychology* 93, 538–547.

Lawless, H. T. (1986). Sensory interactions in mixtures. *Journal of Sensory Studies* 1, 259–274.

MacRae, S. (1987). The interplay of theory and practice in sensory testing. *Chemistry & Industry*, no. 1, 7–12.

McBride, R. L. (1986a). Scale convergence as a validity criterion: a matter of taste. In B. Berglund, U. Berglund & R. Teghtsoonian, eds, *Fechner Day 86*, pp.109–114. International Society for Psychophysics, Stockholm.

McBride, R. L. (1986b). Sweetness of binary mixtures of sucrose, fructose and glucose. *Journal of Experimental Psychology: Human Perception and Performance* 12, 584–591.

McBride, R. L. (1987a). Psychophysics as measurement science, sensory science, food science. *Chemistry & Industry*, no. 1, 25–30.

McBride, R. L. (1987b). Taste psychophysics and the Beidler equation. *Chemical Senses* 12, 323–332.

McBride, R. L. (1988). Taste reception of binary sugar mixtures: psychophysical comparison of two models. *Perception & Psychophysics* 44, 167–171.

McBride, R. L., & Booth, D. A. (1986). Using classical psychophysics to determine ideal flavour intensity. *Journal of Food Technology* 21, 775–780.

McBride, R. L., & Johnson, R. L. (1987). Perception of sugar–acid mixtures in lemon juice drink. *International Journal of Food Science and Technology* 22, 339–403.

Moskowitz, H. R. (1973). Models of sweetness additivity. *Journal of Experimental Psychology* **99**, 88–98.

Pangborn, R. M. (1960). Taste interrelationships. *Food Research* **25**, 245–256.

Pangborn, R. M. (1961). Taste interrelationships. II. Suprathreshold solutions of sucrose and citric acid. *Journal of Food Science* **26**, 648–655.

Pangborn, R. M. (1987). Selected factors influencing sensory perception of sweetness. In J. Dobbing, ed., *Sweetness*, pp. 49–66. Springer-Verlag, Berlin.

Rifkin, B., & Bartoshuk, L. M. (1980). Taste synergism between monosodium glutamate and disodium 5′-guanylate. *Physiology & Behavior* **24**, 1169–1172.

Scott, T. R., & Giza, B. K. (1987). Neurophysiological aspects of sweetness. In J. Dobbing, ed., *Sweetness*, pp. 15–32. Springer-Verlag, Berlin.

PART FOUR
INTERACTIONS OF
THE CHEMICAL SENSES

16

Separating the contributions of smells and tastes in flavor perception

David E. Hornung

*Department of Biology, St Lawrence University,
Canton, New York, USA*

Melvin P. Enns

*Department of Psychology, St Lawrence University,
Canton, New York, USA*

I. Introduction

It is clear from many of the other chapters in this volume that some formidable problems must be solved before an understanding of smell/smell or taste/taste mixture effects is achieved. When mixtures involve the two modalities of smell and taste, the problems become even more complex. That is, questions of the interaction of taste and smell involve all the problems of single modality mixtures as well as problems of how information from two sensory systems is combined.

Basically two experimental approaches have been used to describe taste/smell mixture effects. The first approach focuses on how taste and smell combine to form the sensation of overall intensity or flavor. The results of studies in which this approach has been used have shown remarkable agreement. It is now generally accepted that taste and smell add together to produce the sensation of overall intensity (Fig. 16.1). Although the sum of the estimates of taste and smell is usually higher than the overall intensity of the mixtures, it has recently been observed that how the question of intensity is asked may have some impact on whether the quotient of overall intensity divided by the sum of taste and smell is equal to or lower than 1.0 (Hornung & Enns, 1986; Hornung & Enns, 1987a).

PERCEPTION OF COMPLEX SMELLS
AND TASTES ISBN 0 12 042990 X

285

*Copyright © 1989 by Academic Press Australia.
All rights of reproduction in any form reserved.*

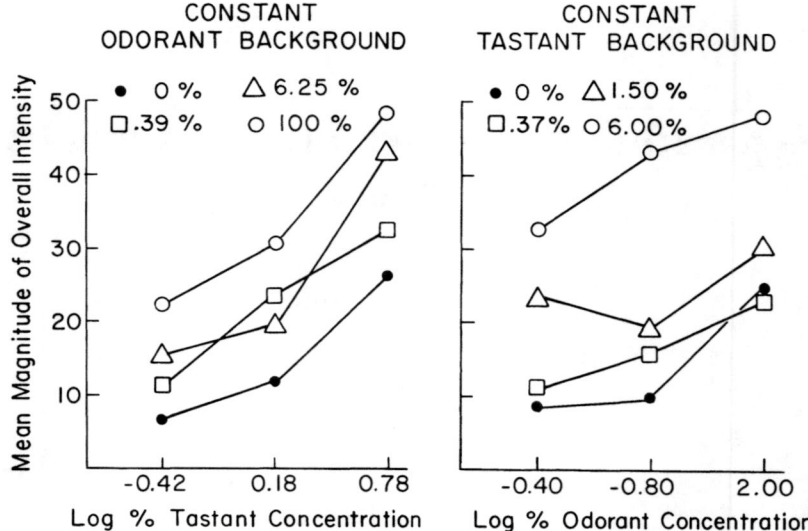

Fig. 16.1. The scaling of the overall intensity of almond extract with four odorant (left plate) and four tastant (right plate) concentrations. Because the two-module delivery system was used, almond extract could serve as both the odorant and the tastant, and the concentrations delivered to the nose and mouth could be varied independently of each other. The mean magnitude estimation (10 subjects × 4 repeats per point) is plotted as a function of the log percentage concentration of almond extract. An analysis of variance indicated that magnitude estimates were influenced by changes in concentration of the odorant and tastant, but not by the interaction of odorant and tastant. Thus, when the magnitude estimations are plotted as a function of the log concentration of the odorant or tastant, parallel lines emerge. Thus taste and smell add together (without statistical interaction) to produce the sensation of overall intensity or flavor. (Reprinted from Enns & Hornung, 1985, with permission.)

The second experimental approach concerns the effect that smell can have on taste and vice versa. The purpose of this chapter is to examine these effects, which are more commonly known as taste or smell confusions. The tendency for people to confuse smell with taste has been recognized for over 150 years. To appreciate these confusions, one need only remember how 'tasteless' food seems when one has a head cold. This example serves not only to suggest the possibility of a smell/taste confusion, but also that these confusions are often resolved in favor of the mouth rather than the nose.

In 1824, Chevreul observed that food stimulates 'touch' receptors in the mouth as well as the taste and smell receptors themselves. Chevreul further suggested that although it was not possible to separate the action of a substance on the touch receptors of the tongue from the action on the taste buds themselves, the sense of smell could be eliminated by pinching the nostrils closed while a substance was examined in the mouth. When the nostrils were closed, Chevreul found that the placement of a piece of camphor gum on the tongue produced only a pricking sensation. Thus, although camphor was thought to have a distinct taste, he concluded that the sensation was a fusion of odor and touch. In the light of what is now known about taste/smell confusions, these observations by Chevreul were remarkably accurate.

Hollingworth and Poffenberger (1917) suggested that a 'very great number' of what are usually perceived as tastes are not tastes at all, but are in fact odors. Further, these authors suggested that the sense organ of smell is so situated that it may be stimulated not only in the 'ordinary way', in which odorant molecules are drawn into the nostrils by a sniff, but also by vapors that pass from the oral cavity to the headspace above the olfactory region.

In support of this position, Hollingworth and Poffenberger reported that when substances were reduced to a similar consistency, it was impossible to distinguish between quinine and coffee or between apple and onion when the nostrils were pinched. In a more practical example they asked, 'How often has the nasty taste of medicine been softened by Chevreul's simple technique of holding the nose'? These authors also suggested that there may be some 'rare' cases in which volatile substances entering the mouth through the nostrils may stimulate the taste buds. In these situations the real taste is mistakenly interpreted as an odor.

Mozell, Smith, Smith, Sullivan Jr and Swender (1969) also demonstrated the importance of olfaction in flavor identification. Human volunteers were unable to identify most common foods (Fig. 16.2) when a stream of humidified air was forced through the external naris (Fig. 16.3). This stream prevented odorant molecules from reaching the olfactory receptors. Since foods could not be identified on the basis of taste, Mozell and his co-workers suggested that olfaction was the critical sense for flavor identification.

II.   Effects of smell stimuli on taste perception

The first study designed to quantify the effect of smell on taste was conducted by Murphy, Cain, and Bartoshuk (1977). Subjects in this study estimated the taste intensity of solutions of ethyl butyrate, saccharin and mixtures of the two solutes. Taste estimates were made with the nostrils open and with the nostrils pinched closed. Although patency of the nostrils did not influence the estimates of the intensity of the taste of saccharin, patency did influence the estimates of the taste of

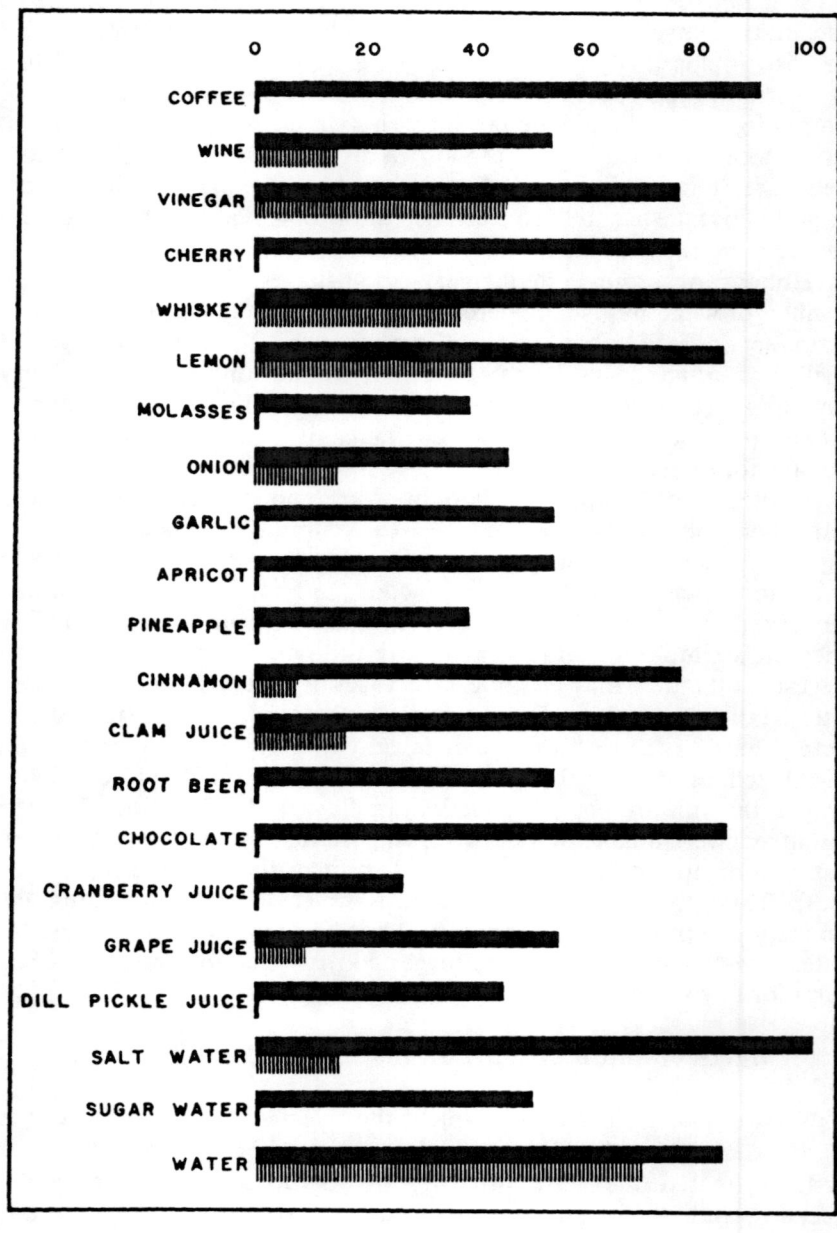

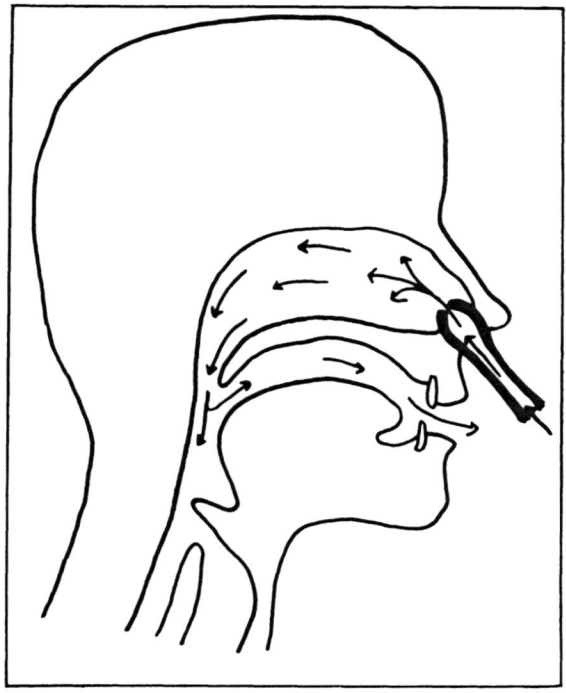

Fig. 16.3. Diagram of the air-flow apparatus used to prevent odorant mo ecules in the mouth from reaching the olfactory receptors. The arrows indicate the direction of air flow. In the experiment, odorant molecules were prevented from entering at the external nares by Teflon noseplugs. (Reprinted from Mozell, Smith, Smith, Sullivan Jr & Swender, 1969, with permission. Copyright 1969, American Medical Association.)

ethyl butyrate and the mixtures, such that the 'taste' magnitudes were higher when the nostrils were open (Fig. 16.4). With the nostrils open, odorant molecules presumably moved on a path around the back of the soft palate to reach the headspace above the olfactory receptors. This movement was most likely accomplished as a result of a number of passive physical processes, including diffusion and convection. These odorant molecules then produced an olfactory sensation, which in turn

◁ **Fig. 16.2.** Percentage of subjects correctly identifying each flavor. The solid bar represents the condition with the nose open, whereas the striped bar represents the condition without nasal chemoreception. The percentages are given across the top of the figure. (Reprinted from Mozell, Smith, Smith, Sullivan Jr & Swender, 1969 with permission. Copyright 1969, American Medical Association.)

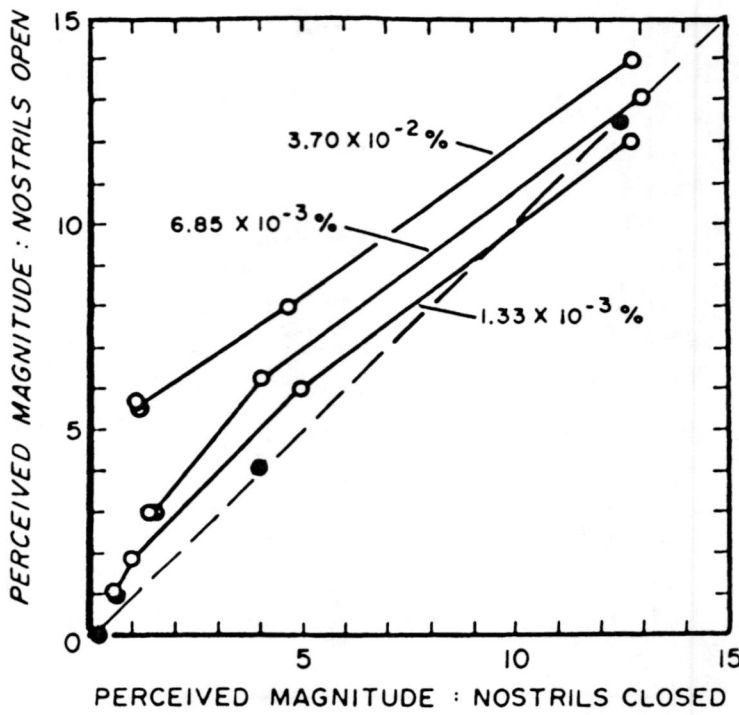

Fig. 16.4. Perceived taste magnitude of mixtures sipped when the nostrils were open versus the taste magnitude when the nostrils were closed. The concentration of the ethyl butyrate is shown. Closed circles represent stimuli with no ethyl butyrate (only saccharin). Diagonal is the line of identity. When the lowest concentration of saccharin was mixed with the highest concentration of ethyl butyrate, the intensity estimates of the taste were five times higher with the nostrils open. (Reprinted from Murphy, Cain & Bartoshuk, 1977, with permission.)

created a taste confusion. As a result, the perceived intensity of the 'taste' of a solution in the mouth was increased.

In a second study, Murphy and Cain (1980) reported that the taste of citral depended on the patency of the nostrils. With the nostrils pinched closed the perceived taste of citral was absent. Thus, they suggested that the entire taste of citral was in fact a taste confusion created by the smell of citral. Further, by pairing citral with either sodium chloride or sucrose, they demonstrated that the confusion created by the citral could increase the perceived intensities of the taste of sugar or salt. However, they were not able to demonstrate that taste affected smell. That is, sucrose and sodium chloride did not alter the intensity estimates of the smell of citral.

Table 16.1. Effects of ethyl butyrate vapor in the nose on the intensity estimates of the taste of water or sucrose solutions

Sucrose concentrations	Water	Ethyl butyrate concentrations		
		Low	Medium	High
Water	0.68	1.11	1.33	1.40
Low	2.94	3.73	3.63	3.63
Medium	4.23	6.67	5.12	10.25
High	6.02	7.94	9.15	10.00

The low, medium and high ethyl butyrate concentrations were 0.01%, 0.04%, and 0.16% v/v r espectively. The low, medium and high sucrose concentrations were 5%, 10% and 20% wt/v respectively.

Thus, Murphy and Cain suggested that the confusion may operate in one direction only.

Recent studies (Hornung & Enns, 1984, 1986) that have employed the two-module delivery system (Fig. 16.5) have proved useful in describing some additional aspects of taste or smell confusions. In one study (Hornung & Enns, 1986), subjects scaled the intensity of distilled water and three concentrations of sucrose with the background of the vapor of distilled water and three concentrations of ethyl butyrate (Table 16.1). This study differed from that of Murphy *et al.* (1977) in that the odorant was delivered to the olfactory receptors via the external nares rather than through the mouth. The taste of distilled water and the taste of sucrose were increased by the smell of increasing concentrations of ethyl butyrate. Thus, a taste confusion was demonstrated when an odorant was presented to the nose via the external nares.

However, this is not to suggest that all smells delivered to the external nares will necessarily create taste confusions. The data from a study of the effects of the smell of instant coffee on the taste of instant coffee showed that different smell backgrounds had no effect on the coffee taste (Hornung & Enns, 1984). Thus, taste confusions appear to be somewhat stimulus specific.

III. Effects of taste stimuli on smell perception

Since smell can at least sometimes affect taste, one might ask if the reverse can also be true. That is, can taste affect the perception of an odorant. In a study (Hornung & Enns, 1986) parallel to the ethyl butyrate/sucrose experiment described above, subjects were asked to scale the intensity of the smell of distilled water and three concentrations of ethyl butyrate paired with the taste of distilled water and three

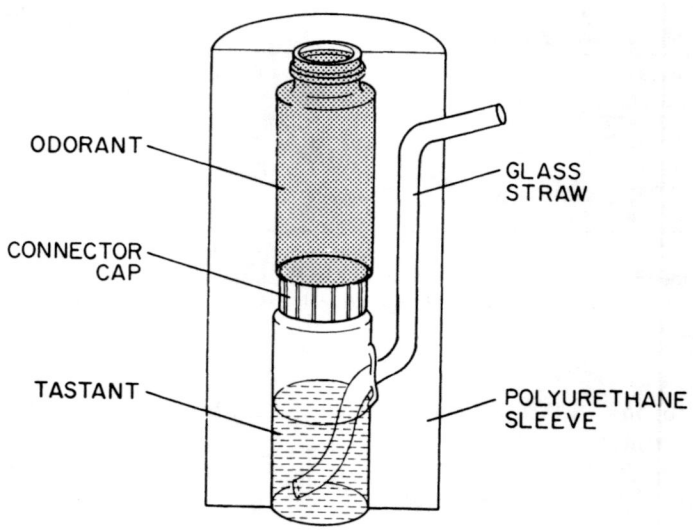

ODORANT

GLASS
STRAW

CONNECTOR
CAP

TASTANT

POLYURETHANE
SLEEVE

Fig. 16.5. The two-module delivery system. Two glass scintillation bottles are fitted into a polyurethane sleeve which covers all but the opening of the top bottle. A glass straw for tasting is inserted through a hole in the side of the lower bottle and extends through the polyurethane sleeve. Thus the concentration of the stimulus delivered to the nose can be varied independently of the concentration delivered to the mouth. (Reprinted from Hornung and Enns, 1984, with permission.)

concentrations of sucrose. The stimuli were again delivered via the two module delivery system. Intensity estimates of the smell of distilled water did not differ significantly when paired with the tastes of distilled water or sucrose. However, estimates of the intensity of the smell of ethyl butyrate were significantly higher when paired with the taste of 10% or 20% sucrose than when paired with the taste of distilled water (Table 16.2). If this represents an example of a smell confusion, this confusion is certainly not as robust or as ubiquitous as that seen with taste confusions.

In a similar example, the taste of a high concentration of almond extract was shown to affect the perceived intensity of the smell of almond extract (Enns & Hornung, 1985). However, lower concentrations of almond extract in the mouth did not affect smell in the same way. Thus, again, the effect of taste on smell seems at best to be a partial one. Since a tastant such as almond also contains an odor, part of the apparent smell confusion may not have been a confusion at all but could perhaps represent a retronasal stimulation of the olfactory receptors (Burdach, Kroeze & Köster, 1984).

Table 16.2. Effects of sucrose in the mouth on the intensity estimates of the smell of water or ethyl butyrate solutions

Ethyl butyrate concentrations	Water	Sucrose concentrations		
		Low	Medium	High
Water	0.61	1.26	1.81	1.61
Low	4.54	6.71	5.60	7.44
Medium	7.01	7.32	9.52	9.25
High	10.16	11.69	11.79	13.10

The low, medium and high ethyl butyrate concentrations were 0.01%, 0.04%, and 0.16% v/v respectively. The low, medium and high sucrose concentrations were 5%, 10% and 20% wt/v respectively.

IV. Mechanisms of smell and taste confusions

To begin to suggest some possible mechanisms by which smell confusions might occur, it was necessary to better describe the phenomenon of referred taste in which odorant molecules in the nose can create a taste sensation. To describe referred taste, subjects smelled one of five chemosensory stimuli (ethyl butyrate, almond extract, vanillin, citral, ethyl alcohol), and then rated the intensity of the taste (Hornung & Enns, 1987b). As part of the data analyses, the slope of the intensity ratings versus concentration was calculated for each stimulus and compared to the slopes (intensity ratings versus concentration) for the respective taste and nasal smell. For each stimulus, the slope describing referred taste was closer to the slope describing smell than it was to the slope describing taste. This observation suggests that referred taste is indeed not a taste at all, but rather reflects mostly the intensity of the olfactory stimulus. Thus, referred taste is perhaps a central (cognitive) confusion.

Since taste confusions might in part be reflected in the perceived intensity of retronasal olfaction, subjects tasted solutions and then rated the intensity of the smell. The slope of the intensity ratings versus concentration for retronasal smell was compared to the slopes for taste and nasal smell. Using the five chemosensory stimuli listed above, the slopes describing retronasal olfaction were somewhat closer to the slopes describing taste than they were to the slopes describing smell. Certainly part of the retronasal smell sensation must have included a direct stimulation of the olfactory receptors as odorant molecules diffused from the mouth toward the olfactory mucosa. However, since the retronasal smell slopes were closer to the taste slopes than they were to the smell slopes, retronasal smell apparently also contains a smell confusion.

Because it has afferents in both the nasal and oral cavities, Murphy and Cain (1980) have suggested that one of the physiological mechanisms for taste and smell confusions may involve a stimulation of the trigeminal nerve. Thus it is possible that as incoming odorant molecules stimulate the trigeminal nerve endings in the nasal cavity, the central nervous system 'confuses' the site of the input and so assigns some of the sensation as coming from the mouth. Likewise, a stimulation of the trigeminal nerve in the mouth could contribute to a smell confusion. Of course the role of the trigeminal nerve in creating confusions is purely speculative. Since, as described above, vanillin, (a non-trigeminal stimulator) can apparently create both smell and taste confusions, it would seem that trigeminal nerve stimulation cannot be the only mechanism for taste and smell confusions.

Recent data have suggested that the influences of taste on smell may affect more than intensity ratings. Burdach *et al.* (1984) compared the effect of nasal versus retronasal olfaction on odor identification. Lemon, rum, ethyl butyrate and amyl acetate in different concentrations were presented nasally and retronasally in a four-alternate, forced-choice detection task. Presenting these solutions nasally or retronasally did not affect the mean detection scores of any of the test solutions. The addition of sucrose to the test solution did not affect the detection of the four smells when presently nasally. However, for all odorants, the presence of sucrose increased the mean detection threshold when the stimuli were presented retronasally. Thus, the results of this study suggest that taste can influence odorant perception.

Another study which has attempted to describe the influences of two modality mixtures on taste and smell qualities is that of Gillan (1983). In that study, subjects rated the intensity of sweetness following a taste of sucrose, saltiness following a taste of sodium chloride, lemon odor following a sniff of citral, and licorice following a sniff of anethol. A sniff of citral or anethol was also paired with a taste of sucrose or sodium chloride. Subjects then rated the intensity of the sweetness, saltiness, lemon odor, and licorice of these smell/taste combinations. Intensity estimates of the quality of the two tastants and two odorants were higher when presented alone than when presented in combination (Table 16.3). These results suggests that a taste sensation can affect the quality of a smell and vice versa.

V.   Conclusions

It should be clear from this chapter that many questions about the nature of the effects of smell on taste and taste on smell remain unanswered.

Table 16.3. Intensity estimates of taste or odor qualities of taste, smell, or mixture stimuli

Modality — quality	Perceived intensity
Taste — sweetness	
Salt	120.0
Salt–anethole	86.2
Salt–citral	77.2
Taste — saltiness	
Sucrose	101.6
Sucrose–anethole	82.5
Sucrose–citral	71.6
Smell — Licorice	
Anethole	103.1
Anethole–salt	62.8
Anethole–sucrose	57.5
Smell — Lemon	
Citral	114.1
Citral–salt	60.6
Citral–sucrose	60.0

Subjects gave the 'standard' stimuli an estimate of 100.
(Reprinted from Gillan, 1983, with permission.)

However, the data that is available from mixtures of smell and taste have implications even to the study of taste/taste or smell/smell mixtures. For example, because smell can have such a dramatic effect on the perception of taste, the possibility of taste confusions needs to be considered when at least one of the tastants in a taste/taste mixture has a olfactory component. Additionally, although the effect is less robust, when olfactory stimuli are placed in the mouth the possibility of smell confusions should also be considered when evaluating smell/smell mixture effects. Because of these types of interactions, a full understanding and appreciation of taste/smell mixtures should parallel an understanding of taste/taste and smell/smell mixture effects.

References

Burdach, K. J., Kroeze, J. H. A., & Köster, E. P. (1984). Nasal, retronasal and gustatory perception: an experimental comparison. *Perception & Psychophysics* **36**, 205–208.

Chevreul, M. E. (1824). *Considerations generales sur l'analyse organique et sur ses applications*. Chez F.-G. Levrault, Libraire, Paris.

Enns, M. P., & Hornung, D. E. (1985). Contributions of smell and taste to overall intensity. *Chemical Senses* **10**, 357–366.

Gillan, D. J., (1983). Taste-taste, odor-odor, and taste-odor mixtures: greater suppression within than between modalities. *Perception & Psychophysics* 33, 183–185.

Hollingworth, H. L., & Poffenberger Jr, A. T., (1917). *The sense of taste.* Moffat, Yard and Company, New York.

Hornung, D. E., & Enns, M. P. (1984). The independence and integration of olfaction and taste. *Chemical Senses* 9, 97–106.

Hornung, D. E., & Enns, M. P. (1986). The contribution of smell and taste to overall intensity: a model. *Perception & Psychophysics* 39, 385–391.

Hornung, D. E., & Enns, M. P. (1987a). Odor/taste mixtures. *Annals of the New York Academy of Sciences* 510, 86–90.

Hornung, D. E., & Enns, M. P. (1987b). Possible mechanisms for the processes of referred taste and retronasal olfaction. *Annals of the New York Academy of Sciences* 510, 375–377.

Mozell, M. M., Smith, B. P., Smith, P. E., Sullivan Jr, R. L., & Swender, P. (1969). Nasal chemoreception in flavor identification. *Archives of Otolaryngology* 90, 367–373.

Murphy, C., & Cain, W. S. (1980). Taste and olfaction: independence versus interaction. *Psychology & Behavior* 24, 601–605.

Murphy, C., Cain, W. S., & Bartoshuk, L. M. (1977). Mutual action of taste and olfaction. *Sensory Processes* 1, 204–211.

17

Mixtures of oral chemical irritants

Harry T. Lawless

Product Evaluation Department, S. C. Johnson & Son,
Racine, Wisconsin, USA

David A. Stevens

Psychology Department, Clark University,
Worcester, Massachusetts, USA

I. Introduction

Experiments on the sensory effects of mixtures of stimuli within a sensory system are conducted for several reasons. The perceived intensities of mixtures may equal those predicted by the addition of the intensities of the components, or be more intense than predicted (synergism), or be less intense (mixture suppression). The conditions under which addition, synergism or suppression is found are unknown. Mixture phenomena are examples of the systematic variance in the operating characteristics of a sensory system which must be described, predicted and explained, if that system is to be fully understood. This information is required both by sensory psychologists who strive to describe and predict the psychophysical characteristics of the system and by sensory physiologists who strive to understand the biological processes that give rise to the psychophysically observed events. The latter workers may use psychophysical data to determine constraints on physiological models and to obtain clues as to the neural organization of the sensory system.

The study of mixtures is not the only way in which the sensory effects of multiple stimuli are examined. Table 17.1 shows three paradigms commonly used in the chemical senses. The paradigms differ primarily along two temporal dimensions — the length of conditioning trials and the duration of the interstimulus intervals. In desensitization experiments, both the conditioning trials and the interstimulus intervals may

PERCEPTION OF COMPLEX SMELLS
AND TASTES ISBN 0 12 042990 X

297

Copyright © 1989 by Academic Press Australia.
All rights of reproduction in any form reserved.

Table 17.1. Time parameters of interaction experiments

Paradigm	Interstimulus interval	Duration of conditioning trial
Desensitization	variable zero–10^4 s	long 10^2–10^4 s
Cross-adaptation	zero to brief zero–10^1 s	brief 10^1–10^3 s
Mixtures	zero	(none)

be very long. In cross-adaptation studies, conditioning trials are relatively short and the interstimulus intervals often approach zero. Mixture experiments may be viewed as a limiting case in which both the duration of the conditioning trials and the interstimulus intervals are zero.

Experiments in which the conditioning trials and interstimulus intervals are very long have been conducted with chemical irritants to study desensitization. For example, Jancso (1960) applied capsaicin topically to one side of the face for repeated trials until the reddening and burning pain were no longer observed. On subsequent trials, ammonia applied to that side of the face elicited no inflammatory reaction and no pain, while a vigorous reaction was obtained to ammonia stimulation on the opposite side. From this experiment we conclude that the neural systems mediating responses to both ammonia and capsaicin stimulation share a common mechanism which can be desensitized with capsaicin.

Since adaptation is rapidly achieved in taste and smell, cross-adaptation experiments in those modalities can employ relatively short conditioning (adapting) trials and brief interstimulus intervals. In such studies, reduced responding to one chemical after adaptation to a different chemical is taken as evidence of a common mechanism of stimulation (Köster, unpublished doctoral dissertation, University of Utrecht (Holland), 1971; McBurney, Smith, & Shick, 1972). Cross-adaptation has not yet been observed with oral trigeminal stimuli, primarily because adaptation itself is difficult to achieve. Taste adaptation is most easily obtained under conditions in which areas of the mouth are stimulated with controlled flowing stimuli or with wetted filter paper swatches which contain the spread of stimulus to limited regions (Gent & McBurney, 1978; Kroeze, 1979). However, when such stimulation is attempted with oral chemical irritants, the sensation grows over time, rather than adapts. Figure 17.1 shows the perceived intensity of irritation (burn) over time during stimulation with a flow chamber (Lawless & Gillette, 1985), and with filter paper swatches (Stevens & Lawless, 1986).

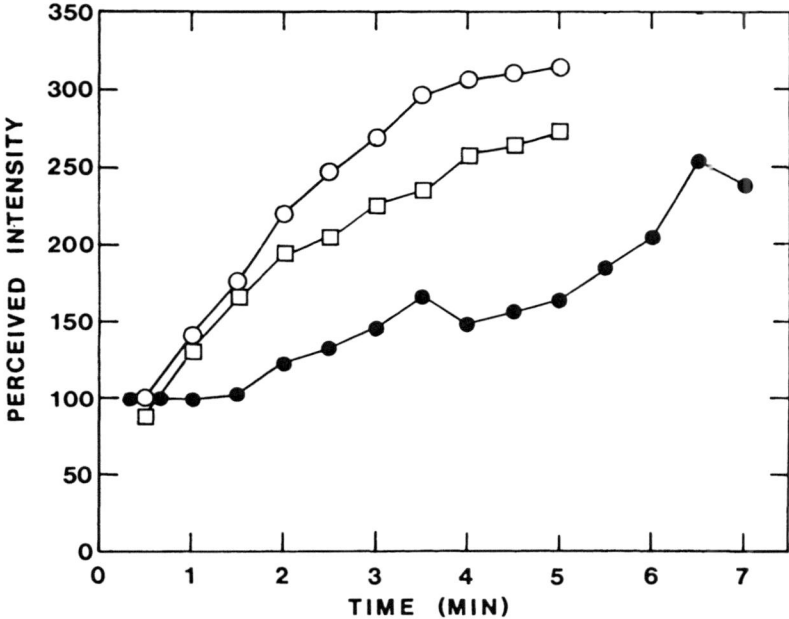

Fig. 17.1. Mean perceived intensities of capsaicin applied to the front of the tongue on filter paper (open circles) or through a flow system (filled circles), and of piperine applied to the front of the tongue on filter paper (squares). Capsaicin data were normalized to a starting value of 100. (Filter paper stimulation from Stevens & Lawless, 1986, and flow system stimulation from Lawless & Gillette, 1985.

Experiments on whole-mouth stimulation have been conducted using the short time parameters characteristic of gustatory cross-adaptation studies. As in the case of the small-area studies, irritation grows over time (Stevens & Lawless, 1987). When one irritant is followed after a brief conditioning trial with either itself or another irritant, perceived intensity increases (see Fig. 17.2). Furthermore, presentation of a *different* second chemical produces a higher intensity of burn than repetition of the same chemical. Although the pattern of adaptation that is normally seen in gustation was not obtained in this experiment, it does provide some clues to the specificity or tuning of the oral trigeminal sense. A simple explanation for the increase in irritation after switching to a different irritant is that some new receptors or fibers are recruited. This explanation implies at least some irritant-specific tuning on the part of oral chemical trigeminal receptors and a higher degree of specificity

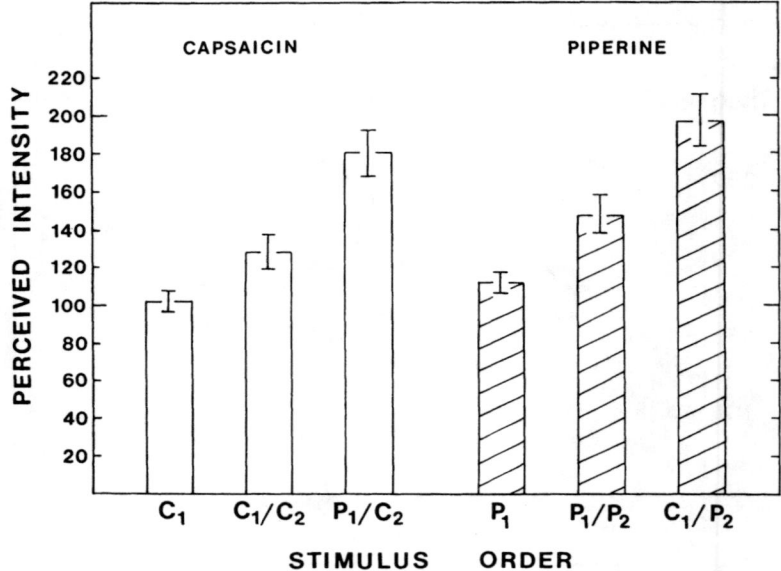

Fig. 17.2. Mean perceived intensities ($\pm$ SEM) of oral irritants as a function of stimulus order. Open bars indicate intensity of capsaicin, hatched bars intensity of piperine. C_1, intensity of capsaicin after water rinse; P_1, intensity of piperine after water rinse; C_1/C_2, intensity of capsaicin after capsaicin; C_1/C_2, intensity of capsaicin after piperine; P_1/P_2, intensity of piperine after piperine; C_1/C_2, intensity of piperine after capsaicin. (Reprinted with permission from Stevens & Lawless 1987. Copyright 1987 Pergamon Press.)

than had been previously suggested by desensitization studies (reviewed in Nagy, 1982) in which responses to all chemical irritants are abolished by capsaicin desensitization.

Turning to the question of mixtures, Kane and Alarie (1978) presented mixtures of airborne respiratory irritants to mice and found evidence consistent with additivity through a common receptor mechanism. To our knowledge, no published studies have reported the interactions among *oral* chemical irritants presented as mixtures. Some anecdotes from the culinary arts suggest the possibility of synergistic interactions. This possibility is supported by the results of sequential stimulation (Stevens & Lawless, 1987), which suggests partial independence of peripheral mechanisms. On the other hand, suppressive or hypoadditive interactions are the rule of thumb in the senses of taste and smell, and the capsaicin desensitization literature suggests enough overlap in neural pathways to provide the substrate for converging systems which might be suppressive. With these possibilities in mind, we undertook the present study of mixtures of capsaicin and piperine, the chemical irritants found in red and black pepper.

II. Method

Seven men and 13 women, aged 18 to 27 years, served as subjects. None had eaten or smoked within an hour of reporting to the laboratory. Informed consent was obtained and a questionnaire on eating habits and body consciousness was administered. Data from this questionnaire are not reported here.

The stimuli were ethanol solutions of capsaicin, piperine and mixtures of them diluted in distilled water. Three concentrations of each stimulus type were used. The concentrations of the stimuli were 0.5, 1.0 and 2.0 ppm capsaicin and 17.5, 35 and 70 ppm piperine. The middle levels of each irritant were found in pilot work to be approximately equally intense. The mixed stimuli consisted of the 50/50 concentrations corresponding to each of these three levels, i.e. 0.25 ppm capsaicin and 8.75 ppm piperine, 0.5 ppm capsaicin and 17.5 ppm piperine, and 1.0 ppm capsaicin and 35 ppm piperine. In the terms of De Graaf and Frijters (1987) these were comparisons of an equiratio mixture type.

The 2.0 ppm capsaicin stimuli were prepared by dissolving 0.8 g of capsaicin (Sigma Grade 1) in 100 mL ethanol and mixing 0.05 mL of this solution into 20 mL distilled water just prior to presentation to the subject. The source solution for the 2.0 ppm stimulus was diluted with ethanol to make the source solutions for the lower concentrations. For the 70 ppm concentration of piperine, 0.84 g of piperine was dissolved in 30 mL of ethanol and 0.05 mL of this was mixed into 20 mL of water immediately before presentation to the subject. The other piperine and mixture stimuli were prepared by using the appropriate dilutions of the ethanol source solutions of piperine or piperine and capsaicin. The amount of ethanol added to water to make the stimuli was constant (0.05 mL ethanol per 20 mL stimulus).

Three irritant stimuli were rated per session for three sessions. The sequences of irritant stimuli were determined by Greco-Latin squares, so that at each session, a subject would rate one sample of capsaicin, one sample of piperine and one mixture, and that each of these stimulus types would be the first, second and third one presented once and only once.

The subjects were instructed in the method of magnitude estimation and given practice by judging the heaviness of visually identical jars weighing 100 g, 200 g, 400 g and 800 g. The subjects then began rating the irritant stimuli for their perceived 'burn'. For each sample, the subject

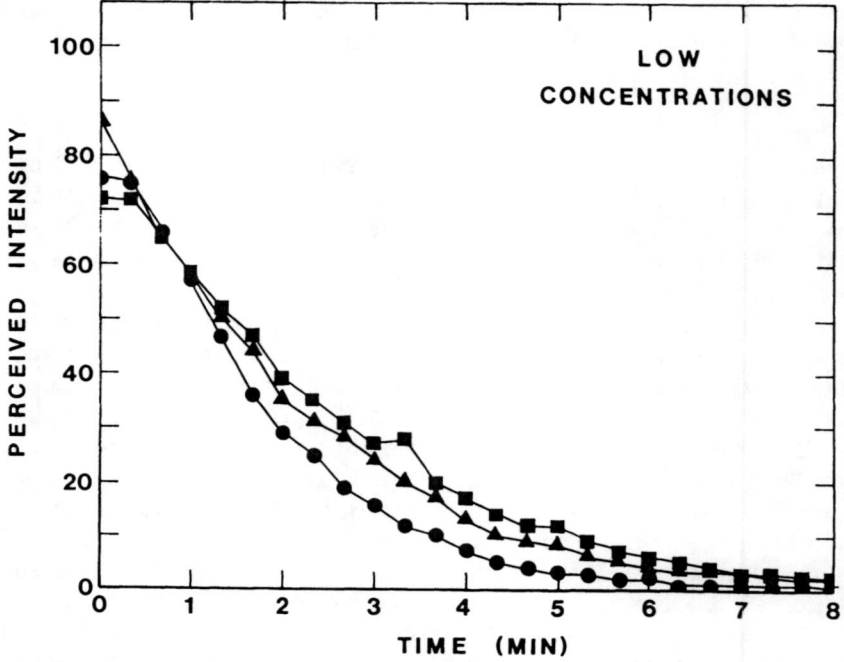

Fig. 17.3. Mean perceived intensities of 0.5 ppm capsaicin (circles), 17.5 ppm piperine (squares) and their 50/50 mixture (triangles).

rinsed with distilled water and expectorated. The 200 g weight was lifted and given a value of 10. The subject was told to use this as a modulus for the judgments of the intensity of oral irritation, and it was made available to lift *ad lib* throughout the experiment. The sample was sipped and circulated in the mouth for 30 seconds and expectorated. A magnitude estimate was made immediately, and every 20 seconds for the next eight minutes, for a total of 25 ratings. After the last rating, the subject rinsed and expectorated and rested until all sensations attributable to the sample disappeared. During this intertrial interval, *ad lib* rinsing was permitted. The intertrial interval lasted a minimum of 15 minutes.

III.   Results

Two sets of analyses were carried out, one on the raw data and one after the data had been normalized by a multiplicative constant for each subject, which set the mean of the first four judgments of each subject for the middle capsaicin concentration equal to 100. Analyses of variance performed on each data set yielded identical patterns (significant effects of time, concentration and concentration by time interactions). Figures 17.3, 17.4 and 17.5 show the mean intensity ratings for the low, middle and high concentrations, respectively.

Separate analyses of variance were carried out on the data from each concentration level to assess whether the mixtures were greater than, equal to or less than the values of the components. For the low and high concentration triads, the mixture intensity was not different from the component intensities and generally fell between the capsaicin and piperine curves. At the middle concentration level, however, the intensities of the mixture were significantly higher than the intensities of the components, especially during early time blocks (stimulus by time interaction, $F(48\ 912) = 2.94$, $p < .01$). Thus the overall pattern of interaction, $F(48\ 912) = 2.94$, $p < .01$). Thus the overall pattern of additivity showed a concentration-dependent synergistic effect, with and a synergistic effect in the middle range.

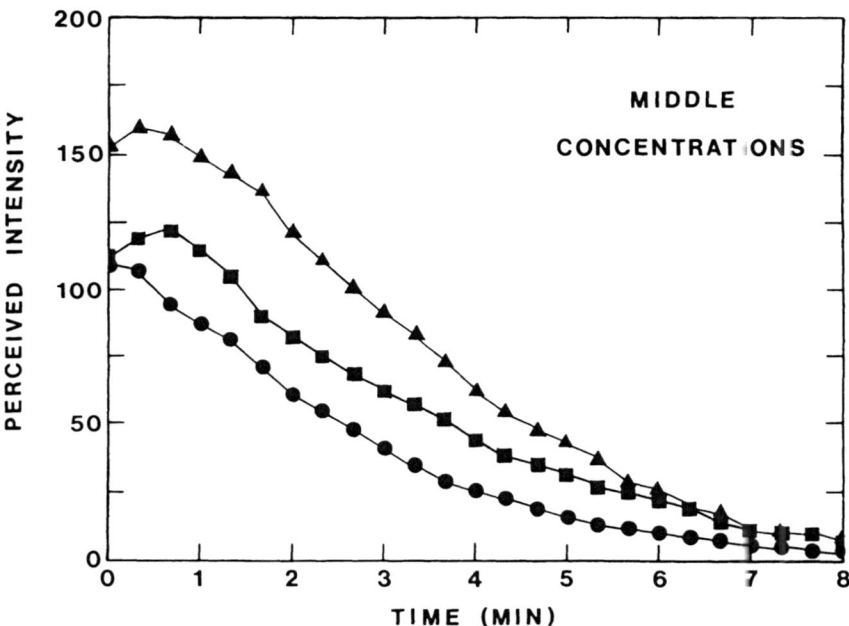

Fig. 17.4. Mean perceived intensities of 1 ppm capsaicin (circles), 35 ppm piperine (squares) and their 50/50 mixture (triangles).

IV. Discussion

The interactions we observed between capsaicin and piperine in mixtures can be described as concentration dependent. Low concentration stimuli show an approximately additive relationship. At intermediate levels interactions of a synergistic nature were observed. This is parallel to the facilitation of oral irritation we observed in the sequential alternation of capsaicin and piperine, at similar concentration levels to those employed in the intermediate mixtures (Stevens & Lawless, 1987). At higher levels, a more additive relationship appears to hold once again. Two questions arise for discussion. First, are concentration-dependent interactions unique to the oral trigeminal chemical sense or are they also seen in mixture studies in other modalities? Second, what kind of physiological model would allow for these interactions?

The question of whether changes occur in the nature of mixture interactions as a function of changes in intensity or concentration has

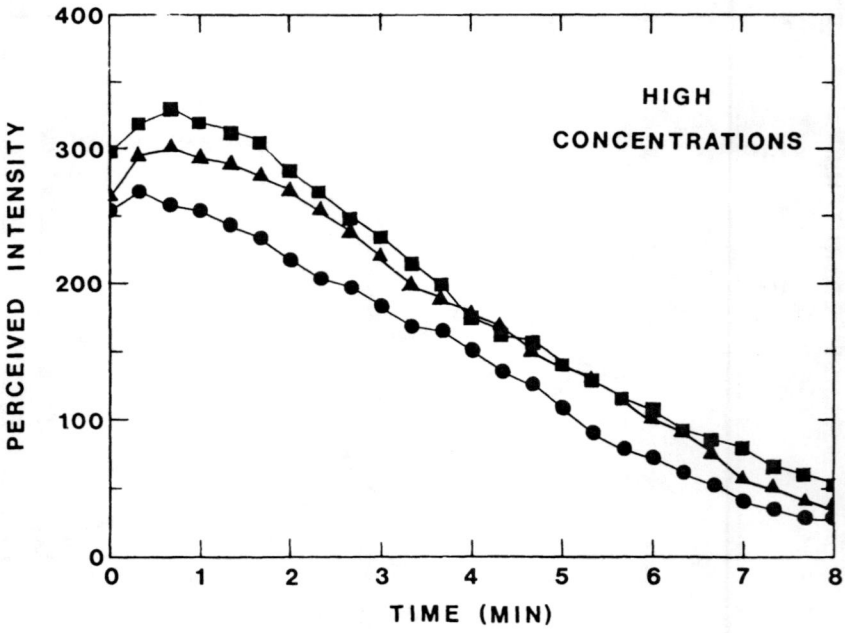

Fig. 17.5. Mean perceived intensities of 2 ppm capsaicin (circles), 70 ppm piperine (squares) and their 50/50 mixtures (triangles).

received little attention. However, one of the most common observations in the sense of taste is mixture suppression, and there is general agreement that this effect is most pronounced at high levels of intensity (Curtis, Stevens & Lawless, 1984; Frank & Archambo, 1986). The intensities of mixtures with lower concentration components may appear more additive or even synergistic, as found, for example, in the sweetener mixtures studied by Curtis *et al.* (1984). A second example of changes in mixture interactions with concentration is in the addition of small amounts of NaCl to sucrose. Because of the sweet taste of NaCl at low levels (Bartoshuk, Murphy & Cleveland, 1978), sucrose sweetness may be enhanced, while at higher levels, mixture suppression is the rule (e.g. Bartoshuk, 1975; Kroeze, 1979). A common anecdote in the sweetener industry is that synergistic interactions are much more likely at low levels. In odor mixtures, facilitation (synergy) is sometimes observed at low levels, with the more common pattern of odor counteraction occurring at higher levels (see, for example, 'Fechner's Paradox' in Gregson, 1986). In summary, intensity-dependent or concentration-dependent interactions are observed in the chemical senses, although the generality of mixture suppression in taste and odor counteraction in olfaction has made them a primary focus of attention in psychophysical research.

What kind of neural organization could produce these data? A simple mechanism explaining these effects is based on Sherrington's (1930) descriptions of neural convergence in spinal reflexes that produces occlusion (analogous to mixture suppression) at high levels of stimulation and synergy at low levels. Our model has the following four assumptions:

1. The neural units responding to capsaicin and those responding to piperine show partial specificity (moderate tuning), as suggested by our results with sequential stimulation (Stevens & Lawless, 1987).
2. For purposes of simplicity, two states exist for these neural units; one below or near threshold, where there is stimulation of the unit but no significant output, and a second state above threshold, at or near saturation (maximal response). These assumptions may be relaxed to provide a more continuous distribution of outputs without damaging the predictions. However, to simplify presentation of the model, two states will be assumed.
3. As stimulus intensity increases, more units are recruited.
4. As stimulus intensity increases, the overlap in response to the two compounds increases. Thus a unit which appears selectively tuned to capsaicin at low levels may also respond to piperine at high levels.

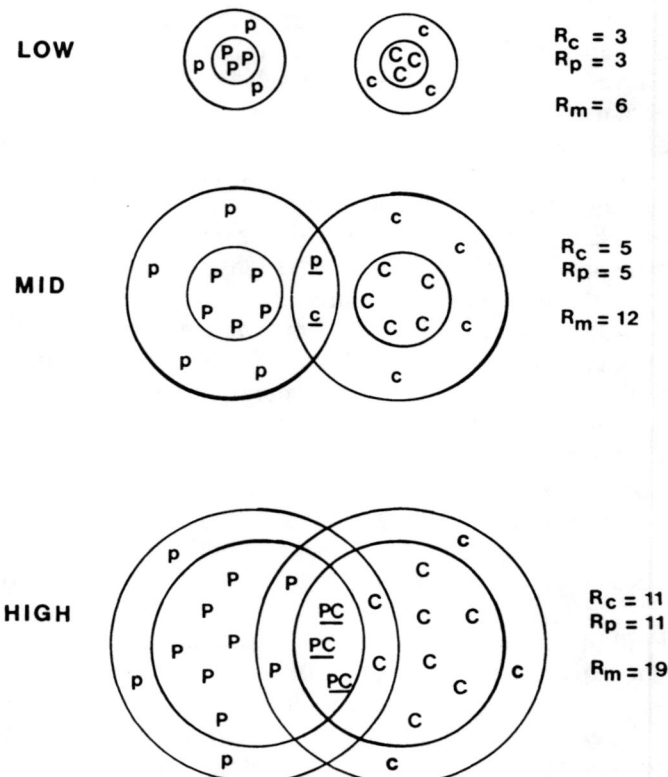

Fig. 17.6. Scheme for intensity dependent interactions (after Sherrington, 1930: *P*, units responding to piperine at or near maximal levels; *p*, units stimulated by piperine at or below threshold; *C*, units responding to capsaicin at or near maximal levels; *c*, units stimulated by capsaicin at or below threshold; *p* and *c*, units stimulated by both capsaicin and piperine, and pushed above threshold, providing hyper-additive response to the mixture due to redundancy; *PC*, units responding to both capsaicin and piperine at or near maximal response, providing hypo-additive response to mixtures due to redundancy; R_c, response magnitude to capsaicin alone; R_p, response magnitude to piperine alone; R_m, response magnitude to the mixtures.

Such a recruitment model will predict relative independence and additivity at low levels. At intermediate levels synergistic interactions will appear, as there is a gain in redundancy due to overlapping stimulation of subthreshold units, pushing them above threshold. At higher levels there is occlusion, due to a loss in redundancy as there is overlapping stimulation of already saturated above-threshold units. These effects are shown in Figure 17.6. Of course, occlusion or suppression was not observed in this study, possibly because high enough concentrations were not used.

Such a recruitment model will also work with multiple output states (rather than the all-or-none output model we described above), provided that the psychophysical function for each neural unit is sigmoical. That is there should be a period of non-responding or baseline responding, followed by a positively accelerated limb, then a negatively accelerating limb approaching an assymptote. The intrinsic flattening of taste functions at high levels is discussed by Frijters, De Graaf and Koolen (1984). Such a relationship is also predicted by the Beidler equation (Fig. 17.7), which has been widely used to describe neural response to chemical stimuli and, more recently, psychophysical responses.

Bartoshuk and colleagues have suggested that the nature of mixture interactions is related to the psychophysical functions of the two components (Bartoshuk, 1975; Bartoshuk & Cleveland, 1977; De Graaf & Frijters, 1987; Frijters *et al.*, 1984). Components with positively accelerated psychophysical functions are predicted to show synergy, while those with negatively accelerated functions should show suppression. However, describing the psychophysical functions for compounds in terms of a single trend (positive or negative acceleration) may

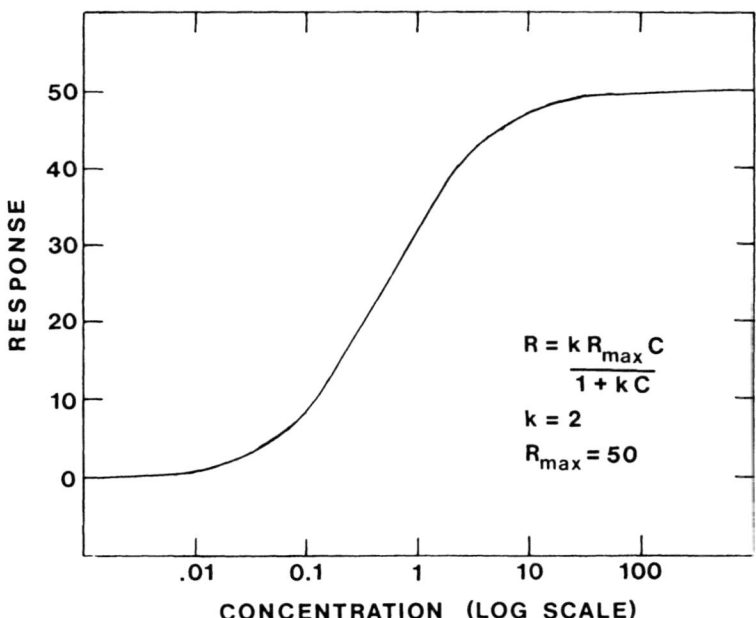

Fig. 17.7. An example of a function from Beidler's equation in semi-log plot

be too simple. As seen in the curve based on Beidler's equation (Fig. 17.7) both trends may occur in the same function. In the case of intensity dependent interactions, we have argued that synergy should be the case at low to intermediate levels, and suppression at higher levels, as is usually found in taste. These observations would be consistent with the theory of Bartoshuk provided that the psychophysical function has zones of different curvature along the concentration continuum.

It is unknown whether occlusion or suppression will be found at high levels with oral trigeminal irritants. It seems reasonable that there are only a limited number of receptors and fibers available, and that these fibers have a maximal rate of response. However, the human observer produces intensity ratings of painful stimuli not only as a function of sensory intensity, but also as a function of the emotional response to pain (Melzack, 1973). Future research should address the nature of responses to trigeminal irritants at high levels, to see if suppression occurs. However, such a study will face the challenge of determining the intensity of irritation separately from the responses engendered by the emotional reactions to pain.

Kane and Alarie (1978) applied the mixture versions of Beidler's taste equations to responses to respiratory irritants and found a good correspondence to predictions based on additivity through a single receptor mechanism (competitive agonism). It is possible that physico-chemical and anatomical factors that govern the access of irritant molecules to receptors play a part in the results observed here. If these factors are concentration dependent, it is possible that a Sherringtonian model would hold under some conditions, but where irritants have equal access to receptors, additivity along the Beidler equation predictions would be observed, as Kane and Alarie found.

Acknowledgments

The authors acknowledge the support of US National Institutes of Health grant NS-20616 and thank William S. Cain for helpful suggestions.

References

Bartoshuk, L. M. (1975). Taste mixtures: is mixture suppression related to compression? *Physiology & Behavior* **14**, 643–649.

Bartoshuk, L. M., & Cleveland, C. T. (1977). Mixtures of substances with similar tastes: a test of a psychophysical model of taste mixture interactions. *Sensory Processes* **1**, 177–186.

Bartoshuk, L. M., Murphy, C., & Cleveland, C. T. (1978). Sweet taste of dilute NaCl: psychophysical evidence for a sweet stimulus. *Physiology & Behavior* **21**, 609–613.

Curtis, D. W., Stevens, D. A., & Lawless, H. T. (1984). Perceived intensity of the taste of sugar mixtures and acid mixtures. *Chemical Senses* 9, 107–120.

De Graaf, C., & Frijters, J. E. R. (1987). Sweetness intensity of a binary sugar mixture lies between intensities of its components, when each is tasted alone and at the same total molarity as the mixture. *Chemical Senses* 12, 113–129.

Frank, R. A., & Archambo, G. (1986). Intensity and hedonic judgments of taste mixtures: an information integration analysis. *Chemical Senses* 11, 427–438.

Frijters, J. E. R., De Graaf, C., & Koolen, H. C. M. (1984). The validity of the equiratio taste mixture model investigated with sorbitol–sucrose mixtures. *Chemical Senses* 9, 241–248.

Gent, J. F., & McBurney, D. H. (1978). Time course of gustatory adaptation. *Perception & Psychophysics* 23, 171–175.

Gregson, R. A. M. (1986). Qualitative and aqualitative intensity components of odour mixtures. *Chemical Senses* 11, 455–470.

Jancso, N. (1960). Role of the nerve terminals in the mechanism of inflammatory reactions. *Bulletin of the Millard Fillmore Hospital of Buffalo* 7, 53–77.

Kane, L. E., & Alarie, Y. (1978). Evaluation of sensory irritation from acrolein-formaldehyde mixtures. *American Industrial Association Hygiene Journal* 39, 270–274.

Kroeze, J. H. A. (1979). Masking and adaptation of sugar sweetness intensity. *Physiology & Behavior* 22, 347–351.

Lawless, H. T., & Gillette, M. (1985). Sensory responses to oral chemical heat. In D. D. Bills & C. J. Mussinian eds, *Recent advances in the characterization and measurement of flavor compounds,* pp. 26–42. American Chemical Society, Washington, DC.

McBurney, D. H., Smith, D. V., & Shick, T. R. (1972). Gustatory cross-adaptation: sourness and bitterness. *Perception & Psychophysics* 11, 228–232.

Melzack, R. (1973). *The puzzle of pain.* Penguin Books, London.

Nagy, J. I. (1982). Capsaicin: a chemical probe for sensory neuron mechanisms. In L. L. Iverson, S. D. Iverson & S. M. Snyder, eds, *Handbook of psychopharmacology,* vol. 15, pp. 185–235. Plenum Publishing, New York.

Sherrington, C. S. (1930). Some functional problems attaching to convergence *Proceedings of the Royal Society of London, Series B* 105, 332–362.

Stevens, D. A., & Lawless, H. T. (1986). Effects of locus of stimulation on human responses to oral chemical irritants. Paper presented at the Psychonomic Society Annual Meeting, November 1986, New Orleans, Louisiana.

Stevens, D. A., & Lawless, H. T. (1987). Enhancement of responses to sequential presentation of oral chemical irritants. *Physiology & Behavior* 39, 63–65.

Index

A page number in italics indicates a reference to a figure on that page. A page number followed by *t* indicates a reference to a table on that page. Figures and tables are not indexed separately if they fall on pages already indexed.

Printed in the United States
107188LV00002B/78/A

9 780120 429905